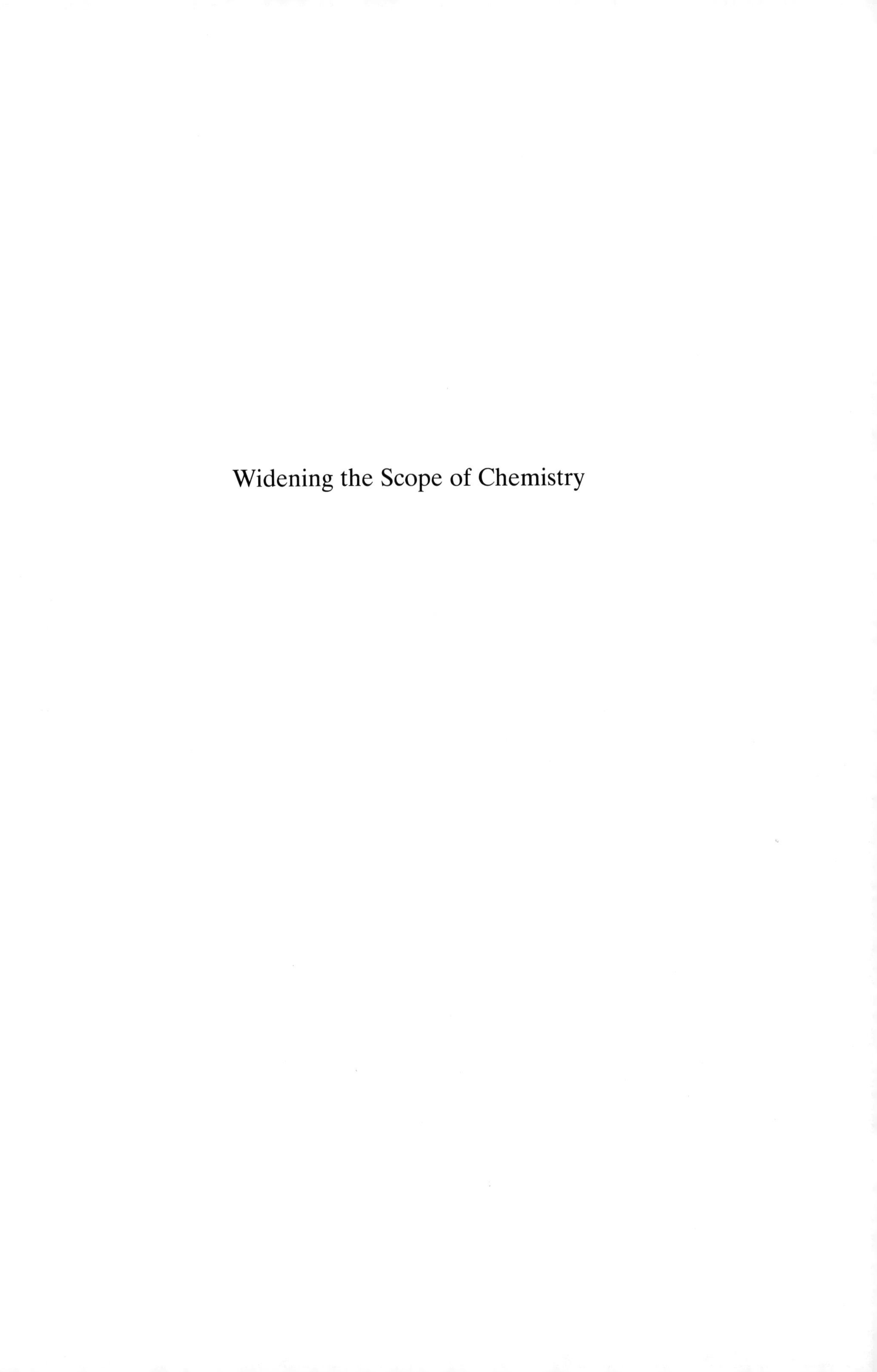

# Widening the Scope of Chemistry

International Union of Pure and Applied Chemistry
in conjunction with
The Chemical Society of Japan

## ORGANIZING COMMITTEE

*Co-Chairmen*

Michinori Ōki
*University of Tokyo*
*Japan*

David J. Waddington
*University of York*
*United Kingdom*

*Secretary General*

John T. Shimozawa
*Saitama University, Japan*

*Executive Members*

Hiroshi Hamada
*Chemical Society of Japan*
Shuichi Hamada
*Science University of Tokyo*
Masakatsu Hirose
*Senshu University*
Haruo Hosoya
*Ochanomizu University*
Shohei Inoue
*University of Tokyo*
Satoshi Niida
*Senior High School attached to Tokyo Gakugei University*
Ushio Sankawa
*University of Tokyo*
Yoshito Takeuchi
*University of Tokyo*
Shozo Toda
*University of Tokyo*

## INTERNATIONAL ADVISORY COMMITTEE

M. Chastrette, *IUPAC-CTC*
H. W. Heikkinen, *North America*
T. R. Hitchings, *Oceania*
R. Isuyama, *Latin America*
G. Keller, *Europe*
J. V. Kingston, *UNESCO*
K. Manunapichu, *FACS*
G. T. Seaborg, *IOCD*

*Organized by*

The Chemical Society of Japan

*Other Collaborating Organizations*

United Nations Educational, Scientific, and Cultural Organization (UNESCO)
Federation of Asian Chemical Societies (FACS)
The Agricultural Chemical Society of Japan
Japan Society of Science Education
Pharmaceutical Society of Japan
The Physics Education Society of Japan

International Union of Pure and Applied Chemistry

# Widening the Scope of Chemistry

Plenary and Invited Lectures
and Selected Contributions presented at the
Eighth International Conference on Chemical Education
held in Tokyo, Japan, 23–28 August 1985

*Edited by*
YOSHITO TAKEUCHI
*The University of Tokyo*
*Japan*

Blackwell Scientific Publications
OXFORD LONDON EDINBURGH
BOSTON PALO ALTO MELBOURNE

Blackwell Scientific Publications
Editorial offices:
Osney Mead, Oxford OX2 0EL
8 John Street, London WC1N 2ES
23 Ainslie Place, Edinburgh EH3 6AJ
52 Beacon Street, Boston, Massachusetts 02108, USA
667 Lytton Avenue, Palo Alto, California 94301, USA
107 Barry Street, Carlton, Victoria 3053, Australia



First published 1987

DISTRIBUTORS

USA and Canada
Blackwell Scientific Publications Inc
PO Box 50009, Palo Alto, California 94303

Australia
Blackwell Scientific Publications
(Australia) Pty Ltd
107 Barry Street, Carlton, Victoria 3053

Printed in Great Britain
at the Alden Press, Oxford

Bound at the
Green Street Bindery, Oxford

British Library
Cataloguing in Publication Data

International Conference on Chemical Education. *(8th : 1985 : Tokyo)*
Widening the scope of chemistry : plenary and invited lectures and selected contributions presented at the Eighth International Conference on Chemical Education held in Tokyo, Japan, 23–28 August 1985.
1. Chemistry—Study and teaching
I. Title II. Takeuchi, Yoshito
III. International Union of Pure and Applied Chemistry
540′.7 QD40

ISBN 0-632-01537-3

Library of Congress
Cataloging-in-Publication Data

International Conference on Chemical Education (8th : 1985 : Tokyo, Japan)
Widening the scope of chemistry.

At head of title: International Union of Pure and Applied Chemistry.
Bibliography: p.
1. Chemistry—Study and teaching—Congresses.
I. Takeuchi, Yoshito, 1934–
II. International Union of Pure and Applied Chemistry. III. Title.
QD40.I536 1985 540′7′1 86–26421

ISBN 0–632–01537–3

# Contents

**Special posters**

# Preface

The roles of chemical education are manifold; the most obvious one is to provide chemists-to-be with the appropriate knowledge necessary for their future occupations. This task is by no means simple, since "chemists" are of various kinds ranging from university professors with the highest academic qualifications to laboratory technicians at small factories. The chemical education of non-chemists should also be important in the light of the relevance of chemistry to society. The scope of chemical education is further diversified since the situation in the developing countries is necessarily different from that of the developed countries.

The Committee on the Teaching of Chemistry (CTC) of The International Union of Pure and Applied Chemistry (IUPAC) has sponsored the International Conferences on Chemical Education (ICCE) to promote the greater development of chemical education at the international level.

The ICCEs have been held in the past in either Europe or the United States, but never in any part of Asia. Some four years ago when we were asked to organize the 8-ICCE, we were delighted and honored to be given this immense task. We felt sure that the marvelous achievements which would be presented by many chemists and chemical educators most actively engaged in this field would generate future challenges for chemical education. Thus, "Widening the Scope of Chemistry" was accordingly chosen as the main objective of the Conference.

We felt that this opportunity was an exciting challenge to display not only international progress in Chemical Education but also our aesthetic talent. This is reflected in the Conference Symbol, designed by Professor H. Sano, Tokyo Metropolitan University and the Vice President of the Chemical Society of Japan. The rim represents a molecular model(space-filling type) of water as seen from the side of the oxygen atom. The design for eight can be regarded either as two hearts or as a young leaf.

The present volume contains fourteen lectures by the principal participants of the Conference in Tokyo during the period of 23 through 28 August, 1985, which well rewarded our expectations. Except for the opening and closing lectures by M. Ōki and G. Pimentel, the twelve lectures have been grouped into four sections based on the major themes of the Conference.

The lectures by S. Sasaki, J. Moore and D. Cabrol were associated with "Chemical Education in the Computer Age", while lectures by J. Malcolm, R. Isuyama and P. Childs focused on "Chemical Education for Fostering Future Chemists of Excellence". The remaining six lectures were designed for more specific fields. Lectures by P. Kelly, Wang-Kui and I. Nakayama were associated with "Chemical Education for Life Science", while those by J. Beek, A. Hofstein and W. Lee were related to "Chemical Education and Industry".

In addition to these fourteen plenary and invited lectures, the opening addresses by our Co-Chairmen, M. Ōki and D. J. Waddington, and the summary of reports of small group discussions by M. Chastrette are included in this Proceedings.

Furthermore, a limited number of general contributions have been selected for inclusion in the Proceedings. Since the total number of general contributions is over 250, it is impossible to accommodate all of them here. Hence, we sent invitations only to those who delivered short lectures (11 lectures) or presented a special poster (43 presentations).

Because of the shortage of time, not all of those who received our invitation could contribute a short paper to this Proceedings. Nevertheless, we believe these additions will provide the reader with a more thorough understanding of chemical education today.

The 8th Conference in Tokyo ended in a success greater than we had anticipated. We were pleased and honored to have been given the opportunity of transforming our enthusiasm into an outstanding Conference week. Finally we would like to express our appreciation for the tireless efforts and diligent work of all who participated in organizing this Conference and to the eminent lecturers for preparing the manuscripts for these Proceedings.

October, 1985
Tokyo

Yoshito Takeuchi
Editor of the Proceedings
Chairman,
The Program Committee, 8-ICCE

# Opening Address

D. J. Waddington

Chairman IUPAC committee on Teaching Chemistry,
Department of Chemistry, The University of York, Heslington, York
YO1 5DD, UK

Ladies and Gentlemen

I have the honor to represent the President of the International Union of Pure and Applied Chemistry in welcoming participants to this Conference.

The International Union is usually known by its initials IUPAC which often sends shivers down the backs of students and probably spells gloom amongst teachers as they struggle to learn the rules of inorganic and organic nomenclature.

However, teachers and students know little of the important role IUPAC plays in coordinating the affairs of chemistry throughout the world. Recently it has initiated CHEMRAWN - Chemical Research For World Needs - a series of meetings to discuss areas of essential concern to mankind, for example, the contributions of chemists in using resources of Organic Raw Materials wisely, and the role of chemists and the World Food Supply. The next will be concerned with the Chemists and the Resources of the Global Ocean.

But now let me turn to another dimension of the world of IUPAC. This is in education. The Committee on the Teaching of Chemistry has many initiatives to encourage good teaching of chemistry throughout the world. Several are specially concerned with development. One of our tasks is to sponsor an International Conference on Chemical Education every two years.

This, the 8th, is being co-sponsored with the Chemical Society of Japan. It is the first in Asia - indeed the furthest east we have been is Poland.

There are two personal pleasures I have had in working with the Chemical Society of Japan.

Professor Ōki, the co-chairman of the Conference, was one of the founder members of IUPAC Committee on Teaching of Chemistry; he sets a splendid example of a distinguished chemist who has involved himself in chemical education at all levels. Professor Shimozawa, the Secretary General of the Conference is, I am delighted to say, one of the 8 members of the present Committee. Both have given distinguished service to our work.

They - and their many colleagues in the Chemical Society of Japan - have worked hard over the last 2 years to make this a splendid feast of activities. Those of us lucky enough to be here will take full advantage of all their work and the generous cooperation of so many organisations.

The Conference brings together teachers from many countries - teachers from the tertiary and secondary sectors - experienced and less so. However, we all learn from each other for we are able, if we listen, to understand each other's problems which, of course, depend on many factors: economic, social and political. I am sure that the scene is set for another very successful meeting.

# Opening Address

Michinori Ōki

Department of Chemistry, Faculty of Science,
The University of Tokyo, Tokyo 113, Japan

Professor Waddington, Professor Tsuruta, distinguished guests, ladies, and gentlemen:

It is my great pleasure to be able to host the 8th International Conference on Chemical Education. To begin with, I should like to extend my warmest welcome to you all, especially to those who flew a long way to participate in this conference from overseas.

As you may know, the IUPAC ICCE has been held mainly in Europe since 1969, the only exception being the fifth conference which was held in Maryland, USA. Thus this conference is the first to be held in Asia. Since chemistry is very important in offering basic materials for the well-being of mankind, chemical education is also important in upgrading the welfare of human beings. Because Asia needs upgrading in its economy, chemistry is one of the musts and should be enhanced in every country. Because of this on one hand, and the idea of introducing the reality in chemical education in Japan, that is far from Europe, to chemical educators of the world on the other, we extended our invitation to hold the 8-ICCE in Japan. It is therefore our great pleasure that we could host the conference here in Tokyo.

As one of the first members of IUPAC CTC, I see that the trend of chemical education is always changing. In the early 60's, the strongest concern was the shortage of good chemists to be recruited by chemical industries, because at that time the petrochemical industry was growing very rapidly. However, as important as this was the world problem of upgrading developing countries. Therefore, international cooperation was another topic which was discussed in an earlier conference. Then came the issue of pollution. Chemical industry was blamed to be the origin of pollution of the environment. It is no wonder that a few conferences were devoted to discussion on the science-and-society problem. And now we are suffering from the shortage of students in chemistry departments of tertiary education. Is this situation unavoidable for chemistry, because of its being basic and because of its nature which could soil the environment?

My answer to this question is "no". And many of you will agree with me in this point. What is wrong then? To solve this problem, we must change our minds in chemical education. That is, we tended to confine ourselves to pure chemistry only when we were teaching chemistry. Chemistry had emerged as a natural science which treated materials. Substances which constituted the organisms were the target of chemical study. However, as chemistry became sophisticated, biological chemistry came to be thought a different discipline in natural science. This is probably because chemistry could not accommodate these new branches of science in education. A similar situation ocurred as polymer science developed. However, as you know, gene technology is essentially the application of organic chemistry because the technique is cleaving DNA at a desired position with the use of an enzyme, separating them by chromatography, and combining these fragments again with the use of enzymes. Why could we not include this in chemical education?

The traditional education in chemistry may have to change at this very time, because of the development of computers. The assessment in chemistry has been carried out on the basis of examining the amount of knowledge students memorized in older days. But the memorization can be better done by computers. What will be the role of good chemists then in coming days? The relationship between industry and chemical education has changed as well, as industry became sophisticated. New attitudes are requested for chemical educators from various changes in society. We need to widen the scope of chemical education.

Education must change as the needs of society change and as natural science develops. However, probably the most important and standing aim in chemical education is how we could raise children so that they can adapt the new era to come. I myself am convinced that the ability of children should not be damaged by the educators. This hope can be seen in the logo of this conference: the letter eight is modified by the design of the seedling. I hope this new bud is raised to become a big tree of chemistry, both as specialists of chemistry and as citizens who understand the nature of chemistry.

As of today, we have 570 registrants from 53 countries. We are sure that through your cooperation this conference will produce important results. Even though it may not be possible to gain something directly useful in your classes from this conference, you may get at least hints and suggestions that you can develop after you go back to your own country. Since we organizers sincerely hope that this conference concludes with the utmost success, please do not hesitate to tell us, if you have anything that you think we can be of help with.

Finally I should like to thank all the members of the organizing committee, especially the general secretary Professor Shimozawa, for their pains-taking cooperation in the preparation of this conference. I also wish to thank our chemical industries for their generous support by donations. Without these, this conference would not have been possible to be open in this manner.

I now declare the 8th International Conference on Chemical Education be open with the hope that you spend a successful and enjoyable conference week.

# Plenary and Invited Lectures

# Chemical education and its surroundings

Michinori Ōki

Department of Chemistry, Faculty of Science,
The University of Tokyo, Tokyo 113, Japan

Abstract In one respect, chemical education is surrounded by other disciplines of natu[illegible] in another, it is strongly connected with soci[illegible], chemical educators have been concerne[illegible]sors and this trend has led chemists to cu[illegible]ces from chemistry. A time has come, h[illegible]tors must think of the relations with oth[illegible]al education has close ties. These aspec[illegible]is on the subthemes of the conference, i.[illegible]chemists, life science, and industry. Som[illegible]roblems which await solution.

INTRODUCTION

Chemistry occupies a central part of natural sciences: it offers necessary materials, and even evidence for correction or modification of theories, for the development of physics, and basic understanding of substances which are important in biological science. Therefore, chemical knowledge is vitally important to other areas of natural sciences and this is why chemistry is called to occupy the central part of science. Despite this situation, chemical education has shown a tendency to confine itself to teaching chemistry which is thought to be fundamental from the viewpoint of chemistry.

The tendency is understandable because the knowledge of chemistry is ever-growing, proportionally to the amount of knowledge increase about the important facts and theories. Within the limited period of time which can be devoted to chemistry teaching, the amount to be taught has naturally some limitations. Thus applications and things which are not basically necessary in understanding chemistry are discarded from the contents of curricula in countries, both developing and advanced.

Our basic concern is "whether the tendency is really suitable for promoting natural science and for increasing the degree of science understanding among citizens?" Some of the previous conferences on chemical education have been devoted to the "chemistry-and-society" problem. This is really an important problem. However, just as important as this is the relation of chemistry with other disciplines of natural science. We believe that, since an interdisciplinary approach is strongly needed in the study of natural science and science education, chemistry in particular should be prepared to lead students to these important areas rather than confining them to basic chemistry. These thoughts led the organizing committee to adopt the theme of 8-ICCE "Widening the Scope of Chemistry".

Under the main theme, we can consider various aspects of chemical education: how can chemistry be used in physics, biology, or earth sciences? It may also be necessary to ask mathematics educators to cooperate with us in teaching chemistry. Taking physics as an example, we have had discussions with physics educators within our country. In the earlier days, the topic of discussion was the syllabus which includes the order of teaching. One of the problems we came across was the placement of atomic structure. In chemistry, we want to use atomic structures, especially the number of electrons, for the explanation of various things such as bonding and ion formation. However, they want to place atomic structure at the very end of study in physics at high school. They say that, after learning mechanics, waves, electromagnetic waves, and others, the atomic structure comes at the very end as a summary of the learning at high school. We did not come to a conclusion and it is all right as long as we are teaching physics or chemistry as separate subjects.

However, it can be a serious problem if we want to teach integrated science. Therefore, the problem must be discussed to find a practical solution among chemists and physicists.

There is another problem which may be mentioned which exists between physics and chemistry. When we teach something, the definition of that terminology is very important. We even give a test on the definition of a term. However, we have found serious problems in this definition between chemistry and physics concerned with the periodic table. We found that physicists look at the table by thinking of the properties of elements. Physicists say that if you are defining metallic elements and non-metallic elements, germanium must be placed in non-metals, which is true if we look at the element of germanium itself. But chemists are trained in such a way from the days of Mendeleev that we must look at the periodic table as indicating the properties of compounds which are derived from the element as well as the element itself. Thus chemists say that germanium is a metalloid. Whenever we classify something into two groups, there occurs a border-line case. Probably this kind of problem can be discussed within ICSU CTS or between IUPAP committee on the teaching of physics and IUPAC CTC. Should we or should we not teach this type of controversial thing at schools? It is clear that there are some topics that should be discussed by the specialists of various disciplines.

Apparently, there are a number of aspects to be discussed. However, because of the limitation of time, we had to choose only one item from these categories, biology. We sincerely hope that relations with other disciplines will be discussed in the forthcoming meetings of this series. Of course, we can not forget the social surroundings because education is affected by the needs of society. On this standpoint, we adopted a subtheme "chemical education in the computer age"; image and computers are changing our daily life and as a consequence they should also modify the way of chemistry teaching in some ways. Fostering good chemists and relations between chemical education and chemical industry are ever-lasting problems. We can not overlook the importance of these problems. We also leave, however, the problems of other social aspects such as population control and agriculture for the forthcoming meetings.

## COMPUTER AGE

Under the subtheme "chemical education in the computer age", almost everyone will think of CAI or CMI. However, the scope of the subtheme is not limited to these areas only, although much effort has been made in the study of CAI. The basic question is, "whereas CAI is good and convenient in the area of memorizing facts and techniques, are there any aspects of chemical education where computers can help educators? Therefore, we should like to discuss various uses of computers in chemical education as well as CAI. When is the best time to introduce electronic hand calculators in chemical education? How can computer graphics change modern chemical education? What are the ways of chemistry teaching when various information is available from the Information Net Service? Can computers be used in the chemical laboratory? The price of computers is getting cheaper more rapidly than anybody has thought, to make it possible to have them in classrooms and at home. We are sure that education which neglects computers and images will become obsolete in a very near future.

Images are well received by the general public but there is some fear of the computer infringing upon one's privacy. This is because the computer is thought to do everything. One of the ways to cope with this situation is to use computers in chemical education and elsewhere, where its use benefits us most. Usually specialists emphasize that a certain methodology is useful in various aspects of chemical education. But my opinion is that we should treat chemical education as a system. If there is more than one way to teach an item, we should evaluate the methods critically to select the best way of teaching it. Therefore, the use of computers in chemical education is not to be taken lightly but should be placed in the most fitting situation.

In our experience, showing molecular models on a screen seems to be a very promising way for teaching stereochemistry to beginners and even for the advanced students. There are varieties of soft ware which can be used for this purpose. If it is connected to calculations such as molecular mechanics, students will be able to see a model which is close to the reality. Even better is that we can rotate the model on the screen so that the best direction is selected. Molecular dynamics can also be seen on the screen which will enable students to understand the dynamic nature of molecules. Computer graphics is certainly a promising area in chemical education. The point which I should like to emphasize here, however, is that educators should carefully think of the good and the bad of the computer graphics in comparison with the traditional molecular models. Here is one of the research problems within education.

## FOSTERING CHEMISTS

"Chemical education for fostering future chemists of excellence" is an everlasting problem. Usually, chemistry teachers teach their students to raise them as successors in chemistry. They demand to increase the contents of curricula to meet the expanding knowledge of chemistry. But the increase has limitations because the time allocated for chemistry teaching is finite. In addition to this problem, the huge increase in the enrollment ratio in higher education among the age group led to the accusation that the existing curricula in chemistry was too difficult. In Japan, we have a national syllabus. This causes the sentiment that scientists are demanding too much in school education. As a result, the contents of the national syllabus in Japan are shrinking. Is this good for promoting science? By contrast, in countries where no national syllabus exists, I hear that the level of science education is not high enough, compared to the national demand. We believe this is a point where we can help each other by talking of our experiences and exchanging ideas.

The other important aspect in this area is how to educate extremely good students. Chemistry is a science which has a very good balance of induction and deduction. The facts we know are too many, however. How much of the facts and theories should we teach in order to make students innovative? Are we not constraining the ability of students by teaching too many things? Some 30 years ago, the idea of the process of inquiry was introduced in the field of chemical education. We believe this idea is still valid today. If there is a problem in this area, it must be the fact that we need highly able teachers in this approach. Although it is said that 5% of students are excellent and become outstanding when they are grown up irrespective of their education, that saying is undermining our responsibility in education. How could we cope with the fear that teaching is deteriorating the ability of students? There are several points to be considered in this area, including teacher education, inservice teacher training, and the balance between memorization and activities of the students' own initiative. We believe that the problems are not confined to school education but extend to college and even to graduate school education.

### a) ASSESSMENT

Assessment of the achievement by the students has been discussed extensively. However, to our belief, there are many points to be improved in the existing system of assessment. Most of the tests examine the knowledge that students acquired. If students memorize the facts, they may get good grades though they may not understand what the content means. Even though there are various efforts made for improving this situation, the progress in this field is not satisfactory. We may have to come back to the very origin of chemical education. What are the essentials in chemistry? Facts are necessary and so is the ability of understanding or grasping the facts as a whole. The number of facts and theories necessary for being successful may differ according to the needs of a person who may be a politician or a scientist. Are we truly identifying the problem in this area? What is the "chemical sense" which is thought to be so important in chemistry?

So-called performance tests are also studied in various countries and many systems have developed. In these systems, mostly technical skills of students are tested. However, the success or the failure in any of the experimental techniques is not important in many areas. Rather more important is the up-grading of skills in a given period of time. This is especially important in primary schools, since children here can be very different in their relative age in a single grade. In the first grade, a child of 6 years old and another who is almost are placed in a single class. Their stage of development is quite different. The older one may be tired if a teacher sticks to the younger, and if a teacher teaches according to the understanding of the older, the younger may not be able to catch up. Although the cause may be different, the same situation can arise even in other grades or schools. Individualized assessment must be developed in this sense.

In our country, the entrance examination to universities and colleges is of grave importance. The problem here is that an all-round ability is required in our system. If everybody prepares well for an entrance examination, it is hard to distinguish performance in languages, natural science, or social science. The subject which gives a difference is mathematics. Thus it is said that the results of the mathematics tests determine whether the applicant is successful or not. Is this good for selecting the students in chemistry? There is no doubt that mathematics plays an important role in physical chemistry but it is still a debated question how much mathematics is needed in organic chemistry.

b) TEACHERS

The problem of teachers cannot be forgotten if we want to improve the situation of chemical education. Even though excellent curricula may be produced, if the teachers who really contact with children at schools are not aware of the essentials of the curricula, the curricula may not be so effective as they should be. Computers may reduce some parts of the burden of teachers, but the subject of teaching must be taken by the teachers. It is especially so, if one thinks that education is the interaction between teachers and children. We will discuss the problems by classifying them into primary school teachers, high school teachers, retraining, and college teachers.

Education in primary schools is carried out usually by a single teacher for all the subjects. This is believed to be good from the standpoint of the interaction between teachers and children and I believe this viewpoint should be supported. However, it causes problems as well. The most important point from the standpoint of chemical education is that many teachers teach chemistry at primary schools without satisfactory training in chemistry. As a result, teachers tend to say that chemicals are dangerous and to avoid them as far as possible. This is of course not good for chemistry. Sufficient retraining in chemistry is needed. To cope with this problem, specialists in science are stationed in schools. They handle matters much better, but the trouble is they tend to teach beyond the level of primary schools: the latter is not good as well.

We in Japan adopted the way of teacher training developed in the USA after the Second World War. As a consequence, general universities can offer qualifications for teaching in high schools, provided that the students take certain courses in addition to the basic courses in education. It may not be common worldwide, but this caused trouble in our country. The origin is that in Japan, education in a special field is so intense that students are not usually able to attend courses of other disciplines. Consequently, in many universities and colleges, it is common practice that, for qualification, a certain course in chemistry, for example, is read as a course in biology. This is necessary because students get qualifications as a science teacher but not as a chemistry teacher. At lower secondary schools however, a teacher is required to teach what we call the first area which is really a combination of physics and chemistry. A new teacher may even be asked to teach the second area which is the combination of biology and earth sciences. It is no wonder that new teachers get frustrated under these circumstances. When the national syllabus is changed, these teachers are forced to study new contents which they have not learned in universities.

To cope with these problems, teacher retraining is absolutely necessary. It is fortunate for us to have diligent teachers on average. We have established science centers in every prefecture for retraining. They are functioning very well, although they do not give a very extensive, lengthy courses but rather short ones. The short courses are convenient for teachers to participate in as a refresher. In addition, the retraining of science teachers is operated by local authorities which enables teachers to leave schools more easily. Thus they can visit the science centers to attend the retraining courses more often.

However, it will not be fair to mention only the optimistic side of the retraining system. It is said that good teachers often visit science centers. Thus those who have been trained well attend the courses though they may not be required to do so. On the other hand, those who do really need to attend the refresher course do not visit the science center. Our system gives an average attendance ratio among all teachers which can be favorably compared with those in other countries.

Because of these problems, teacher training is one of the big concerns of people who are interested in science education in Japan. The modification of the system of training teachers in Japan is discussed within our Ministry of Education. The basic idea is to introduce internship. This will give candidates ample time for experience in teaching and give them time to think whether they are really suited to the profession. From the standpoint of chemical education, this system is also welcome because it gives time for the candidates to study science, chemistry in particular, which is needed for the teaching of chemistry. Courses leading to the master's degree have been established in various institutions and they are working well. But we cannot expect that every teacher finishes the master course as a requisite: they should be leaders in the field when they got the degree.

The biggest problem in the training of teachers is, however, that of university professors. Indeed, in Japan, university professors are the only people who can teach students without any qualification. For school teachers, our law requires that they should obtain some qualifications by attending courses which are specially designed for the teaching profession. But the university professors are chosen because they publish research papers on chemistry, for example. Their accomplishment in teaching is scarcely counted when one is considered for selection or promotion. Thus university people tend to write research papers on chemistry but not on education. This tendency is extended even to teacher training colleges and universities. In order to promote research in education methodology and in science education itself, we may have to change our basic attitude toward the evaluation of the work in chemical education.

This does mean that our average chemical educators are trained so that they can conduct research in chemistry in some ways but not in a manner that they conduct research in science education. It is easy for everyone to repeat what they have learned from their teachers. Our science teachers are thus tending to teach science rather than what is suitable at the levels of the children: they tend to teach things which are difficult to understand for children. Unless we change this situation in the very near future, I am afraid that the parents' worry that we teach too much in schools cannot be eased and, as a consequence, we may have to admit the general feeling that science is too difficult to teach in low grades in primary schools.

The situation in university chemical education is not good either in Japan. It is said that the level of our average first year students in universities is about one year higher than that in the US, thanks to the severe competition in entering universities. However, after four years, when they finish their courses as undergraduates, the levels of the students both in Japan and in the US become almost equal, and when they finish the course at the graduate school the US students are on an average more advanced than ours. There should be various reasons for this, but one of the big reasons is the less endeavor toward education of Japanese university professors than the US professors. Unless we cope with this problem, our chemistry cannot become of very high quality and highly innovative new chemists cannot be expected.

Of course, there is no royal road in teaching graduate students. Some will prefer to be taught everything to find out his way of thinking. Some will say that they find a better way by being taught only part of the necessary things. Indeed, history teaches us that the originality of young people has often been better than a very careful thought of a senior person. "Which is better" depends on the personality and the level of development of the student. However, this does not mean that the superviser does not need to think of education. Again, if one says that the development of the chemical sense in the graduate school should depend on individual capability, that means we are giving up the effect of education at this stage.

## LIFE SCIENCES

The reason why "chemical education for life sciences" has been chosen as a subtheme of this conference was mentioned above. Due to the effect of modern biology teaching developed in fifties and sixties, chemical substances such as sugars, proteins, and nucleic acids are introduced at very early stages of school education in Japan, although the structures of them are not explicitly shown. We believe this tendency prevails in other countries as well. However, in chemistry, we teach organic chemistry at senior high schools in rather a primitive way: structural theory may not be included in the curricula. Therefore, complex molecules such as nucleic acids are beyond the scope

of organic chemistry in high schools today. Is it not possible to compromise in teaching these substances from both sides?
I recall serious discussions among chemists and biologists when a new department, biophysics and biochemistry, was established in our Faculty of Science, at The University of Tokyo. The discussion focused on whether biochemistry should be taught at undergraduate level: some insisted that biochemistry should form a department which is involved in research and graduate education only. The reason for this opinion is that, at the undergraduate level, the amount of knowledge is not large enough to do research in biochemistry. Consequently, our Department of Biophysics and Biochemistry offers only a limited number of courses to their undergraduate students who choose various courses in other departments like chemistry, physics and biology.

Bioorganic chemistry and bioinorganic chemistry have emerged in recent years and these could fill the gap between chemistry and biology to some extent. Bioorganic chemistry and bioinorganic chemistry are imitating the biological system to reproduce the activity of the latter. Thus they are chemistry. They believe that by enhancing the sophistication they use in the chemistry, they can approach organisms more and more precisely. But the viewpoint of biologists seems to be different: one biologist told me that this kind of approach had failed. Probably biologists were in some haste and chemists pretended that they could imitate organisms very effectively in the very near future. My opinion is that it takes time to satisfy both sides of the people. It is possible that the two sides could come to an agreement and education can be worked out accordingly.

Another point which has connections both with chemistry and biology is photosynthesis. We had a chance to work with biologists to write a text book. When we came to the point of photosynthesis, we felt differently. The biologists insisted that the chemical equation for the photosynthesis should be:

$$6\ CO_2 + 12\ H_2O \longrightarrow C_6H_{12}O_6 + 6\ H_2O + 6\ O_2$$

Their point was that, if we think of the mechanisms of the photosynthesis, we must admit that 12 molecules of water are used and six new water molecules are formed. Thus the above is the correct equation. However, chemists argued that since we do not teach mechanisms of the reaction and we teach that, if common molecules appear on both sides of a chemical equation those should be eliminated, the correct chemical equation is:

$$6\ CO_2 + 6\ H_2O \longrightarrow C_6H_{12}O_6 + 6\ O_2$$

From the educational viewpoint, the opinion of chemists seems to be more acceptable than that of biologists, but probably we should discuss the problem between chemists and biologists.

## INDUSTRY

"Chemical education and industry" is another topic we should not forget. In older days, the result of school education was useful in industries. However, sophistication of industries has brought changes in demands in industry. I recall the first ICCE (1) which was held in Frascati, Italy, in 1969. The theme of the conference was university chemical education. A person from industry suggested that university professors made students dislike industries by teaching too pure theories and too simplified models in universities. They aroused a lot of discussion, but the same person expressed his desire that students should be trained in basic fields so that they do not hesitate to enter into alien fields: necessary techniques can be taught in industry itself. My question is then, "can chemical education in schools and colleges become directly useful in industry?" If the answer is no, what are the roles of vocational schools and professional schools? Since chemical knowledge is needed not only in chemical industry but also other industries as well, it should be worthwhile for these problems to be discussed by chemical educators at this opportunity.

I always feel that more industrial materials should be included in school chemistry. By saying this, I am not insisting that the method of preparation of some metals which are useful in special application should be taught. They may become obsolete in a near future. Rather I am pointing out that the materials we find in our environment are familiar to children. They should be given knowledge at least of the best way of dealing with it. I am thinking of things like plastics. Although plastic and its products are very abundant in our surroundings, I have not heard that the ways of handling it is included in primary school education. Indeed, this way of teaching will make children more familiar with industry and informal talks by teachers about industry, which is not necessarily put in curriculum, will arouse interest among children.

Our industry people often say that more industrial materials should be placed in the school curricula. But it may not be wise because the newest materials will become obsolete in days to come. However, I agree that the newest knowledge often attracts the interests of young people. Therefore, we should not overlook the role of this kind of knowledge in recruiting young people to chemistry or forming a group of sympathizers of chemical industry. This may be best accomplished by publishing subsidiary text books which contain a lot of illustrations and pictures. From our experience, it may be wise to publish books of this sort at two levels, primary school and high school. The latter is readable even for the general public and will play some role in making the public feel that chemical industry is indispensable.

Since chemistry provides materials necessary for the development of various tools and instruments, chemical knowledge is indispensable for those who work in industries other than chemistry. One of the examples was the case of polychlorinated biphenyl. Although this problem became unveiled when we had a tragedy of Kanemi oil disease, it could be wrong in other ways as well. Polychlorinated biphenyl was a very convenient material in the sense that it is heat-stable, it is a fluid in a large range of temperatures, and it does not conduct electricity. But nobody thought of the danger when it was exposed to the environment at that time. Even after the prevalence of the Kanemi oil disease, some thought it was a special case, and the polychlorinated biphenyl which was enclosed in a transformer case was safe: they forgot what would happen when the transformer had to be discarded.

Another example of this sort is the handling of mercury. Mercury is indispensable in dry cells because it gives overvoltage which is good for practical use. However, this fact had not been taught in any school curricula. And suddenly the use of mercury in cells became an issue. I do not think that this is the way to get sympathizers of industry. Workers in industry and the general public should have been informed that mercury is necessary in dry cells because its action makes the cells useful. Fluorescent lamps pose the same problem in discarding them. Yet people are less aware of the fact that they contain more mercury than necessary. Some day I am afraid that the industry will be attacked because of the mercury problem unless we teach people about the nature of the convenient electric tool.

It has been mentioned so far only about the industry concerned with electricity. But the situation is not confined to the electric industries. The automobile industry, which is the biggest machine producer now, is also very closely connected with chemistry, because it uses various chemical products. It may also be necessary for workers in this industry to understand that burning oil in engines produces nitrogen oxides. Even in trade companies, it has become necessary to recruit people who have some knowledge of chemistry, because the salesman may have to respond to the claims which customers raise. Chemistry is now the basic knowledge in all kinds of industry in as far as they treat materials at any stage of their activity.

## ENVIRONMENTAL PROBLEMS

The chemical knowledge remaining in average adults in Japan is very little to our regret, although we teach chemistry in high schools and colleges. This is one of the reasons why we had very severe problems of pollution: our citizens failed to identify correctly what was at the center of the problem. Today the problems seem to be forgotten but have potential of recurrence. It is natural that we teach chemistry in a way that stresses that pollution should be minimized by industry as well as individuals. However, this is putting emphasis on negative sides. If mankind forgets positive attitudes, I am afraid that he shall decline. Chemical education has responsibility to foster sympathizers rather than opponents of chemical industry. Since there is a possibility in a near future that citizens vote pro or con to construction of chemical firms in their locality, this type of chemical education should urgently be explored.

The problem of environmental issues is that they have been discussed, at least in Japan, mainly from the stand-point of biology. The pollution of a river may be determined by the fish which can live in a given area. Thus by examining the species of fish at a certain place of a river, they claim that the river is polluted to a certain extent. The problem here is that this type of assessment is the recognition of an existing condition and does not imply any of the efforts which are trying to improve the situation, both from the producer's side and the citizen's side. Since pollution is caused by foreign materials in a given area, it is a problem of chemistry: chemistry must be involved in discussing pollution or environmental problems.

The point which arises from chemistry is that a minute amount of foreign materials is not avoidable if there is chemical activity at a place. The chemical industry is very concerned with the environmental problem these days and avoids contamination of the surroundings by every possible means. Yet it is possible that they pollute the surroundings to an extent which is not recognizable by the present methods of analysis. This is very close to the situation whereby we contaminate our environment unconsciously by our personal activities. Although we may continue to pollute, we must live. When a nuclear power plant is to be installed, there is always an argument whether the local government should accept the offer of a company. People who are against it argue that we are already suffering from cosmic radiation and any excess of radiation will cause damage to the population. Others who wish to promote the installation may say that, since we are already suffering from the cosmic ray radiation, only a minute amount of increase in radiation does not harm us to a measurable extent, and because our energy resources are running out, a nuclear power plant is a must at present. Although chemistry cannot alter their opinions, at least chemistry can offer them information which is needed for their judgement. At least, it is the role of chemistry and chemical education that they provide people with correct data and a basis on which people can make their decision on the problem. The problem posed to chemical educators is then, what are the suitable ways of introducing these things to the people? And at what age and how?

It may be worthwhile to mention here the recycle of materials. Since Japan is a very poor country as far as the natural resources are concerned, it is a good place to develop the system of teaching the recycling. Collecting cans in the field is carried out on the basis that wastes damage the environment but it should be done from the standpoint of recycling. In this sense, only quoting as cans is not doing any good. If cans of iron and aluminum are mixed, they may not be useful. However, if aluminum cans only are collected as such, it will greatly reduce the use of energy in producing new aluminum lots and hence foil. Similarly, iron cans collected as such can be used in blast furnaces without much trouble and reduce the amount of coke necessary for reduction of iron ore. To be able to do this collection of wastes, the general public has to have some knowledge of chemistry and by doing this, they will feel that they are contributing to the preservation of energy and natural resoures.

At the same time, we must ask people who work in industries that they should make materials easily identified and easily separable from one component to another which is not necessarily made of the same material. I strongly feel this, especially in looking at polymers which are used in various places. Polystyrenes, polyacrylates and polypropylenes may be recycled if they are separately collected from the waste. But this separation is practically impossible at present, because they are not identifyable without special instruments, and composite materials are used. Again, people can feel that they are contributing to the preservation of natural resoures, if they could do so. This point must be worthwhile to consider for the industry people in the future. The materials are made today to satisfy the users for their own purpose, that is the usage, but we may have to consider the case when it is discarded.

## LOCALITY OF EDUCATION

As has been mentioned above, education is implemented according to the needs of societies. That means the education which is best suited in the locality is different from country to country. I recall the situation of Pilot the Project which was held in Bangkok, Thailand, in the early sixties by the sponsorship of UNESCO. There, Professor Strong who was a chairman of CBA (2) and Professor Watton who chaired the curriculum development in Australia headed the project. As a whole the project was successful in as far as it produced some new materials which cost very little and some new films which can be used in classrooms. However, those kinds of material have limitations in that they may not be applicable to some parts of the world. It seemed to me that instead of only one center of such projects, at least several of them should be set up, though they may be small, to meet the needs of the locality.

I must confess that, although the CHEM Study (3) was received with enthusiasm in Japan, and Japanese translations have been published, the CHEM Study as a system has not been accepted. This is of course not suggesting that it was not usable, however. Because of the limitation of time allocated for chemistry in high schools, the text book is too thick in addition to the fact that in Japan, text books of high schools must be authorized by our government after inspection. As a result, many of the experiments developed in the CHEM Study and the ways of introducing a topic are adopted in this country but not the whole of it. I am afraid that this practice may be ruining the whole spirit of the CHEM Study. I must stress that there is no universal text book nor universal teaching materials because of the difference in background of each country.

Education is a human activity to which everybody can express opinions. The needs and levels of chemical education are affected by the background of students: demands of society are different from country to country. Therefore, it may not be possible to find a unique answer to a problem which prevails worldwide. Yet I believe that discussing some aspects in chemical education by scientists from all over the world will give individuals strong impact, and will produce some thoughts which are useful when they go back to their home countries.

REFERENCES

1) Proceedings of the International Symposium on University Chemical Education, Pure & Appl Chem., 22, 3-211 (1970)
2) Chemical Bond Approach Project, "Chemical systems", Webster Division, McGraw Hill, St. Louis (1964)
3) Chemical Education Materials Study, "Chemistry —— an Experimental Science", W. H. Freeman, San Francisco (1960).

# Chemical education in the computer age

Shin-ichi Sasaki

Toyohashi University of Technology, Tempaku, Toyohashi 440, Japan

Abstract - Human beings are now reaching the age when they expect the computer to have creative and associative abilities, through the course of the progress of computers from the first, second, third and fourth generation ones. In these circumstances, it will be very important to tell correctly the functions of the computer to students on chemistry courses. For this purpose, the brilliant features of the computer, the enumeration of all the possibilities for given information, prediction of the future by analyzing or synthesizing the information given and so on, are described referring to the author's results of computer chemistry studies up to date. In addition to the above, that education by CAI becomes a key element of chemical education, and that the CAI should be introduced into classrooms, taking many things into consideration is very important.

## INTRODUCTION

Fifty years have now passed since the advent of a computer in this world. In the early days of its appearance, the computer was said to be a calculating machine for performing a mass of operations rapidly and exactly. Even at present this function is indeed the greatest characteristic of the computer. As is well-known, however, through the course of its progress from the first, second and third generation computers to the present age of the fourth generation ones it has become a tool, in addition to the above-mentioned function, for a data base to retrieve wanted information freely from a mass of memory, and further has become a tool called artificial intelligence which can forecast the future by synthesizing and analyzing various sorts of information and list up all the probabilities corresponding to the given information. Moreover, the advent of the fifth generation computers is now thought possible, and human beings are reaching the age when they can expect the computer to have creative and associative abilities.

Under these circumstances, it is extremely important to teach correctly the functions of the computer to students on chemistry courses. The author would like to describe the brilliant features of the computer, referring to some examples.

## COMPUTER EDUCATION FOR STUDENTS ON CHEMISTRY COURSES

One of the typical data bases is Chemical Abstracts. In 1945 the thickness of the book of Abstracts was less than 30cm, but in 1980 it has reached more than 200cm, and the number of abstracts of papers collected from all over the world is more than 500 thousand. It is quite difficult to find out manually the information from these enormous abstracts. These days, however, the document required (an abstract of the contents, the name of the author and that of the journal, etc.) can be obtained in a moment only by inputting the key words into a computer. It must not be forgotten that these levels of performance have been achieved by many efforts in the integration of the nomenclature of substances, the expression and manipulation of chemical structure within a computer, the establishment of algorithms for various logical operations and the like.

When a chemist majoring in organic synthetic chemistry tried to prepare a new compound in the past, he used to determine the course of synthesis by making the most use of his knowledge on synthetic chemistry as well as to investigate the literature from the past, and to push ahead the study work along the course selected. At present, however, when a chemist intends to synthesize a new substance and inputs the structural formula of it into a computer, the computer will tell him how to obtain the substance by combining one thing with another. This is a result brought about by a gigantic data base of chemical reactions and the computer logic which combines these reactions freely.

The author is preparing the following 3 systems as a tool for facilitating studies in chemistry: the first is a chemical structure determining system; the second is a data base of chemical spectra; and the third is a molecular design system.

## Chemical structure determining system

When a man wants to determine the chemical structure of an unknown substance, he measures the spectra of the substance, combines the substructures presumed from the spectra with each other and constructs the full structure, or he degrades the unknown substance chemically into smaller fragments, determines the structures of these fragments and combines them together. In these cases abilities of association and presumption possessed by human beings are made the best use of and the correct structure is finally constructed in most cases through trial and error. Once a man obtains a plausible answer, however, he becomes extremely lazy in considering successively the further possibilities which may exist. Thus, he makes an unexpected failure, being convinced that the first possibility is the only one. It often happens that a publication of the determination of a chemical structure is later found to be an error and replaced with another publication of a corrected chemical structure.

Now, if we want to make a computer determine chemical structures, we must have a logic suitable for using a computer. A computer can store quite a huge amount of memory. Then, cannot we make it remember all the possible structural formulae? If we can do so, we only need to confirm which formula among the memories of the computer corresponds to the given sample.

The number of organic compounds known to date is said to be about 6 million. If the number of structural formulae is limited to those of the 6 million compounds, we can make a computer memorize them. But these 6 million compounds are substances obtained in the past. If we consider additionally substances to be found in the future and those uncertain to be found but possible to exist chemically, the number of the structural formulae will be an enormous one. For example, for the molecular formula $C_{40}H_{82}$ only, 6200 billion structural formulae are possible (Fig. 1). When only one molecular formula includes 6200 billion structural formulae, all the possible structural formulae can be said to be approximately infinite, which can never be memorized even by a computer.

| n | 6 | 7 | 8 | 9 | 10 |
|---|---|---|---|---|---|
| $C_nH_{2n+2}$* | 5 | 9 | 18 | 35 | 75 |
| $C_nH_{2n}$ | 25 | 56 | 139 | 338 | 852 |
| $C_nH_{2n-2}$ | 75 | 222 | 654 | 1894 | 5495 |
| $C_nH_{2n+2}O$ | 32 | 72 | 171 | 405 | 989 |
| $C_nH_{2n+3}N$ | 39 | 89 | 211 | 507 | 1238 |
| $C_nH_{2n}O$ | 211 | 596 | 1864 | 4745 | 13373 |
| $C_nH_{2n-2}O$ | 745 | 2564 | 8608 | 28162 | 90769 |

* n = 40 62,491,178,805,831

Fig. 1. Number of constitutional isomers.

Therefore, we need to consider the problem in another way, that is, to describe the way of enumerating the structural formulae corresponding to a molecular formula from a chemical viewpoint.

For example, there is the molecular formula $C_6H_6$. The formula $C_6H_6$ means that the single molecule includes 6 C's and 6 H's and that naturally C is tetravalent and H is monovalent. At the same time, as $C_6H_6$ is included in the classification of hydrocarbon, it means that the structural formulae corresponding to $C_6H_6$ are composed of any of the parts of $CH_3$, $CH_2$, CH and C combined in certain numbers (Fig. 2).

Then, as the first step to create the structures corresponding to $C_6H_6$, the calculation is

$C_6H_6$ → How many structural formulas?

6 × C + 6 × H in the molecule

$C^{IV}$ , $H^{I}$

$CH_3$ $CH_2$ CH C

Fig. 2. Meanings of $C_6H_6$.

$C_6H_6$

No. 1 $CH_3$ × 2 C × 4

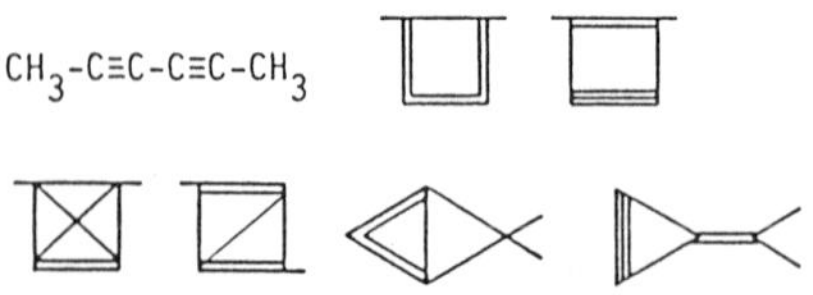

$CH_3$-C≡C-C≡C-$CH_3$

(7)

Fig. 3. Possible structures consistent with $CH_3$×2, C×4.

done to determine which of, and how many of the parts of, $CH_3$, $CH_2$, CH and C can constitute $C_6H_6$. One of the possibilities is as shown in Fig. 3, and from the set of the parts, 7 structures are possibly generated.

Further, the 2nd possibility is as shown in Fig. 4, where 6 structures come from the set of 6 CH's. The sets of the parts corresponding to $C_6H_6$ including Nos. 1 and 2 amounts to 114, and from the all sets up to No. 114 a total of 217 structures can be constructed (Fig. 5).

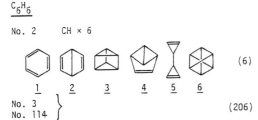

$C_6H_6$

No. 2 CH × 6

1 2 3 4 5 6 (6)

No. 3 }
No. 114 } (206)

The number of structural formulas 217

Fig. 4. Possible structures consistent with CH×6. The number of structures for $C_6H_6$.

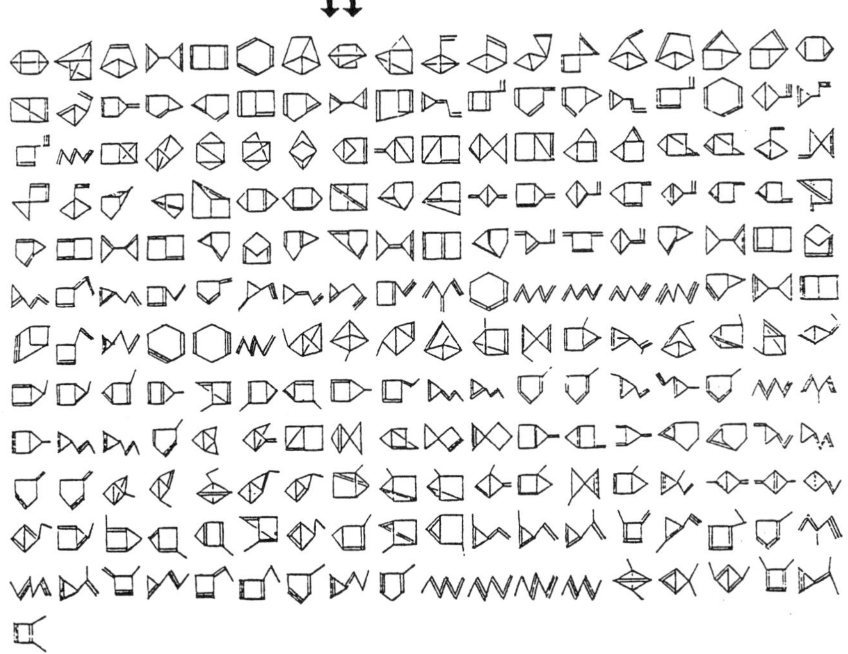


Fig. 5. 217 structures for $C_6H_6$.

The author does not think that all of these 217 structures can exist chemically. Here, it is only said that the total possible structures corresponding to the information of $C_6H_6$ is 217. However, one should not claim that such a structure can never exist from a chemical point of view.

For example, in the case of above No. 2 set (Fig. 4), the computer tells us that when the molecular formula is $C_6H_6$ and it is composed of the CH part only, the possible structures will be 6. Before 1960, that is before Dewar benzene 2 was found, structure 1, namely, benzene only was known, and the majority of people thought that 2, 3, 4, 5 and 6 structures could never exist from a theoretical point of view in chemistry. Now, it is well-known owing to the progress of photo-chemistry that all the structures other than 6 in fact exist.

Referring to the above discussion, let us think about a structure-determining method of unknown substances for computers.

If we know the molecular formula of an unknown substance, it will be known automatically from the molecular formula from what parts the structure of the substance is made up, and all the possible structures of the substance can be generated on the basis of those parts. From the molecular formula only, however, 217 structures are possible for $C_6H_6$ and a huge number of 6200 billion of structures are probable for $C_{40}H_{82}$, and it is indeed almost impossible to pick a structure up among them. Fortunately, an unknown substance has its spectra at peculiar values, which can easily be measured. Therefore, can we not somehow forecast the existence of any parts from these spectra?

On the other hand, let us select from the viewpoint of organic chemistry a number of parts with which any structure may be constructed. In the case of the author, the number of parts is 189 (Table 1, 2), and the authors examined, by referring to the data and literature in the past, where NMR chemical shifts for carbon and hydrogen occur for each part, and prepared a correspondence list. Thus, a correlation table between the 189 parts and NMR chemical shift values attributable to each of them were made and stored in a computer (Table 3).

The parts selected for the structure determination system.

TABLE 1.

| # | substructure | adjacent group | # | substructure | adjacent group |
|---|---|---|---|---|---|
| 1 | TERT-BUTYL- | (O)* | 51 | CH3-(CH): | (A)(P)** |
| 2 | TERT-BUTYL- | (Y) | 52 | CH3-(CH): | (Q)(Q) |
| 3 | TERT-BUTYL- | (K) | 53 | CH3-(CH): | (Q)(T) |
| 4 | TERT-BUTYL- | (D) | 54 | CH3-(CH): | (T)(T) |
| 5 | TERT-BUTYL- | (T) | 55 | CH3-(CH): | (C)(O) |
| 6 | TERT-BUTYL- | (C) | 56 | CH3-(CH): | (C)(A) |
| 7 | GEM-DIMETHYL- | (O) | 57 | CH3-(CH): | (C)(Y) |
| 8 | GEM-DIMETHYL- | (Y) | 58 | CH3-(CH): | (C)(K) |
| 9 | GEM-DIMETHYL- | (K) | 59 | CH3-(CH): | (C)(D) |
| 10 | GEM-DIMETHYL- | (D) | 60 | CH3-(CH): | (C)(T) |
| 11 | GEM-DIMETHYL- | (T) | 61 | CH3-(CH): | (C)(C) |
| 12 | GEM-DIMETHYL- | (C) | 62 | :CH. | (O,O,O) |
| 13 | CH3-(C)- | (O) | 63 | :CH. | (O,O,A) |
| 14 | CH3-(C)- | (Y) | 64 | :CH. | (O,O,P) |
| 15 | CH3-(C)- | (K) | 65 | :CH. | (O,A,A) |
| 16 | CH3-(C)- | (D) | 66 | :CH. | (O,A,P) |
| 17 | CH3-(C)- | (T) | 67 | :CH. | (O,P,P) |
| 18 | CH3-(C)- | (C) | 68 | :CH. | (A,A,A) |
| 19 | ISO-PROPYL | (O) | 69 | :CH. | (A,A,P) |
| 20 | ISO-PROPYL | (A)** | 70 | :CH. | (A,P,P) |
| 21 | ISO-PROPYL | (Y) | 71 | :CH. | (Q,Q,P) |
| 22 | ISO-PROPYL | (K) | 72 | :CH. | (Q,T,T) |
| 23 | ISO-PROPYL | (D) | 73 | :CH. | (T,T,T) |
| 24 | ISO-PROPYL | (T) | 74 | :CH. | (C,O,O) |
| 25 | ISO-PROPYL | (C) | 75 | :CH. | (C,A,O) |
| 26 | CH3O- | (O) | 76 | :CH. | (C,O,P) |
| 27 | CH3O- | (Y) | 77 | :CH. | (C,A,A) |
| 28 | CH3O- | (K) | 78 | :CH. | (C,A,P) |
| 29 | CH3O- | (D) | 79 | :CH. | (C,Q,Q) |
| 30 | CH3O- | (T) | 80 | :CH. | (C,Q,T) |
| 31 | CH3O- | (C) | 81 | :CH. | (C,T,T) |
| 32 | CH3- | (Y) | 82 | :CH. | (C,C,O) |
| 33 | CH3- | (D) | 83 | :CH. | (C,C,A) |
| 34 | CH3- | (T) | 84 | :CH. | (C,C,Y) |
| 35 | CH3CO- | (O) | 85 | :CH. | (C,C,K) |
| 36 | CH3CO- | (Y) | 86 | :CH. | (C,C,D) |
| 37 | CH3CO- | (K) | 87 | :CH. | (C,C,T) |
| 38 | CH3CO- | (D) | 88 | :CH. | (C,C,C) |
| 39 | CH3CO- | (T) | 89 | -CH2- | (O)(O) |
| 40 | CH3CO- | (C) | 90 | -CH2- | (O)(Y) |
| 41 | CH3CH2- | (O) | 91 | -CH2- | (O)(K) |
| 42 | CH3CH2- | (Y) | 92 | -CH2- | (O)(D) |
| 43 | CH3CH2- | (K) | 93 | -CH2- | (O)(T) |
| 44 | CH3CH2- | (D) | 94 | -CH2- | (Y)(Y) |
| 45 | CH3CH2- | (T) | 95 | -CH2- | (Y)(K) |
| 46 | CH3CH2- | (C) | 96 | -CH2- | (Y)(D) |
| 47 | CH3-(CH): | (O)(O) | 97 | -CH2- | (Y)(T) |
| 48 | CH3-(CH): | (O)(A) | 98 | -CH2- | (K)(K) |
| 49 | CH3-(CH): | (O)(P) | 99 | -CH2- | (K)(D) |
| 50 | CH3-(CH): | (A)(A) | 100 | -CH2- | (K)(T) |

TABLE 2.

| # | substructure | adjacent group | # | substructure | adjacent group |
|---|---|---|---|---|---|
| 101 | -CH2- | (D)(D) | 151 | -O- | (K)(T) |
| 102 | -CH2- | (D)(T) | 152 | -O- | (K)(C) |
| 103 | -CH2- | (T)(T) | 153 | -O- | |
| 104 | -CH2- | (C)(O) | 154 | -CO- | (O)(O) |
| 105 | -CH2- | (C)(Y) | 155 | -CO- | (O)(Y) |
| 106 | -CH2- | (C)(K) | 156 | -CO- | (O)(K) |
| 107 | -CH2- | (C)(D) | 157 | -CO- | (O)(D) |
| 108 | -CH2- | (C)(T) | 158 | -CO- | (O)(T) |
| 109 | -CH2- | (C)(C) | 159 | -CO- | (O)(C) |
| 110 | CH2= | (C)(C) | 160 | -CO- | (Y)(Y) |
| 111 | METHYLENEDIOXY | | 161 | -CO- | (K)(Y) |
| 112 | TROPOLONE | | 162 | -CO- | (K)(K) |
| 113 | Y-OH | | 163 | -CO- | (D)(Y) |
| 114 | Y-H | | 164 | -CO- | (D)(K) |
| 115 | CYCLOPROPENONE-H | | 165 | -CO- | (D)(D) |
| 116 | T-H | | 166 | -CO- | (T)(Y) |
| 117 | -CH=<KETENE> | | 167 | -CO- | (T)(K) |
| 118 | -CH=<OLEFIN> | | 168 | -CO- | (T)(D) |
| 119 | -OCHO | (O) | 169 | -CO- | (T)(T) |
| 120 | -OCHO | (Y) | 170 | -CO- | (C)(Y) |
| 121 | -OCHO | (K) | 171 | -CO- | (C)(K) |
| 122 | -OCHO | (D) | 172 | -CO- | (C)(D) |
| 123 | -OCHO | (T) | 173 | -CO- | (C)(T) |
| 124 | -OCHO | (C) | 174 | -CO- | (C)(C) |
| 125 | -OH | (O) | 175 | O=C= | |
| 126 | -OH | (D) | 176 | T | |
| 127 | -OH | (C) | 177 | Y | (O) |
| 128 | COOH | (O) | 178 | Y | (Y) |
| 129 | COOH | (Y) | 179 | Y | (K) |
| 130 | COOH | (K) | 180 | Y | (D) |
| 131 | COOH | (D) | 181 | Y | (T) |
| 132 | COOH | (T) | 182 | Y | (C) |
| 133 | COOH | (C) | 183 | C | (O) |
| 134 | -CHO | (Y) | 184 | C | (Y) |
| 135 | -CHO | (K) | 185 | C | (K) |
| 136 | -CHO | (D) | 186 | C | (D) |
| 137 | -CHO | (T) | 187 | C | (T) |
| 138 | -CHO | (C) | 188 | C | (C) |
| 139 | -CHO | (CH) | 189 | D | |
| 140 | -CHO | (CH2) | | | |
| 141 | CYCLOPROPENONE : | | | | |
| 142 | :C= <KETENE> | | | | |
| 143 | :C= <OLEFIN> | | | | |
| 144 | =C= <KETENE> | | | | |
| 145 | =C= <ALENE> | | | | |
| 146 | FURAN(O) | | | | |
| 147 | -O- | (K)(K) | | | |
| 148 | -O- | (K)(O) | | | |
| 149 | -O- | (K)(Y) | | | |
| 150 | -O- | (K)(D) | | | |

* means the adjacent atom or functional group, they are saturated oxygen (O), aromatic carbon (Y), carbonyl carbon (K) olefinic carbon (O), acetylenic carbon (T) and saturated carbon (C), respectively.

** A:←O-CO-
P: Y K D T
Q: Y K D

| No. | Components | | $^1$H-NMR(PPM) | | $^{13}$C-NMR(PPM) | | | |
|---|---|---|---|---|---|---|---|---|
| 16 | $(CH_3)_2CH-$ | (CS) | 1.20 | 0.50 | 28.9 | 13.4 | 40.7 | 18.2 |
| 132 | $CH_3CO-$ | (O ) | 2.50 | 1.80 | 24.2 | 17.7 | 174.0 | 165.8 |
| 196 | $>CH_2$ | (N ) | 5.60 | 1.10 | 75.5 | 25.7 | | |
| 197 | $>CH_2$ | (O ) | 6.10 | 2.30 | 88.6 | 43.2 | | |
| 198 | $>CH_2$ | (Y ) | 5.40 | 1.70 | 60.4 | 6.6 | | |
| 199 | $>CH_2$ | (CD) | 6.20 | 0.50 | 58.0 | 11.9 | | |
| 208 | -CH< | (N ) | 7.30 | 1.10 | 96.9 | 27.1 | | |
| 209 | -CH< | (O ) | 7.70 | 1.80 | 111.1 | 41.3 | | |
| 211 | -CH< | (CD) | 4.40 | 0.80 | 75.8 | 15.4 | | |
| 274 | -CH= | (N ) | 9.60 | 6.60 | 184.5 | 91.6 | | |
| 277 | -CH= | (CD) | 9.00 | 4.50 | 165.0 | 90.1 | | |

TABLE 3. Correlation table between the parts (called components in the present system) and appearance range of chemical shifts.

Molecular Formula $C_9H_{14}O$

| $^1H$ NMR | Position | Area | $^{13}C$ NMR | Position | Intensity | Multiplicity |
|---|---|---|---|---|---|---|
| 1 | 344.6 | 45 | 1 | 24.4 | 1679 | Q |
| 2 | 342.5 | 40 | 2 | 28.3 | 4549 | Q |
| 3 | 339.6 | 25 | 3 | 33.5 | 895 | S |
| 4 | 128.2 | 239 | 4 | 45.2 | 2380 | T |
| 5 | 124.3 | 228 | 5 | 50.8 | 2119 | T |
| 6 | 114.8 | 342 | 6 | 125.4 | 2494 | D |
| 7 | 60.1 | 712 | 7 | 159.9 | 1084 | S |
| | | | 8 | 199.2 | 861 | S |

Fig. 6. Data of an unknown.

Next, it supposes that there is an unknown substance, its molecular formula is $C_9H_{14}O$ and NMR spectra for it we measured as shown in fig. 6. As the index of hydrogen deficiency for $C_9H_{14}O$ is 3, it can not be an aromatic compound. Accordingly, it is not necessary to take into consideration the parts related to aromatic compounds among the 189 parts. Similarly, as only one oxygen atom is in the molecular formula $C_9H_{14}O$, those parts having 2 or more oxygens such as COOH and -OCHO groups except ethereal, alcoholic and carbonyl groups may be excluded. In this way, only by checking up the molecular formula $C_9H_{14}O$ can the 189 parts be reduced to 108. After that, the 108 parts are compared with the spectra. This procedure is called spectral analysis, in which the parts having chemical shifts corresponding to the NMR spectra of the unknown substance are selected by referring to the above-mentioned correlation table. Through this selection process, although its details are omitted, 29 parts remain finally (Fig. 7).

When the author was first given a sample to determine its structure, he faced the 189 parts. Since the 189 parts were selected so that structure possible to exist in this world may be constructed with them, it can be said that the author faced at that time an infinite number of structures. After that, however, the molecular formula of the unknown sample has been determined, its spectra have been measured and compared with the 189 parts, and at last the number of parts has been reduced to 29 (Fig. 8).

One only needs to make up structures using these 29 parts. The answer should certainly be included in these structures. Then, how to make up the structures? One can according to the same principle as in the case described above, $C_6H_6$. That is, it is necessary at first to

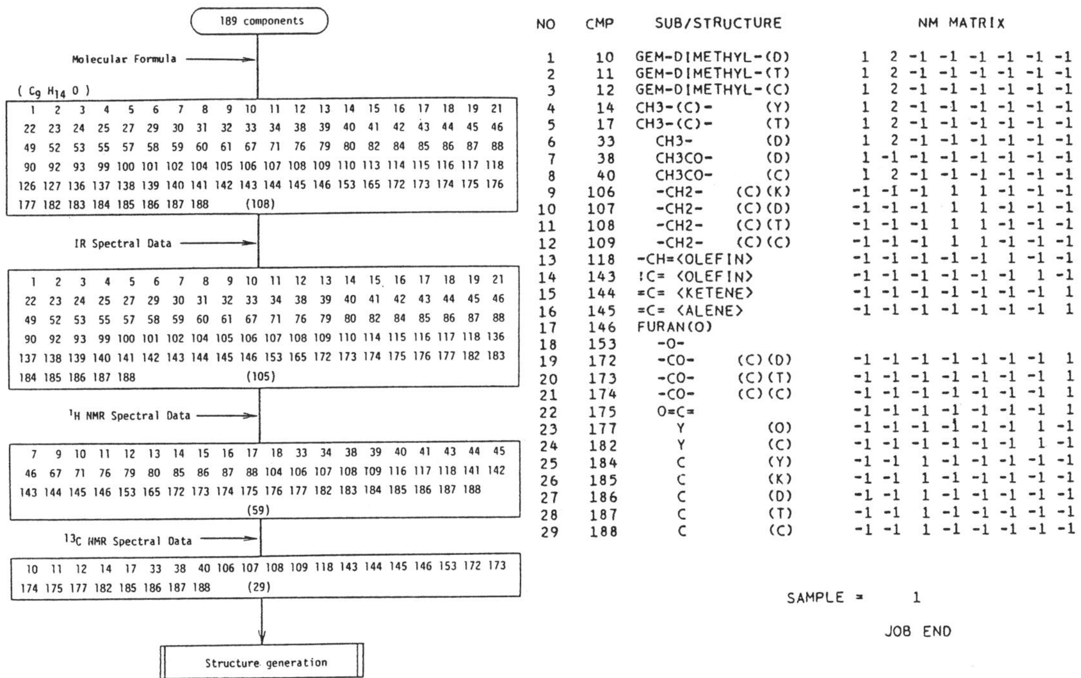


| NO | CMP | SUB/STRUCTURE | | NM MATRIX |
|---|---|---|---|---|
| 1 | 10 | GEM-DIMETHYL-(D) | | 1 2 -1 -1 -1 -1 -1 -1 |
| 2 | 11 | GEM-DIMETHYL-(T) | | 1 2 -1 -1 -1 -1 -1 -1 |
| 3 | 12 | GEM-DIMETHYL-(C) | | 1 2 -1 -1 -1 -1 -1 -1 |
| 4 | 14 | CH3-(C)- | (Y) | 1 2 -1 -1 -1 -1 -1 -1 |
| 5 | 17 | CH3-(C)- | (T) | 1 2 -1 -1 -1 -1 -1 -1 |
| 6 | 33 | CH3- | (D) | 1 2 -1 -1 -1 -1 -1 -1 |
| 7 | 38 | CH3CO- | (D) | 1 -1 -1 -1 -1 -1 -1 -1 |
| 8 | 40 | CH3CO- | (C) | 1 2 -1 -1 -1 -1 -1 -1 |
| 9 | 106 | -CH2- | (C)(K) | -1 -1 -1 1 1 -1 -1 -1 |
| 10 | 107 | -CH2- | (C)(D) | -1 -1 -1 1 1 -1 -1 -1 |
| 11 | 108 | -CH2- | (C)(T) | -1 -1 -1 1 1 -1 -1 -1 |
| 12 | 109 | -CH2- | (C)(C) | -1 -1 -1 1 1 -1 -1 -1 |
| 13 | 118 | -CH=<OLEFIN> | | -1 -1 -1 -1 -1 1 -1 -1 |
| 14 | 143 | !C= <OLEFIN> | | -1 -1 -1 -1 -1 -1 1 -1 |
| 15 | 144 | =C= <KETENE> | | -1 -1 -1 -1 -1 -1 -1 1 |
| 16 | 145 | =C= <ALENE> | | -1 -1 -1 -1 -1 -1 -1 1 |
| 17 | 146 | FURAN(O) | | |
| 18 | 153 | -O- | | |
| 19 | 172 | -CO- | (C)(D) | -1 -1 -1 -1 -1 -1 -1 1 |
| 20 | 173 | -CO- | (C)(T) | -1 -1 -1 -1 -1 -1 -1 1 |
| 21 | 174 | -CO- | (C)(C) | -1 -1 -1 -1 -1 -1 -1 1 |
| 22 | 175 | O=C= | | -1 -1 -1 -1 -1 -1 -1 1 |
| 23 | 177 | Y | (O) | -1 -1 -1 -1 -1 -1 1 -1 |
| 24 | 182 | Y | (C) | -1 -1 -1 -1 -1 -1 1 -1 |
| 25 | 184 | C | (Y) | -1 -1 1 -1 -1 -1 -1 -1 |
| 26 | 185 | C | (K) | -1 -1 1 -1 -1 -1 -1 -1 |
| 27 | 186 | C | (D) | -1 -1 1 -1 -1 -1 -1 -1 |
| 28 | 187 | C | (T) | -1 -1 1 -1 -1 -1 -1 -1 |
| 29 | 188 | C | (C) | -1 -1 1 -1 -1 -1 -1 -1 |

SAMPLE = 1

JOB END

Fig. 7. 189 parts reduce to 29 by input of data of the unknown substance (Fig. 6).

Fig. 8. The members of the surviving 29 parts.

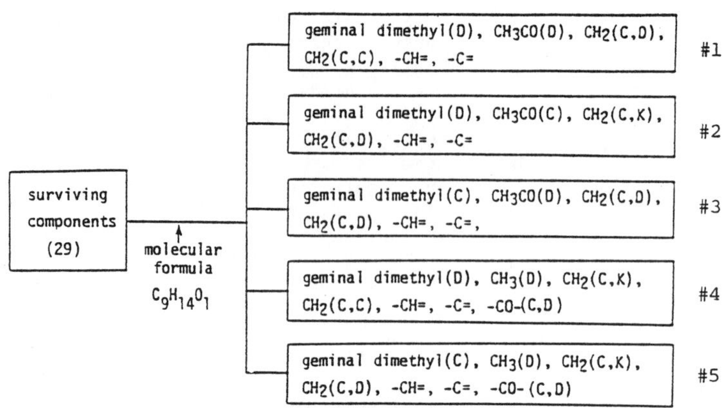


Fig. 9. Sets of parts corresponding to $C_9H_{14}O_1$.

calculate which of the 29 parts, and how many of the selected parts, can constitute $C_9H_{14}O$ in total. In this case, 5 possible sets of the parts are obtained (Fig. 9). Then, structures are to be made up for each set using the parts within it. From Nos. 1, 2, 3, 4 and 5 sets of the parts respectively, 3, 1, 2, 3 and 3 structures, namely 12 structures in total are constructed and outputted (Fig. 10).

Of these 12 structures those strange ones are included may doubt their actual existence, but any of the 12 structures has the same degree of plausibility as far as they are calculated by the computer system by referring to the molecular formula and spectra of the sample. Therefore, one should not conclude at this stage which is good and which is bad among them. It is only possible with the addition of further information to approach the single correct answer.

The author thinks that young students will understand through this example the role of computers in chemical studies. If a computer is given a certain logic, it will enumerate all the possibilities corresponding to the given information. One must consider what information should further be given to the computer in order to select the only one correct answer among these possibilities. This is applied to not only the case of the structure determination but also all the cases of prediction work by computer. Thus, one will be engaged in a higher intellectual production on the basis of the results supplied by the computer. The advent of computers will make human beings increasingly successful in such intellectual activities.

| selected components | set of components #1 | #2 | #3 | #4 | #5 |
|---|---|---|---|---|---|
| 10 | 1 | 1 | 0 | 1 | 0 |
| 12 | 0 | 0 | 1 | 0 | 1 |
| 33 | 0 | 0 | 0 | 1 | 1 |
| 38 | 1 | 0 | 1 | 0 | 0 |
| 40 | 0 | 1 | 0 | 0 | 0 |
| 106 | 0 | 1 | 0 | 1 | 1 |
| 107 | 1 | 1 | 2 | 0 | 1 |
| 109 | 1 | 0 | 0 | 1 | 0 |
| 118 | 1 | 1 | 1 | 1 | 1 |
| 143 | 1 | 1 | 1 | 1 | 1 |
| 172 | 0 | 0 | 0 | 1 | 1 |
| 189 | 1 | 1 | 1 | 1 | 1 |
| generated structures | 3 | 1 | 2 | 3 | 3 |

13

Fig. 10. Structures generated from the five sets of parts.

It has been said in the past that the realization of artificial intelligence (AI) or the expert system may deprive human beings of their jobs. But AI is a tool for human intelligence and the expert system is also a tool for experts. AI and expert system will make human beings more and more active intellectually, by which science and technology will be developed increasingly.

Data base of chemical spectra (2)

A data base of chemical spectra, if it is designed deliberately, can function not only as a tool for identification of substances but also as a tool for chemical education.

The $^{13}$CNMR data base prepared by the author stores about 10 thousand units of data, each of which includes the name of the substance, its structural formula and spectrum, the assignment of each carbon atom to which chemical shifts correspond, measuring conditions of the

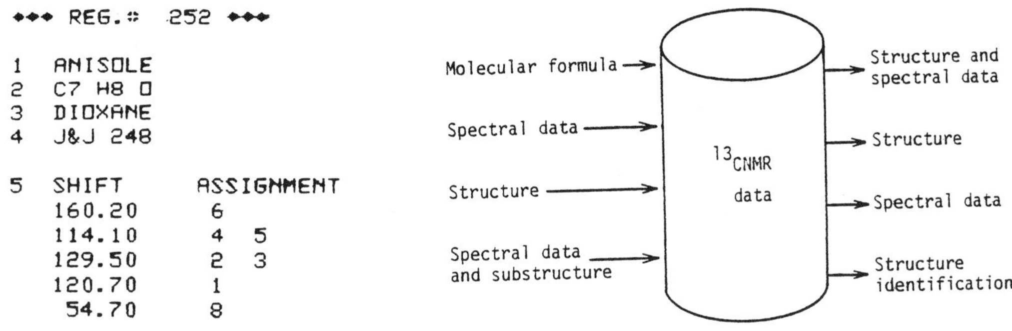


6 MOLECULAR STRUCTURE

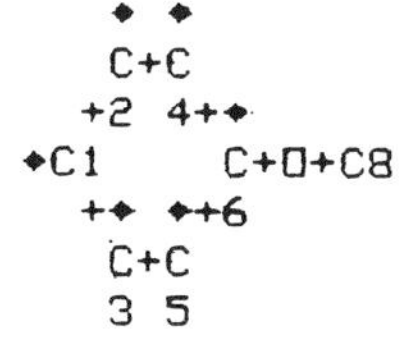


COMMAND ? END
* JOB END *

Fig. 11. Data set in $^{13}$CNMR data base.

Fig. 12. Usual method of use of the data base.

spectrum and literature (Fig. 11).

The usual method of use of the data base is as shown in Fig. 12.
Input the spectrum of an unknown substance and obtain the data set corresponding to the spectrum.
Input a structure and obtain the spectrum corresponding to it, and vice versa.
Input the spectrum and substructures of an unknown substance with each other, and identify the unknown substance.
By employing so-called inverted files, one can also perform the following work.
Determine under what circumstances the carbon is found which shows the chemical shift of 120-130 ppm.

Retrieve carbons having the chemical shift of 120-130 ppm from the data base, collect those found under the same circumstances and represent them in a histogram (Fig. 13). Although the carbon of 120-130 ppm is usually regarded to be unsubstituted aromatic carbon, carbons in terminal methylene, acetylene and some others may appear in the location of 120-130 ppm depending on their circumstances. This fact suggests, as a matter of interest, that a simple thinking of '120-130 ppm = aromatic carbon' should certainly be cautionary. On the contrary, the chemical shifts of certain carbons in a structure are obtained. For example, when t-butyl is connected to an aromatic ring, chemical shifts of the quarternary carbon, and methyl carbon of t-butyl and the aromatic carbon to which t-butyl is directly connected are outputted as shown in Fig. 14.

According to the figure, chemical shifts of the above-mentioned carbons do not differ so greatly depending on what is connected to the remaining parts of the aromatic ring. If the

```
COMMAND ? PSTRUCTURE
  SET # ? 2
   SIZE ?
 HISTOGRAM FOR TARGET CARBON..... Y(ES)? Y

  >C<  ;    2 **
 =CH2  ;    2 **
 =CH-  ;   29 *****************************
  =C<  ;    7 *******
  #C-  ;    3 ***
 =C=O  ;    3 ***
 =AH-  ;  364 **************************************************
              **************************************************
              **************************************************
              **************************************************
              **************************************************
              **************************************************
              **************************************************
              **************
  =A<  ;   31 *******************************
# AND A INDICATE TRIPLE BOND AND AROMATIC CARBON, RESPECTIVELY

STATISTICS FOR ; TOTAL= 441, KIND= 8 FOR 120.00-130.00 PPM
```

Fig. 13. Properties of carbons appearing at 120-130 ppm. #C, AH and A stand for acetylenic, unsubstituted aromatic and substituted aromatic carbons, respectively.

```
COMMAND ? STRUCTURE C3 C Y Y1 Y1 C3 C3 1 2 3 4 5 0 2 6 0 2 7
        ENTRY   3
        SET #   7

COMMAND ? SSPECTRA
  SET # ? 7
VERTEX # ? 1 2 3
  HISTGRAM ONLY [NO] ?

      * REG.#     SHIFT (PPM) FOR VERTEX #  1
         339       29.50
         426       31.40        } methyl carbon
         429       29.70

  STATISTICS FOR ;   3 SHIFTS FOR VERTEX #  1

      * REG.#     SHIFT (PPM) FOR VERTEX #  2
         339       34.40
         426       34.20        } quarternary carbon
         429       34.40

  STATISTICS FOR ;   3 SHIFTS FOR VERTEX #  2

      * REG.#     SHIFT (PPM) FOR VERTEX #  3
         339       136.10       } aromatic carbon
         426       148.00         connected to t-butyl
         429       135.60

  STATISTICS FOR ;   3 SHIFTS FOR VERTEX #  3

 COMMAND ? END
```

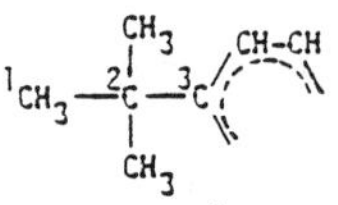

Fig. 14. Chemical shifts of methyl, quarternary and aromatic carbon to which the substituent connects in t-butylbenzenes.

```
COMMAND ? CPEAK 40.20 0.10
        ENTRY  20
        SET #   1

COMMAND ? PSTRUCTURE 1 1 2

              * SUBSTRUCTURE (UP TO ALPHA POS. )

        1 C1= 40.30 PPM(REG.#    2)

            BR
             \
              C1-C--
              / 2
            BR

        2 C1= 40.20 PPM(REG.#   64)

            O          /
             =        /
             C2-C-C3
              / 1  \
             /      \

        3 C1= 40.20 PPM(REG.# 216)
```

Fig. 15. Atom or atomic group at α-position of 40.00 ppm carbon (1)

```
        * SUBSTRUCTURE (UP TO BETA POS.)

  1 C1= 40.30 PPM(REG.#    2)

      BR
       \
        C1-C-BR
        /  2
      BR

  2 C1= 40.20 PPM(REG.#   64)

                O
                =
      O        C5
       \      / \
        C2-C-C3   \
        =  1 !
        O    S

  3 C1= 40.20 PPM(REG.# 216)

      N-C-C-C--
        1 3 4
```

Fig. 16. Atom or atomic group at β-position of 40.00 ppm carbon (1)

difference should be mentioned, it can be observed that the chemical shift of the methyl carbons is affected a little by the way of the connection, but the chemical shift of quarternary carbon is not affected at all.
Further, what surrounds the carbons of 40.00 ppm? One can detect atoms or atomic groups of and positions of these carbons and display them (Fig. 15,16). Carbon of 40.00 ppm is surrounded by two bromines and one carbon, or connected to a carbonyl and one carbon (Fig. 15). Also, the carbon is surrounded by two bromines and C-Br linkage, or connected to COO and CSCO, or one nitrogen and C-C linkage (Fig. 16).

In these ways, the data base can be used as a tool for training students in teaching NMR and also as a means to easily obtain the materials for NMR analysis by simply operating the terminal without resorting to any excessive memorization. Brains of human beings should be used for thinking rather than for simple memorization.

Molecular design system (3)
If speaking only about herbicides and insecticides, 1800 compounds were obtained by synthesis and the like by 1956, among which one compound was commercialized, but 12,000 compounds were to be synthesized by 1977. Thus, it is expected that by 1997 possible compounds obtained by

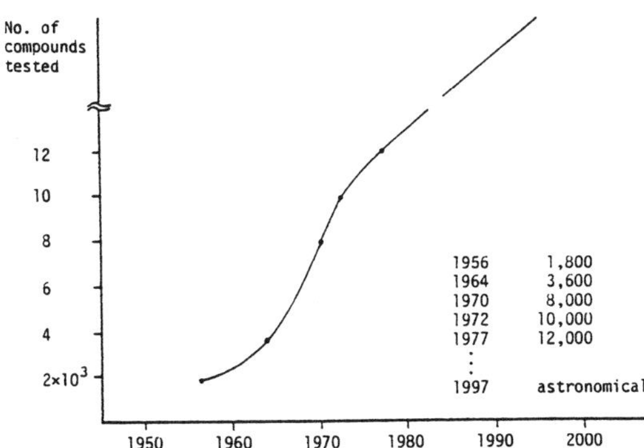


Fig. 17. The number of compounds tested to obtain one product to be commercialized.

synthesis will amount to astronomical figures (Fig. 17).

It is necessary to change this present situation to stop unreasonable syntheses of substances and develop a system which can make computers forecast exactly the relevancy between biological activity and structure of substances. If one can make a computer tell whether a structure is appropriate to show certain biological activity, then one needs only to synthesize a substance having such a structure. Of course, the computer will tell us the procedure of this synthesis and also confirm whether the substance synthesized through the procedure has, in fact, the structure expected.

The system TUTORS being prepared by the author is a system for this purpose. The system includes techniques to determine the correlation between structure and activity such as the techniques of pattern recognition and regression analysis, the method of abstracting the longest common substructures from many chemical structures having the same activity, and moreover the methods of calculating molecular orbital and molecular mechanics (Fig. 18).

Using these techniques complementarily, the system will show up fragments having certain activities and make up a structure by connecting them with each other. The structure obtained should be an ideal type of the structure to show the activity.

If such a so-called molecular design system is completed and the appropriate structures to show certain activities are suggested at A of Fig. 19, we can synthesize the compound having the structure through synthetic route prediction at B. Further, we can confirm whether the substance synthesized through the procedure has in fact the structure expected by structure determination step C. Although this total system for molecular design is not yet completed to the final stage, it will soon come into the world as a tool of providing a biologically active compound, and contributing to the development of science and technology as well as to the welfare of human beings.

How can we in Japan train the talent which can develop such a high level of computer chemistry contributing to welfare of human beings, or those who do not make the system themselves but can understand and use it?

It can hardly be said of Japanese universities that the training of students who major in chemistry for the use of computers has been conducted enthusiastically. Several professors have dared to neglect or despise computers. This is because not all people could keep up

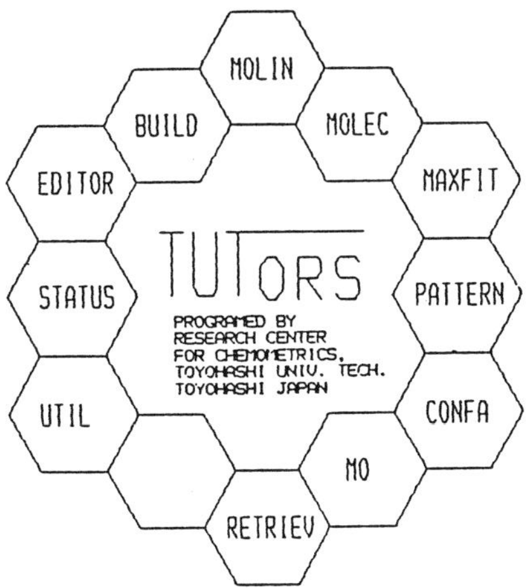


Fig. 18. The concept of molecular design system, TUTORS.

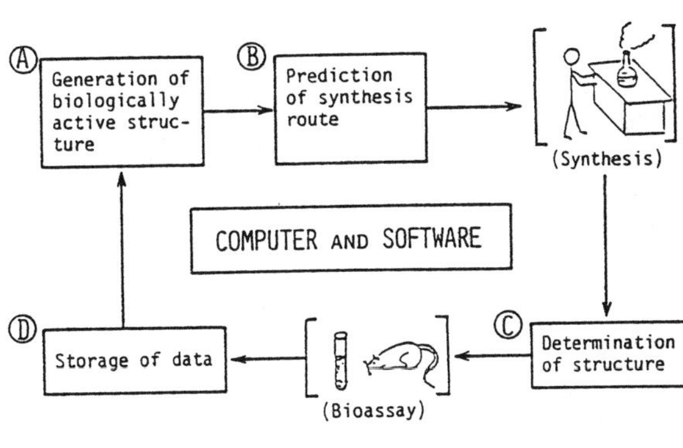


Fig. 19. The total system for molecular design.

Hückel MO calculation

Learning program for configuration of valence electron by electron dot model

Estimation of molecular formula by mass spectrum

Program for analysis of infrared spectrum

Simulation of gas chromatogram

Program for synthetic design of organic compound

Program for preparation and retrieval of reagent data base

Fig. 20.

Detection of asymmetric center and the R/S configuration of molecule

Program for three-dimensional display of molecule

Generation of substituted isomers by Polya method

Calculation of some kinds of topological indices

Generation of Morgan name

Program for multiparameter regression analysis

Program for pattern recognition technique

Fig. 21.

Exercises assigned to students for computer education.

with the rapid development and progress of computers, which might have been, in a sense, a natural consequence.

But this situation seems to have changed greatly with the boom of microcomputers and personal computers, especially among young people. It can be said that the computer 'allergy' of some adults is countered by the interest among young people and has become less problematic.

In computer training, it is necessary not only to teach computer languages, but also to assign various exercises for using computers in chemistry.

Huckel's MO calculation, the technique of pattern recognition, estimation of molecular formula, etc. assigned to students are to be especially important among many exercises (Fig. 20,21). In this university, those students who have learned this so-called computer chemistry as well as general basic chemistry, proceed to the special education courses of inorganic chemistry, respectively, and not many students participate in the studies of computer chemistry itself.

## CHEMICAL EDUCATION BY COMPUTER

Now, the author would like to speak a little of computer aided instruction (CAI) in chemistry. CAI in chemistry and studies on it are especially prevalent in the U.S. Notices offering a program for chemical education are always found in Journal of Chemical Education. PLATO of Illinois University is especially famous as a large-scale system. The system has been transferred to personal computers to use PLATO more economically.

In Japan, the CAI system designed by Shimozawa and his group (4) by selecting Koyamadai High School as the field of experiment, has made a great contribution to improving chemical education in high schools. Yoshimura (5) is also making a study on dry laboratory by personal computers independently of Shimozawa, and has achieved hopeful results. Two figures are picked up from their floppy discs, one (Fig. 22) shows how to handle a balance, the other

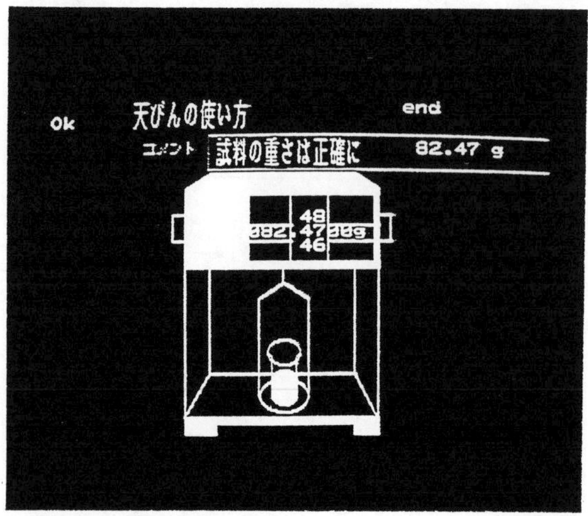


Fig. 22. Simulation of chemical experiment (Yoshimura).

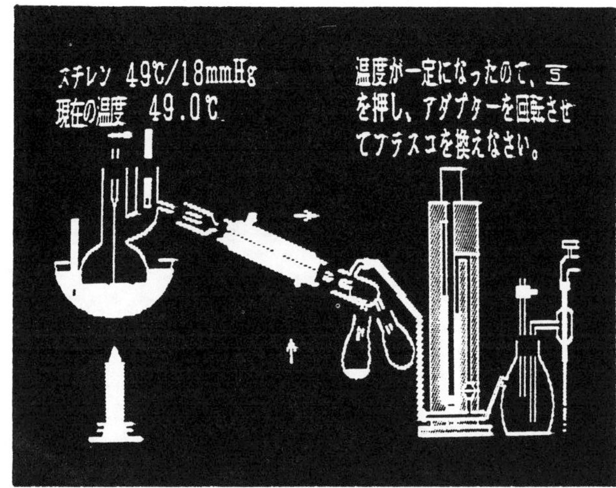

スチレン 49℃/18mmHg
現在の温度 49.0℃
温度が一定になったので、
を押し、アダプターを回転させ
てフラスコを換えなさい。


Fig. 23. Simulation of chemical experiment (Shimozawa, Kuroishi).

(Fig. 23), the procedure of vacuum distillation.

Education by CAI is becoming a key element of chemical education. Education is, however, an extremely high mental activity, and there still are many difficulties even for the application of advanced computer technology. Is it fair to say that for successful examples of simulation by CAI, one can only refer to the computerized system of language laboratory and flight simulation for pilots? At any rate, we have not yet been able, in the field of chemical education, to replace lessons of real teachers by CAI. However, we must be encouraged to use CAI as a complementary means of teaching, taking into consideration its necessity and effectiveness. It will be quite effective as a tool to bring together the knowledge learned in study oneself, as preparation and reviews of one's lessons.

It is rather dangerous, however, to readily introduce a CAI program designed by others. Just such a CAI program which is designed with good coordination between teachers and software makers, and in which an ideal of education is understood, should be operated in classrooms. Of course, the best situation is that the teacher himself is the maker of the software. Essentially, education is not composed only of logic, but ambiguity and unexpectedness are important elements. These elements can only be ensured by real teachers, and should never be neglected by mechanization under the title of CAI.

Finally, it should not be forgotten that chemistry is a science of experimentation. Those who have once learned chemistry should be capable of remembering vividly the smell or other various properties of a substance when told the name of it, and should have an ability to convert easily one substance into another. There is no chemistry without experimentation. It must be kept in mind that CAI is only a complementary tool to aid students in learning chemistry and arouse their interest in chemistry.

## REFERENCES


1. K. Funatsu et al., Automated Structure Elucidation System CHEMICS, Colloquium Spectroscopium International XXIV Garmisch - Partenkirchen, (1985), Fresenius Zeitschrift fur Analytische Chemie (in press).
2. H. Abe et al., $^{13}C$- and $^{1}H$-NMR Spectra Collections as a Base for the Retrieval System SPIRES in Computer Supported Spectroscopic Data Bases , ed. J. Zupan. Ellis Horwood Ltd., Int'l, Publ. (1985) (in press)
3. Y. Takahashi et al., Research on the System for Molecular Design, TUTORS, Proceedings of VIIth International Conference on Computers in Chemical Research and Education Garmisch-~Partenkirchen, (1985) (in press).
4. J.T. Shimozawa, Int'l. News Letter Chem. Edu., No. 19, 5 (1983).
   J.T. Shimozawa, Y. Kuroishi, Chem. Edu. 31, 189-192 (1983); J. Sci. Edu. 8, 131-138 (1984).
5. T. Yoshimura, J. Assoc. Pers. Compt. Chem., 7, 3-6 (1985).

# Steps towards achieving the computer's potential

John W. Moore

Department of Chemistry, Eastern Michigan University, Ypsilanti, MI 48197, USA

Abstract - Recently there has been considerable effort in the U.S. to incorporate computers and computer-related materials into the undergraduate curriculum in chemistry. Project SERAPHIM, sponsored by the National Science Foundation, Science Education Division, exemplifies what is currently being done. By describing SERAPHIM's activities I will give an overview of the steps being taken to include in the curriculum the many ways that chemists use computers.

The computer's ability to calculate rapidly, to manipulate symbols quickly and without error, to simulate other devices and processes, and to present results of its operations in graphical and other readily understood forms have led to the incorporation of computers in all aspects of chemical research, chemical instrumentation and analytical laboratories, and the chemical industry. These same capabilities have also aroused considerable interest on the part of chemistry teachers with regard to applications of computers in chemistry classrooms. The computer is a tool that chemists everywhere use in their daily work and as such is worthy of inclusion in the chemistry curriculum. The computer is also a medium for transmission of information and knowledge and consequently ought to be used to teach chemistry. I believe that computers can and should be used in these two roles throughout the program of studies of all chemistry students; any student who is not being taught about computers and being taught with computers is being shortchanged. Many others throughout the world share such beliefs and are working hard to live up to them.

Examples of the importance currently attached to the inclusion of computers in the chemistry curriculum in the United States are provided by a recent survey done by the American Chemical Society (ACS) and by the 1983 Guidelines for Undergraduate Programs in Chemistry published by the ACS's Committee on Professional Training (CPT). The survey results are summarized in Table 1, where it can be seen that computer training is thought to be nearly as important as the core chemistry curriculum, and more important than training in calculus and statistics. The CPT Guidelines call for the inclusion of computer topics in

TABLE 1. Results of ACS Survey of Members Regarding Undergraduate Chemistry Curriculum

| In current undergraduate curricula in chemistry this area is very important: | | My undergraduate work in this area was very useful: | |
|---|---|---|---|
| ORGANIC CHEMISTRY | 88% | ORGANIC CHEMISTRY | 73% |
| ANALYTICAL CHEMISTRY | 82% | ANALYTICAL CHEMISTRY | 69% |
| PHYSICAL CHEMISTRY | 74% | PHYSICAL CHEMISTRY | 54% |
| INORGANIC CHEMISTRY | 73% | INORGANIC CHEMISTRY | 54% |
| WRITING AND SPEAKING | 73% | WRITING AND SPEAKING | 52% |
| COMPUTER SCIENCE | 70% | COMPUTER SCIENCE | 49%[a] |
| CALCULUS/STATISTICS | 62% | CALCULUS/STATISTICS | 45% |

[a]Percentage of those who did undergraduate work in computer science.

all chemistry courses, both lecture and laboratory, and give specific suggestions about ways that computers can become part of the curriculum. Some excerpts from the guidelines indicate the flavor of what is proposed: "Principal changes from the previous (1977) edition ... are increased emphases on computer literacy, information retrieval, self-instruction programs, and basic inorganic chemistry." "Laboratory work should give students the ... competence to: use and understand modern instruments, particularly ... laboratory computers [and to] analyze data statistically and assess the reliability of results. Based on these recommendations, both from chemists themselves and from the American Chemical Society, there is considerable effort currently in the United States to include computers effectively in the curriculum.

The activities of Project SERAPHIM (sponsored by the Science Education Division of the U.S. National Science Foundation) exemplify what is currently being done to involve computer-based activities in the chemistry curriculum more effectively. SERAPHIM is:

- Collecting and disseminating information and materials regarding computer-based chemistry
- Training teachers to use computers effectively in chemistry classes
- Maintaining communication and professional relations among teachers via a national computer network and conference, CHYMNET
- Preparing computer interfaces and laboratory exercises that allow teachers to use computers easily, effectively, and inexpensively in the laboratory
- Considering the question "what do students need to learn about computers" and preparing materials that will aid teachers in teaching about chemical applications of computers
- Designing and creating new computer-based materials that:
  - Bring simulated industrial experiences into the classroom
  - Relate chemical science to environmental and other studies
  - Allow students to explore chemistry and develop creative approaches to chemical problems

In all of these activities we have involved high school and college chemistry teachers directly in designing, evaluating, and using our materials, and we have sought advice from chemists in all areas: teaching, research, government, and industry. I will describe Project SERAPHIM's efforts in some detail as examples of what we think needs to be done in order to improve chemistry students' training and knowledge through the use of computers.

Project SERAPHIM has collected more than 300 computer programs for Apple II, IBM PC, Commodore, Atari, and Radio Shack microcomputers, and we have distributed to chemistry teachers throughout the U.S. and the world, more than 12,000 diskettes containing software. Programs are distributed first of all to give other persons ideas about how computers can be used effectively; in addition, the programs provide an inexpensive way for teachers to get started using computers in chemistry classes. We also distribute written material of several kinds: tutorials for program authors that will help them to write more effective programs; reviews and evaluations that aim to improve the quality of our programs and of commercially available software; descriptions of how computers can be used effectively in chemistry classes; and information about sources of computer materials. In the last category we publish a Software List that includes more than 500 different computer programs that can be used for chemistry instruction and are available from commercial and other suppliers. Thus as a first step we have collected and can distribute to teachers a wide variety of computer programs and other materials that will help the teachers to incorporate computers into their classes. Table 2 summarizes the clearinghouse/distribution activities of the project.

TABLE 2. Clearinghouse/Distribution Activities of Project SERAPHIM

| SOFTWARE | | | WRITTEN MATERIALS | | |
|---|---|---|---|---|---|
| Computer | Disks | # Distrib. | Type | Number | # Distrib. |
| Apple | 26 | 11 400 | Information | 13 | 6 800 |
| Atari | 1 | 11 | Author | 18 | 1 970 |
| Commodore | 7 | 150 | Review | 13 | 240 |
| IBM PC | 3 | 540 | Text | 5 | 620 |
| Radio Shack | 5 | 160 | Laboratory[a] | 10 | 300 |

[a]Distribution began in July 1985.

To aid in distribution of materials we have involved many Regional Distribution Centers, both within the United States and abroad. Each of these centers provides computer disks to teachers in its local area, keeps records of how many disks have been distributed and to whom, and reports once a year to Project Headquarters. Some centers are doing even more: translating programs, preparing disks for distribution, holding workshops in their areas, and so on. Currently there are International Distribution Centers at the following locations: North York, Ontario, CANADA; Nice, FRANCE; Berlin, GERMANY; Schwerte, GERMANY; Italian Chemical Society, Bologna, ITALY; Nairobi, KENYA; Johannesburg, SOUTH AFRICA; Ljubljana, YUGOSLAVIA. Certainly Project SERAPHIM would be interested in cooperating and collaborating with any other countries or projects that are concerned with the problems of incorporating computers into the curriculum.

To make teachers aware of our materials and to train teachers to use them we have developed a Workshop Leader Training Program. During the summer of 1984, 26 high school chemistry teachers were trained as workshop leaders. Ten more have been trained this summer. These teachers designed and produced an introductory workshop whose purpose is to bring other teachers to the level where they know enough to begin to use computers in their chemistry classes. The SERAPHIM Workshop Leaders prepared a 50-page booklet that is given to each workshop attendee and contains simple information about how to arrange a classroom or laboratory to use computers effectively, how to evaluate computer programs to decide whether they will enhance chemistry instruction, and where materials are available. Each workshop provides a hands-on session during which teachers (many of whom may not have used computers before) can sit down at a computer and evaluate both commercial software and Project SERAPHIM software. Each workshop leader has agreed to organize and operate at least four workshops during the period 1984-1986; some have already done three and have many more planned. By this method we expect to be able to reach nearly every chemistry teacher in the country and introduce him or her to computers within the next two years. The map in Fig. 1 indicates where Regional Distribution Centers and SERAPHIM Workshop Leaders are located within the United States.

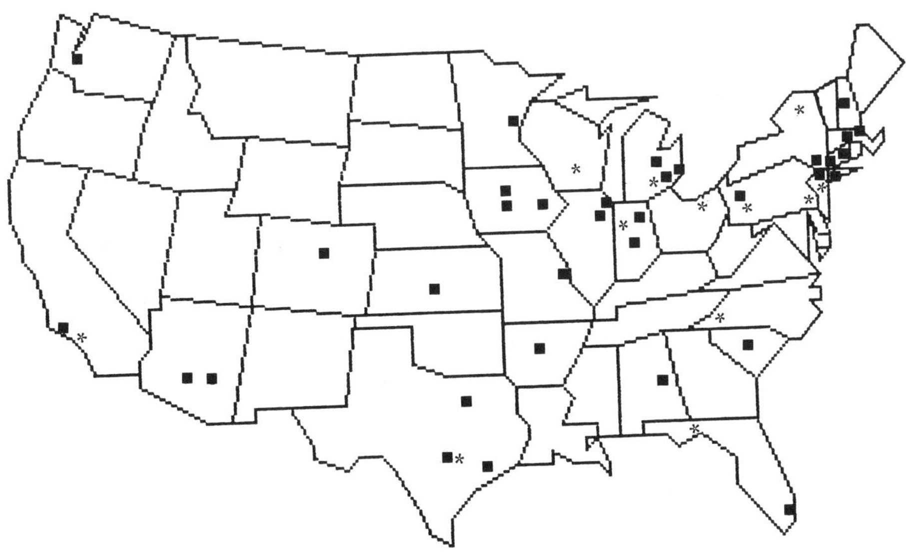

Fig. 1. Location of SERAPHIM Workshop Leaders (■) and SERAPHIM Regional Distribution Centers (*) within the United States.

Since our workshop leaders are widely separated geographically, we have instituted a novel and effective way of maintaining communication among them: a computer network and conference that we call CHYMNET. Fig. 2 indicates schematically how the computer network and conference work. Each workshop leader has been given a modem to use with his or her (or a school's) microcomputer. The teacher can call a local (no charge) telephone number to reach a nationwide network, called TELENET, from which a large computer at the University of Michigan can be reached. The mainframe computer runs a program called CONFER that allows each workshop leader to transmit or receive a <u>message</u> to or from any other workshop leader, enter an <u>item</u> that can be read and responded to by all other workshop leaders, or post a <u>bulletin</u> that everyone will see the next time they use the computer conference. Messages are like private letters--they are only seen by the person or persons to whom they are sent. Bulletins are public messages, like those posted on a bulletin board or kiosk; they are intended to be seen once and reacted to immediately. Items are like bul-

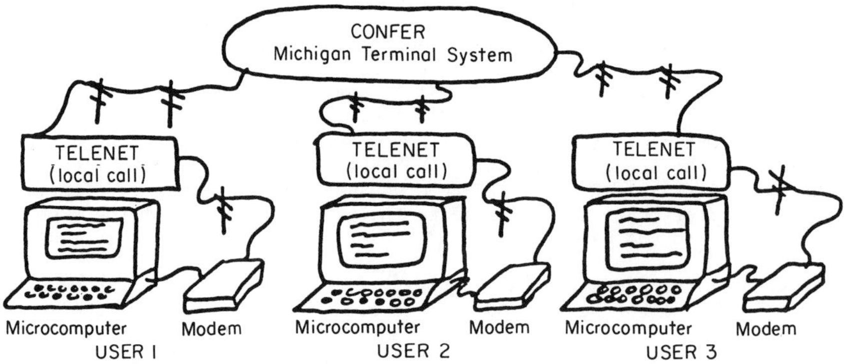


Fig. 2. Schematic diagram of CHYMNET computer network and conference.

letins, but persons who read them can submit <u>responses</u>. A response is attached to the item, where it can be seen by other persons in the conference. The other conferees can in turn respond to the responses they read, or to the original item. In this way a continuing discussion can take place on the subject of the item. Since a computer handles the communication it is not necessary for everyone to be using the conference at the same time--whenever a workshop leader signs on he or she can get all the new messages, items, responses, and bulletins that all the other conferees have sent. This means that high school chemistry teachers can communicate effectively even though their daytime hours are very busy. Most of our workshop leaders communicate during evenings or weekends, which also corresponds to times when network charges are lowest. Communication via CHYMNET allows teachers all across the country to exchange information about how to organize workshops better, how to teach chemistry better, new material that becomes available from publishers or others, curriculum improvement in chemistry, and many other subjects. Teachers who formerly felt isolated are now able to communicate about their problems and successes, and about the subject of chemistry. They have been brought more effectively into the profession of chemistry by CHYMNET.

Project SERAPHIM is also working hard to prepare materials that will aid teachers in using computers in chemical laboratories. An optional portion of our introductory workshops involves simple experiments that can be done using very inexpensive equipment interfaced to the game ports of Apple and other microcomputers. This summer we have developed a more advanced interfacing workshop for secondary school chemistry teachers in which three items

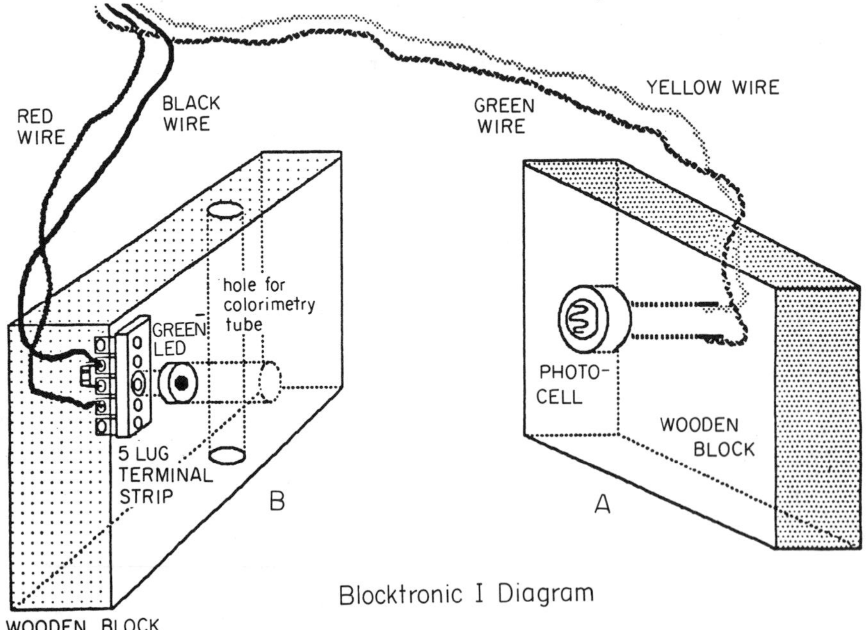


Fig. 3. Schematic diagram of the photometer (Blocktronic I) constructed from wood blocks, light-emitting diode, and CdS photocell; this device can be connected directly to microcomputer

of equipment will be constructed. These are: (1) a thermistor-based temperature sensor, (2) a photometer based on a light-emitting diode and CdS cell, and (3) an adapter box that allows the thermistor or photometer to be connected to the game port without opening the computer to gain access to the electronics. A schematic diagram of the photometer is shown in Fig. 3. During the SERAPHIM Fellows Malcolm Rasmussen (Australian National University), Patricia Barker (Minisink Valley High School, New York), and Kenneth Hartman (Ames Senior High School, Iowa) have developed software to read and analyze data from each device. They have also designed several experiments that are easily done in the laboratory and make effective use of the powers of the computer. the cost of each interface (exclusive of the computer) is less than $10, and parts for constructing the devices are readily available at electronics stores (such as Radio Shack) throughout the United States. The low cost of these devices is possible because they make use of an analogue/digital converter that is built into the computer--the computer's game port. Therefore an expensive analogue/digital converter need not be purchased. Table 3 summarizes the laboratory materials we currently have available.

TABLE 3. Equipment and experiments developed for laboratory computing

| EQUIPMENT: | | EXPERIMENTS: | |
|---|---|---|---|
| | Thermistor | | Thermistor Calibration |
| | Colorimeter | | Heat of Reaction |
| | Game-port adapter | | Photochromic Kinetics |

The three experiments that have been developed so far are intended as examples of the advantages of using a computer to collect and analyze data. The first, Calibration of a Thermistor, is quite simple, involving a comparison of thermistor readings collected by a computer with temperature readings from a mercury thermometer. However, in the process of calibrating the thermistor, students learn several things: (1) in quantitative work the instruments and apparatus we use should be calibrated; (2) there are limits on the range of readings that can be taken--due to a limit on the resistance readings of the computer game port and the particular thermistor chosen, temperatures below about $8^{\circ}$ C cannot be measured; and (3) instruments may be more sensitive to an experimental variable in one part of their range and less sensitive in another--due to the characteristics of the computer/thermistor combination a one degree rise in temperature causes a much larger change in computer reading at $10^{\circ}$ C than at $80^{\circ}$ C. We believe that these lessons are essential for students who will often use computer/instrument combinations in their professional work.

The Heat of Reaction experiment involves measuring the heat of neutralization of acid and base, but with an added feature that would not be possible without the computer's aid. Because the computer can measure temperatures rapidly, it is possible to carry out the reaction as a titration instead of a batch process. (In the usual experiment stoichiometric quantities of acid and base are mixed in a calorimeter and the heat of reaction is calculated from the temperature rise.) By adding 10%, 20%, 30%, etc. of the stoichiometric quantity of acid to base and measuring the temperature increase upon each addition, and then adding 10%, 20%, 30%, etc. of the stoichiometric quantity of base to acid and similarly measuring temperature increases, it is possible to collect in about 20 minutes the data needed for a Job's Method determination of the stoichiometry of the reaction as well as to determine the heat of reaction. The computer handles collection of the data and prompting of the students, so that the experiment is not difficult to perform and can be completed in less than an hour if only one or two heats of reaction are to be measured. If time permits, it is possible to study diprotic and triprotic acids and to learn more about the details of Job's Plots for such species. (See ref. 1 for details of the type of information that can be obtained from this type of experiment.)

The Photochromic Kinetics experiment was originally suggested by Dr. James George of DePauw University. In it, mercury(II) dithizonate is exposed to bright light and undergoes a change in color from pale orange to blue. When the light source is removed the blue material decays back to orange and the process is first order. However, the process is rapid enough that it is almost impossible for students to collect the concentration versus time data needed to determine the rate constant. Since a computer interfaced to a Blocktronic I colorimeter can collect data many times per second it becomes possible to measure concentration/time data over a range of temperatures above and below room temperature. The computer can analyze the data to determine rate constants and to obtain activation parameters, yet the experiment itself, which involves exposing a sample to light and then placing it into the colorimeter, is quite simple. Thus the computer has expanded the range of data that can be collected and made it possible to incorporate into the curriculum a simple experiment in which kinetic studies are done by collecting concentration/time data--the way most research on kinetics is carried out.

We are also developing written and computer materials to show teachers and students how to use general purpose software (spreadsheets, equation solvers, graphical and statistical analysis programs, symbolic mathematics programs, databases, etc.) for analyzing laboratory and other data. For example, one of our SERAPHIM Fellows is working on spreadsheet templates that facilitate computation of nutritional value of various meals. This is designed to accompany a written unit on food and nutrition that is being developed by the American Chemical Society's Chemistry in the Community (ChemCom) project (also sponsored by the Science Education Division of the National Science Foundation). Our aim here is to make it easier for teachers to show students how to use programs that the students will surely encounter after graduation and will use to carry out much of their day-to-day work.

Also in conjunction with the American Chemical Society's ChemCom project we have developed new computer programs to accompany written lessons. These are aimed at forcing students into decision-making situations, often with respect to science/society questions. The titles of programs produced for this purpose are listed in Table 4. Two examples should

TABLE 4. Computer-based Decision-Making Exercises for Scientific/-Technical/Social Decisions

| Program | Author | Program | Author |
|---|---|---|---|
| DALTON | Bendall | REFINERY | Susskind |
| SULFURIC ACID | Dalton/Newman | LAKE STUDY | Whisnant |
| WAQUAL | Estell | BCTC | Whisnant |
| DESIGN-A-DRUG | Meisenheimer | MINERAL RESOURCES | Whisnant |
| RUTHERFORD | Rittenhouse | | |
| OCTANE | Susskind | SEPARATIONS | Whisnant |

give an idea of our goals. To accompany written materials on energy resources we have developed a tutorial and simulation of the processes involved in refining of petroleum. The student first learns about the processes and then is allowed to operate an oil refinery. The latter requires decisions about the purchase of crude oil from several sources, decisions about how much to refine, and decisions about what new equipment to install in the refinery. More than one student can use the program at the same time, and competition with others sharpens students' perceptions of the problems encountered every day in the chemical industry. Figure 4 shows the screen display at the beginning of the tutorial

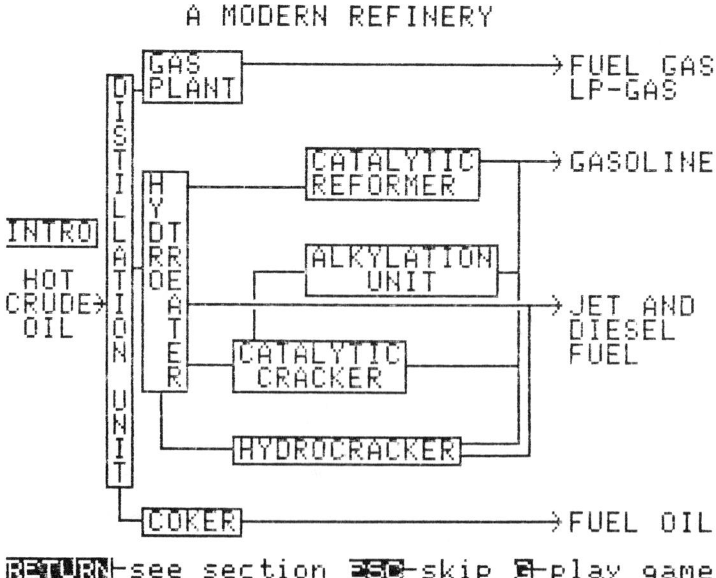


Fig. 4. Schematic of the refinery processes that also serves as the main menu for REFINERY.

portion of the program. In addition to indicating the relationships among the different processes that would go on in a refinery this screen serves as a menu from which a student can select which process he or she wishes to learn about. Once a process has been selected, students receive a tutorial that usually includes animation of the process involved; for example, in Fig. 5 is shown a screen display from the tutorial on distillation. On the actual computer screen the vapors bubbling up through the distillation tower actually

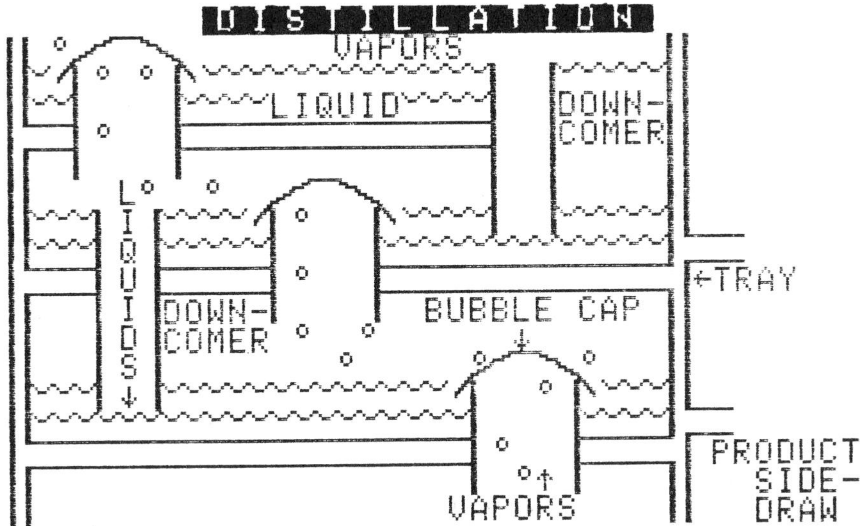


Fig. 5. Screen display from the distillation tutorial--program REFINERY.

move upward and liquid flows back down the tower through the downcomers. This animation accompanies a discussion of separation of petroleum by distillation. Student reactions to this program have been very enthusiastic, and those who have completed the simulation by running their refineries successfully through the year 2020 or so, indicate that they have learned a great deal about the processes and problems of a modern refinery.

A second decision-making simulation involves a study of fish dying in a lake. Students can collect and analyze water samples, observe symptoms of the dying fish, obtain information about toxicity of various substances from a simulated library, consult a simulated colleague, and design controlled experiments to resolve questions about the cause of the fish mortality. All this can be done in a few hours' time and without having to develop laboratory skills (which can happen later if students are really science oriented), but students are introduced to the scientific method in a realistic way and they also learn a good deal about the complexity of environmental problems. Two example screen displays are shown in Fig. 6. Each represents an instrument being used for a simulated analysis of lake water. The program tells students a little bit about the operation of the instrument, but it does not attempt to explain the principles on which the instrument is based. Students obtain digital read-outs of concentrations, and these can then be compared to information obtained from books and journals in the simulated library. An important feature of the program is that different students obtain somewhat different results and find different references in the library. This means that initially different students may come to different conclusions regarding the source of the problem--only after some scientific discussions have gone on and controlled experiments have been done in simulated laboratory fish tanks can everyone agree on the nature of the problem. this simulates the differences that often occur among scientists before enough data have been obtained to fully define a problem, and it may help non-scientists to understand how it can happen that scientists have honest differences of opinion regarding complicated environmental problems.

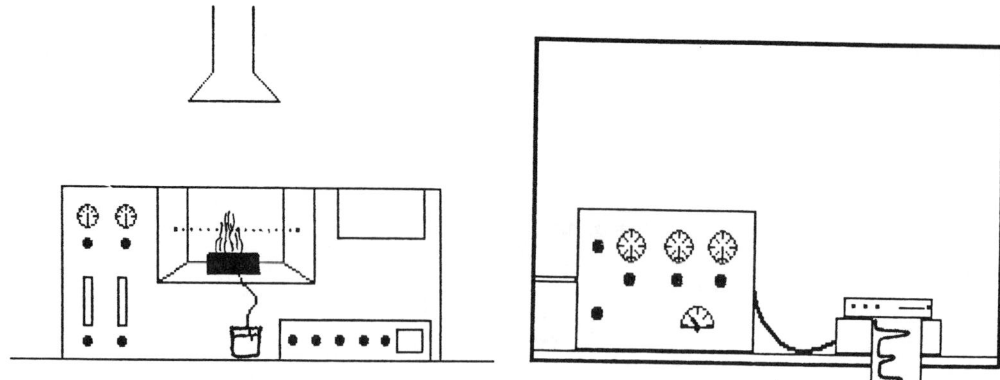

Fig. 6. Two screen displays from Lake Study: a. Analyzing for metals by atomic absorption spectrophotometry; b. Analyzing for pesticides by gas-liquid chromatography.

In order to obtain ideas and provide direction for our efforts, Project SERAPHIM has sponsored and organized numerous workshops, symposia, and meetings to attempt to chart the future of computer applications in chemistry education. One of these meetings, Powwow I, was held in May 1984, and the ideas generated at that time have been developed much further by now. In May 1985 we held Powwow II, at which five concepts were developed in detail by groups of experts convened specifically for that purpose. Two of the ideas developed at these meetings will be described here to indicate the kinds of new approaches we are working on.

The first idea involves using a computer to simulate an instrument. Instrument simulations are currently being developed for NMR, IR, and HPLC, and many others have been proposed as being useful at various points in the curriculum. The first of these was programmed by Paul Schatz (University of Wisconsin) as pre-laboratory preparation for students who needed to run NMR spectra as part of their organic laboratory work. The principal screen display from Schatz's program (which is available from a commercial source, not from Project SERAPHIM) is shown in Fig. 7. The computer behaves, insofar as possible, exactly as the real instrument would. By using the computer simulation to learn how the

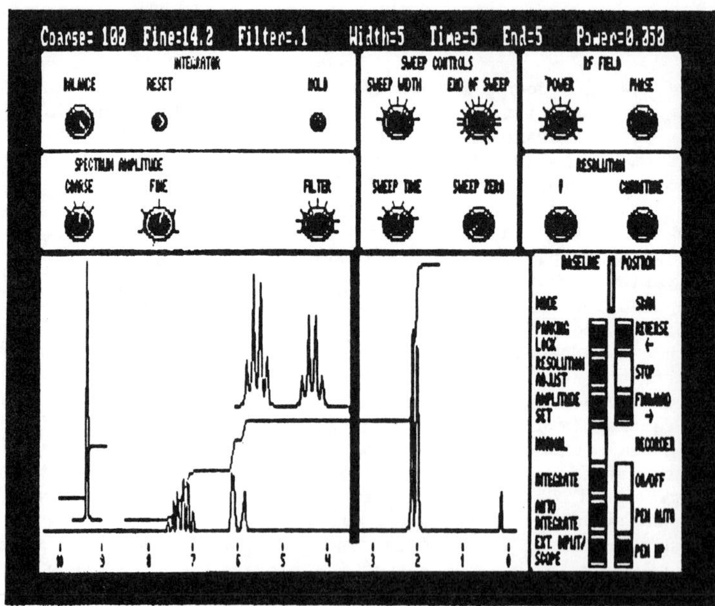


Fig. 7. Screen display from NMR Simulator, by Paul Schatz, University of Wisconsin.

controls of the real instrument work students are able to use the instrument more effectively and to run their samples more rapidly. More recently a high school chemistry teacher has submitted an IR simulator to Project SERAPHIM, and SERAPHIM Summer Fellow Robert Rittenhouse has produced an HPLC simulator. A screen display from the latter program is shown in Fig. 8. Since an HPLC consists of several modular parts, the screen will allow

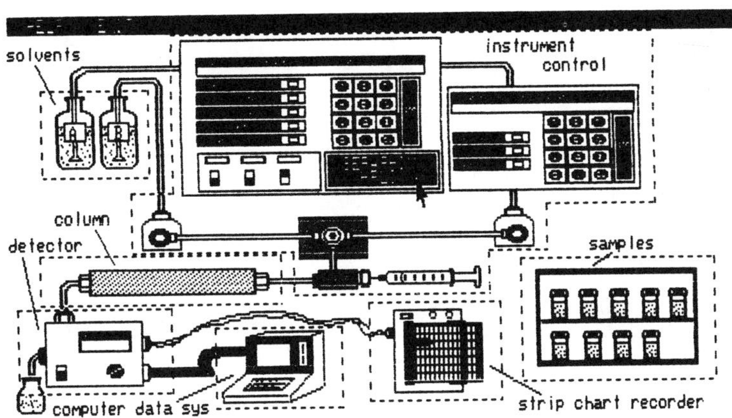


Fig. 8. Screen display from HPLC Simulator; selecting an instrument module (within dotted lines on screen) allows adjustment of parameters controlled by that module.

students to select each part of the instrument and make necessary adjustments. When the program is complete such selections will be made using a mouse or other pointing device. Underneath this easy-to-use exterior lies a computational kernel that can accurately simulate the chromatograms that would be obtained from samples containing a variety of substances individually or in mixtures. The chromatograms vary as they would in real life when eluting solvent composition changes, when elution time changes, when sample size changes, and so on. Thus the computer behaves very much the same as a real HPLC would.

Other advantages of instrument simulations are that the same computer can simulate more than one instrument, that a much greater range of samples and instrument settings can be explored, that problems of down-time and instrument repair are much less troublesome, that all students may learn the same instrument at the same time, even though only a single instrument is available, and that simulations may often be used to design experiments before these are tried on the real instrument. In some cases, such as secondary schools and smaller colleges, an instrument simulation can make it possible for students to learn about equipment that is far too expensive to be purchased for actual laboratory use. Thus instrument simulations will be able to broaden greatly the experience of many students with respect to modern ways of doing chemistry.

As another example of Project SERAPHIM's pioneering in development of new types of computer programs I will describe KC? Discoverer, a program that we hope will allow more effectively than by non-computer, methods for the development of creativity on the part of students. KC? Discoverer is designed to allow students to explore the properties of the chemical elements and make discoveries about relationships among those properties. The program currently includes a database of about 25 properties for each of the 106 elements together with means for sorting, listing, finding, and graphing the data. Table 5 summarizes the data and processes included in the program. Several examples will illustrate how KC? Discoverer might be used by students or teachers.

TABLE 5. Processes and data available in KC? Discoverer.

| | | | |
|---|---|---|---|
| PROCESSES: | FIND elements | DATA: | Physical properties |
| | LIST properties | | Chemical properties |
| | SORT numeric data | | Atomic properties |
| | GRAPH numeric data | | Practical information |
| | Select periodic group | | |
| | Advise student | | |

First, suppose we want to find all the elements that are gases at room temperature and normal atmospheric pressure. KC? Discoverer includes in its data the state of matter for each element, and so we can FIND all elements whose state is "g" for gas. Fig. 9 indicates how this would be done: the user chooses FIND from the main menu. This produces the

```
     F -- Find all elements that have a particular property

     G -- Graph one property against another

     L -- List on the screen (and optionally on a printer) values of
          several properties

     S -- Sort the elements in order of increasing value of a property

     T -- use the periodic table to sort and graph

     ? -- obtain help in using the program


Find, Graph, List, Sort, Table or Help : (F/G/L/S/T/?)
```

Fig. 9. Main menu for KC? Discoverer program.

```
     Property that can be searched :

     1. name                                  Reaction with :
     2. symbol                                  6. air
     3. structure                               7. water
     4. state(s,l,g)                            8. acid
     5. appearance                              9. base

     Type the number of property to search ( 1 ~ 9 ) : 4

     Search value ? g
```

```
  NAME:            Hydrogen            ATOMIC NO:        1
  SYMBOL:          H                   ATOMIC WT:        1.008
  STATE(s,l,g):    g                   MELTING PT:       13.9 K
  DENSITY:           0.071 g/cm**3     BOILING PT:       20.1 K
  ATOMIC VOL:        14.1  cm**3/mol   COV.RADIUS:       32.0 pm
  1st ION.ENG:     1318.0  kJ/mol      HT OF FUSION:      0.1 kJ/mol
  2nd ION.ENG:     No data.            HT OF VAP:         0.4 kJ/mol
  3rd ION.ENG:     No data.            ELEC.AFFIN:       73.0 kJ/mol
  4th ION.ENG:     No data.            ELECTRONEG:        2.2
  POLARIZABILITY:    0.4  A**3         ELEC.CONDUCT:   No data.
  STRUCTURE:       No data.
  RX w AIR:        burns explosively
  RX w WATER:      none
  RX w ACID:       none
  RX w BASE:       none
  APPEARANCE:      light, colorless gas, no taste or smell when pure

 Search next ?(y/n)
```

Fig. 10. Using KC? Discoverer to find all elements that are gases.

secondary menu shown at the top of Fig. 10. This indicates all the properties that the program can find. Next the user selects "state" (number 4) to find elements according to their state of matter, and indicates "g" as the state to search for. After a brief pause the user is given a complete screen display of all the data for the first element that is a gas: hydrogen. By answering "y" to the Search next? prompt, the user would then see helium, nitrogen, oxygen, fluorine, etc. Once all elements had been found, their positions in the periodic table would be shown as in Fig. 11.

PERIODIC TABLE OF THE ELEMENTS

group:

| 1 | 2 | 3 | 4 | 5 | 6 | 7 | 8 | 9 | 10 | 11 | 12 | 13 | 14 | 15 | 16 | 17 | 18 |
|---|---|---|---|---|---|---|---|---|---|---|---|---|---|---|---|---|---|
| H | | | | | | | | | | | | | | | | | He |
| Li | Be | | | | | | | | | | | B | C | N | O | F | Ne |
| Na | Mg | | | | | | | | | | | Al | Si | P | S | Cl | Ar |
| K | Ca | Sc | Ti | V | Cr | Mn | Fe | Co | Ni | Cu | Zn | Ga | Ge | As | Se | Br | Kr |
| Rb | Sr | Y | Zr | Nb | Mo | Tc | Ru | Rh | Pd | Ag | Cd | In | Sn | Sb | Te | I | Xe |
| Cs | Ba | La | Hf | Ta | W | Re | Os | Ir | Pt | Au | Hg | Tl | Pb | Bi | Po | At | Rn |
| Fr | Ra | Ac | | | | | | | | | | | | | | | |

Elements that have been found in state have "G" value. <CR>

Fig. 11. Periodic table display from KC? Discoverer program indicating all elements found to be gases at room temperature and ordinary pressure.

As another example, suppose we wanted to graph the difference between second and first ionization energies versus atomic number. The sequence of commands needed is illustrated in Fig. 12. After choosing GRAPH from the main menu (Fig. 9) the user selects the variable to plot on the x-axis; for the y-axis it is possible to combine two items of data by adding, subtracting, multiplying, or dividing. Then the data are graphed, as in Fig. 13.

Property that can be graphed :

1. atomic number
2. atomic volume
3. electric conductivity
4. polarizability
5. electronegativity
6. 1st ionization energy
7. 2nd ionization energy
8. 3rd ionization energy
9. 4th ionization energy
10. atomic weight
11. density
12. melting point
13. boiling point
14. covalent radius
15. heat of fusion
16. heat of vaporization
17. electron affinity

Type the number of property on X-axis ( 1 ~ 17 ) : 1
Calculate any of the two properties on Y-axis ? (y/n) y
Type the number of first property (1 ~ 17) : 7
Type the number of second property (1 ~ 17) : 6
Choose operator ( + , - , * , / ) : -

Fig. 12. Secondary menu indicating properties that can be graphed by KC? Discoverer.

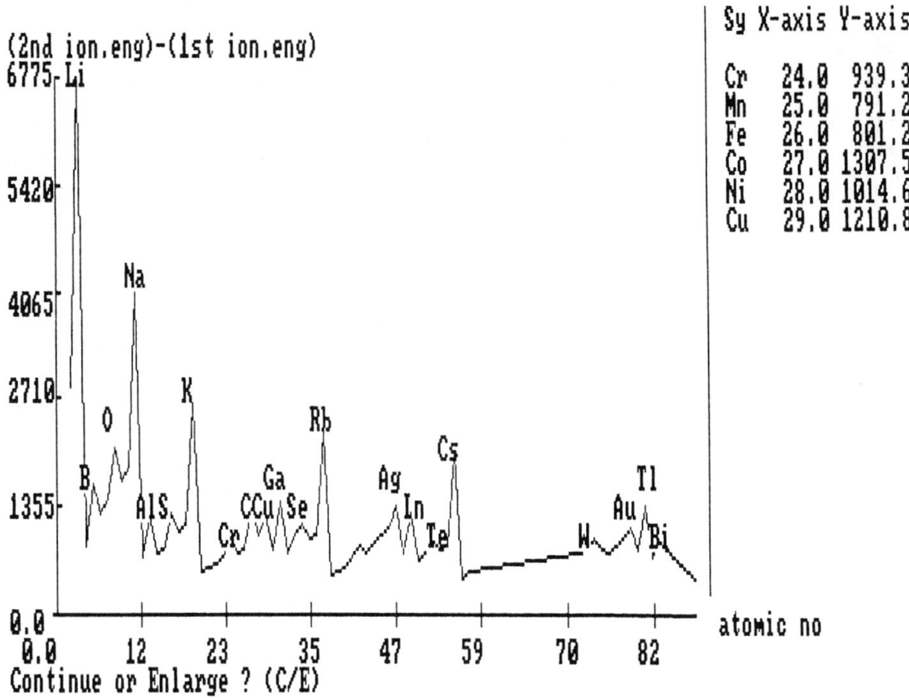


Fig. 13. Graph of difference between second and first ionization energies versus atomic number produced by KC? Discoverer. Note that exact values of plotted points may be displayed at upper right of graph.

Once the graph appears it is possible to add grid lines, to display the x-axis and y-axis values for any element by specifying the symbol, and to enlarge the graph to study more carefully any individual region. An enlarged graph is shown in Fig. 14.

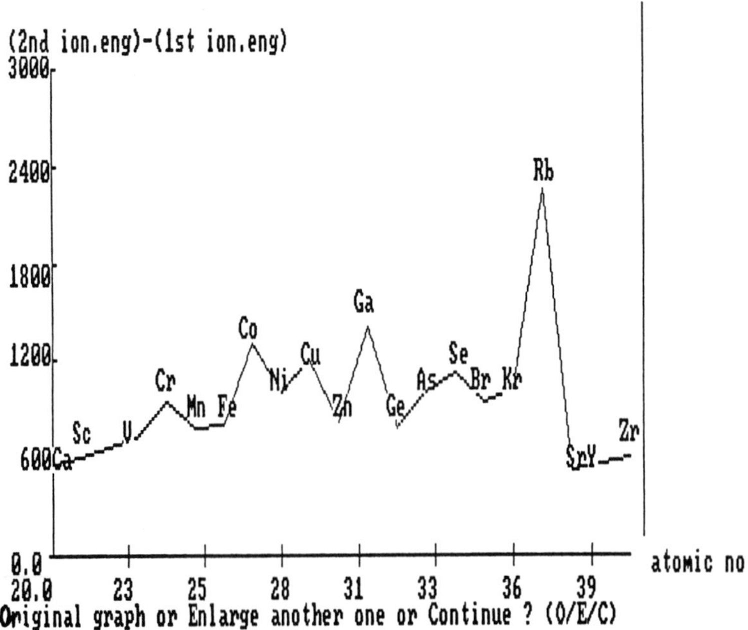


Fig. 14. Enlargement of graph shown in Fig. 13

Another very useful feature of KC? Discoverer **is** the selection of elements by periodic group. In this mode it is possible to sort or to graph information about one or more columns in the periodic table. The groups to be used are selected from a periodic table on computer screen as shown in Fig. 15, and an example graph is shown in Fig. 16.

PERIODIC TABLE OF THE ELEMENTS

| group: 1 | 2 | 3 | 4 | 5 | 6 | 7 | 8 | 9 | 10 | 11 | 12 | 13 | 14 | 15 | 16 | 17 | 18 |
|---|---|---|---|---|---|---|---|---|---|---|---|---|---|---|---|---|---|
| H | | | | | | | | | | | | | | | | | He |
| Li | Be | | | | | | | | | | | B | C | N | O | F | Ne |
| Na | Mg | | | | | | | | | | | Al | Si | P | S | Cl | Ar |
| K | Ca | Sc | Ti | V | Cr | Mn | Fe | Co | Ni | Cu | Zn | Ga | Ge | As | Se | Br | Kr |
| Rb | Sr | Y | Zr | Nb | Mo | Tc | Ru | Rh | Pd | Ag | Cd | In | Sn | Sb | Te | I | Xe |
| Cs | Ba | La | Hf | Ta | W | Re | Os | Ir | Pt | Au | Hg | Tl | Pb | Bi | Po | At | Rn |
| Fr | Ra | Ac | | | | | | | | | | | | | | | |

Choose another group to graph on "melting pt" ( 1 ~ 18 ) : 10

Fig. 15. Periodic table display from KC? Discoverer showing selection of groups 1, 2, 6, and 10 (being selected) for graphing. Graph is shown in Fig. 16.

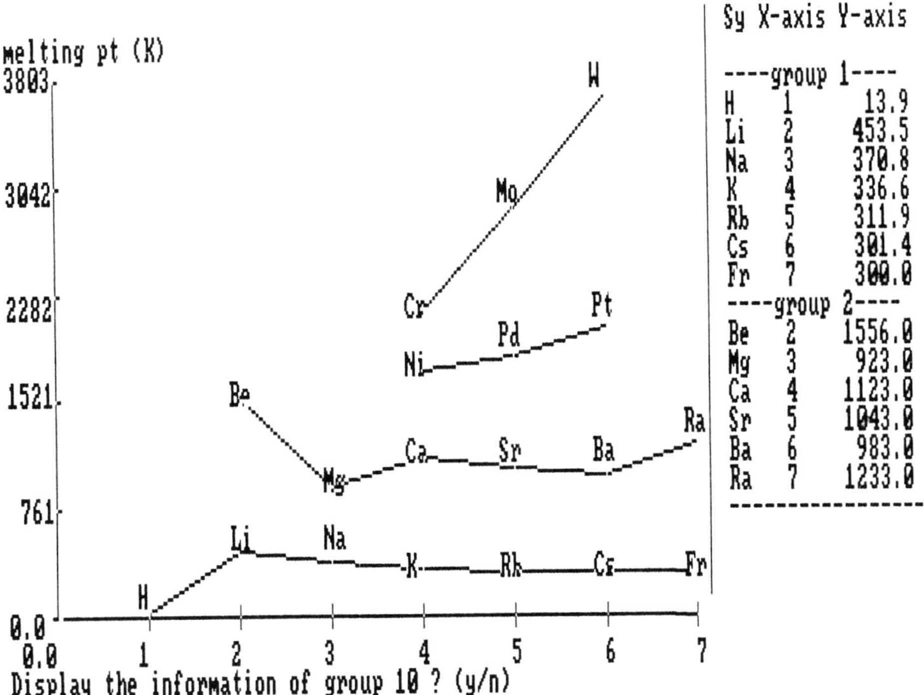


Fig. 16. Graph of melting point versus period number for periodic groups 1, 2, 6, and 10 as produced by KC? Discoverer.

Two other aspects of KC? Discoverer are worthy of mention. Students may be given research-like assignments to be completed with the help of the program. Table 6 summarizes some of the types of questions that students might be asked to answer based on information

TABLE 6. Example questions that might be posed for students to answer using KC? Discoverer.

Find--all elements that are gases
--the best electrical conductor
--the worst thermal conductor
--a replacement for some element for some application

Compare density of an atom with density of the element--for all elements

List all elements that would form diatomic, nearly non-polar, hydrides

A mineral contains a metal ion $M^{3+}$, whose radius is 78 pm; suggest other metals that are likely to be present

included in the database. To aid students in such problem solving we are constructing a knowledgable counselor within the program to advise students who are unsure about how they might proceed. The K and C in KC? Discoverer's name derive from Knowledgable Counselor, and the question mark character is to be used by students to ask for help from the counselor. This feature of the program has not yet been completed, but we have concluded that the help provided by the counselor needs to be specifically designed for the question the student is trying to answer. SERAPHIM Summer Fellow David Whisnant has prepared a program for entering help messages and for presenting them at appropriate stages in a student's progress toward a solution of a problem. In the near future we will be testing further the efficacy of these messages.

Currently, data are represented by words or numbers in KC? Discoverer, but eventually we hope to use videodisc technology to provide sight and sound for properties such as appearance and chemical reactions of the elements. Even without video the database is fascinating to chemists--so many interesting ideas can be tried out so quickly and so many new relationships can be explored that many persons become almost addicted to exploring the data. We are currently trying the program out with students to see what their reactions will be; we expect that this approach will make it possible for much more descriptive chemistry to be assimilated by students in a much shorter time.

The activities of Project SERAPHIM involve a broad spectrum of chemists and chemistry teachers from throughout the United States and elsewhere in the world. It is impossible to list here the many persons and organizations that have contributed to our efforts and should be thanked for their help, but the work I have described above is indeed a cooperative effort. I hope that by summarizing it here I have given a good indication of the current status of computer-based chemistry education and have stimulated you to think further about how best to use computers in teaching chemistry.

## REFERENCE

1. D. W. Mahoney, J. A. Sweeney, D. A. Davenport, and R. W. Ramette, J. Chem. Educ. 58,730-731 (1981).

# Some examples of the use of artificial intelligence techniques in chemical education

Daniel Cabrol

Centre de Recherches Pédagogiques
et de Rénovation Didactique en Chimie.
Université de Nice, F06034 Nice Cedex, France

Abstract - Recent successes in developing efficient expert systems, and the announcement of the Japanese "fifth generation" computer project may lead to over optimistic expectations from the field of artificial intelligence. In this contribution we critically consider the possible impact of developments in this field on chemical education. Building intelligent teaching systems is nowadays possible, in particular if they focus on the teaching of highly specialized skills. However, this requires considerable computer resources and a huge amount of work. Consequently, this approach is not expected to lead to numerous concrete applications in the short -term. The use of artificial intelligence techniques to build mixed initiative systems seems more appealing. By contrast with the expository instructional paradigm, this alternative approach emphasizes learner control over the teaching material, and learning by doing or by discovering. Two examples of this approach are presented. The potential of using object oriented and logical declarative languages such as SMALLTALK and PROLOG is briefly discussed.

## INTRODUCTION

Until recently artificial intelligence was almost unknown outside the small community of academic researchers. The announcement of the Japanese MITI "fifth generation" computer project followed by similar programs in Europe (ESPRIT), Britain (ALVEY) and the United States (MCC & DARPA) have pushed some of the ideas of artificial intelligence into the public view.

In a very wide sense, artificial intelligence can be defined as "the science of getting computers to do things which, if done by humans, would be said to be intelligent" (ref. 1). Examples of tasks which fit this definition can be found in very different areas, such as playing chess, diagnosing medical problems, elucidating a chemical structure, prospecting minerals or translating from one language to another. Obviously, teaching must be considered as an intellectual activity, probably one of the human activities which should require a very high level of intelligence. Therefore, the application of artificial intelligence to the design of computer—assisted instruction / learning systems should be extremely valuable.

After many years of research which produced results of limited practical interest, such as the chess-playing programs, artificial intelligence research has now provided highly practical expert systems. These are specialized "problem solving programs that solve substantial problems generally conceded as being difficult and requiring expertise" (ref. 2). Expert systems performance was achieved only by specialization. Successful systems which can compete with human experts rely mainly on the domain of specific knowledge and strategies rather than on general heuristic methods. For this reason they are often called knowledge based expert systems. One of the first major expert systems, DENDRAL was used in the field of Chemistry (ref. 3, 4). DENDRAL can derive the structure of an organic molecule from its mass spectra, nuclear magnetic resonance and other experimental data. For this task the system surpasses most human experts. Several other systems, the most famous two being MYCIN (ref. 5) for medical diagnosis, and PROSPECTOR (ref.6) for geological prospecting, demonstrate the effectiveness and economic importance of such systems. Expert systems like these are the results of substantial effort extended over many years. The pioneering work that was done to develop these systems makes it somewhat easier to create similar expert systems today. However, the effect of media amplification combined with potential commercial implications may lead to some unrealistic expectations. In this paper we will critically consider the possible uses of artificial intelligence techniques in chemical education.

## COMPUTERS IN EDUCATION

The use of computers in education during the last twenty-five years has proved that computers can make instruction more effective. Two major types of approaches have been explored : computers as tutors, and computers as tools. A great deal of knowledge about the effectiveness and limits of these approaches, and about usable strategies, languages and systems for computer assisted instruction has been accumulated during this time (ref. 7, 8). The microcomputer revolution provided a boost and favored the widespread use of computers in education, but at the same time it caused a regression of the average quality of educational software. The reasons for this regression were: 1/ the limitations of both the hardware and programming facilities initially available, 2/ the change of people involved, who were no longer specialists in this field but primarily teachers and new computer users, 3/ the change in objectives from research projects to practical in-class applications. In spite of these dificulties, the best educational software now available on microcomputers has reached the mean quality of software produced on main-frame computers a decade before.

## COMPUTERS AS TUTORS

The uses of computers as tutors, relying on behaviorist theories, evolved rapidly from linear programmed instruction to sophisticated systems which try to adapt themselves to the needs of individuals. However, the fundamental limitation of using computers as tutors remains. It lies in the fact that the program lacks knowledge both in chemistry itself and in education. In most applications, the program does not have the capacity to answer questions or solve problems which have not been explicitely taken into account by the software author. As a result, the dialog between the system and the student is one-way with respect to questioning. The computer generates questions or problems and the student gives the answers. The direction of this pseudo-dialog is not reversible. The student cannot ask his own questions or submit his own problems because they cannot be anticipated. Although fundamental, this limitation is not the only one. In most common cases, the program flow is controlled by the analysis of the last response provided by the student. In more elaborate programs, the branching is also based on a series of flags and counters which have been raised or incremented according to previous answers. Dedicated systems and languages, such as the PLATO system and the COURSE-WRITER language, have been developed to facilitate the author's task in writing such programs. Notwithstanding these facilities, managing a large number of indicators becomes rapidly cumbersome. As a result, with a few exceptions, current tutorial programs have very primitive teaching strategies.

To overcome these limitations, it has been suggested that one use artificial intelligence techniques to build intelligent computer tutors. It seems quite natural to admit that a computer tutor should have some knowledge in the subject matter it is supposed to teach, and to have some kind of expertise and flexibility in its teaching strategy. Thus it seems promising to combine an expert system of the subject matter, a teaching expert system, a model of the learner and a natural language interface (Fig. 1). This approach has been suggested as early as 1970 by Carbonnel (ref. 9).

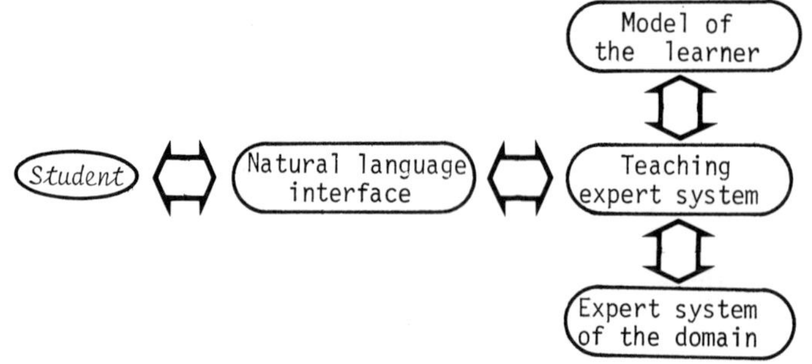


Fig. 1
General structure of intelligent tutoring systems (adapted from ref. 10)

A typical expert system comprises at least three parts : a knowledge data base, a workspace and an inference engine. The knowledge data base comprises symbolic representations of facts, models, rules and other particular knowledge related to the domain. The workspace holds the description of the current status of the problem to be solved. The inference engine is a piece of software to process symbols that embody the knowledge of the system. The inference engine selects facts and rules from the knowledge base and applies them to the facts in the workspace to deduce new facts. An expert system must also include a user interface to allow the user to define the problem under consideration. The credibility of such a system is achieved only if it can explain and justify the path followed and the rules applied to reach the solution. Thus an explanatory sub-system is sometimes added to expert system. This is absolutely essential for educational purposes. Differences in architecture are common, but the idea of separation between the knowledge data base and the part which does the processing seems fundamental.

Writing such systems is a complex task probably not feasible for most teachers. Conventional programming languages are not well suited for this purpose. The most commonly used language for manipulating symbolic representation of knowledge is LISP. For building a new expert system, it may be worthwhile starting from an existing system, or using specialized programing tools, rather than starting from scratch. Several tools derived from existing systems are available for building expert systems, one of the most powerful inference engines being OPS5 (ref. 19). Choosing the appropriate tool for building a particular system is a difficult task. Most of these tools coming from research projects need large computers; however, some less complex development systems especially designed for medium-sized microcomputers are now offered on the software market (ref. 20). In spite of some shortcomings and limitations of these first generation expert systems builders, the direction has great potential for creating valuable tools for students. However, even with more efficient expert systems builders, the amount of knowledge required by the user for conceiving a useful system, except for very simple cases, may be considerable. Thus, dedicated problem solvers and expert systems for chemical education will more likely be provided by educational software distributors rather than written by the individual teacher. Very few examples of problem-solvers and expert systems for students can be found nowadays. The two programs decribed below share the "partner" philosophy in the sense that the user and the program both have a part of the work to do.

The first example deals with organic synthesis. While general programs for computer assisted synthesis design which are available for research need main-frames and are not affordable for teaching purposes, the program "micro-synthese" (ref. 21, 22) uses an Apple II microcomputer. This program uses a retro synthetic method to help the user to design a synthetic route for a given molecule. The user selects the target molecule he wishes to obtain. For this purpose he simply draws on the computer screen the structural formula of the target molecule. Then the program searches through a list of organic reactions for those which can give the target molecule as a product. For each selected reaction, it displays the possible precurssor. The user has the responsibility of retaining or discarding the proposed precursors. Each precursor retained creates a ramification on the synthetic tree (Fig. 2). The process is repeated recursively using each retained precursor as a new target molecule. The search is ended when no new precursor can be proposed by the program, or when the user is satisfied with the starting materials. The synthetic tree can be dispayed on the screen.

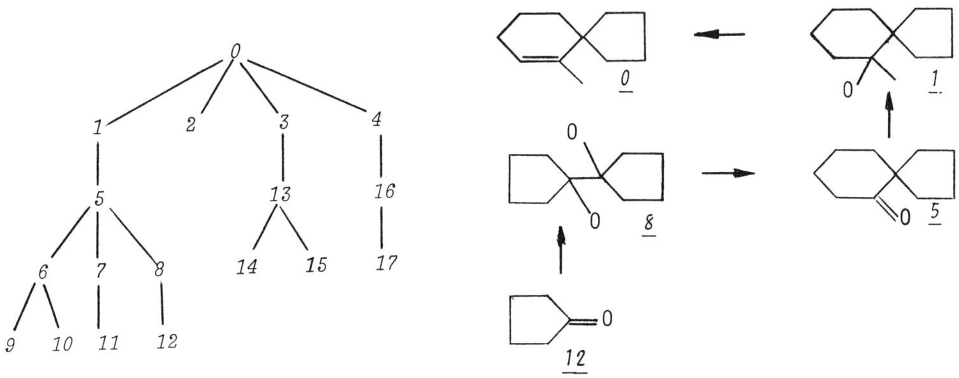


Fig. 2: The synthetic tree leading to 0. A new synthetic route 12-8-5-1-0

The expert of the domain generates and solves problems, answers questions, and analyzes the students' responses. The teaching expert controls the dialog with the student, decides when and how to step in, and provides hints and explanations. The model of the learner is a data base representing the knowledge and skills of the learner. This model is consulted and updated by the teaching expert. The natural language interface accepts input from the learner and converts it into the appropriate internal representation. It also does the inverse job, producing in plain English the messages coming from the teaching expert.

Some systems built on this general scheme are described in reference 11. One of the most successful applications of these techniques is SOPHIE (ref. 12, 13). This program provides the student with an interactive learning environment for electronic trouble-shooting. SOPHIE incorporates an expert system in trouble-shooting which can answer student's questions and make suggestions, a coach which applies the teaching strategy, and a powerful natural language interface, but the model of the learner knowledge has not been well developed. Among other interesting examples are ACE : a system which analyses complex explanations (ref. 14) in the interpretation of nuclear magnetic resonance spectra, and GUIDON (ref. 15) which is a case method tutor for medical diagnosis training.

Despite the impressive capability of these and related systems, the actual implementation of intelligent computer tutors remains problematic. These systems work on large machines and cannot be adapted for microcomputers that are currently available in schools. With the fast increase in the power of microcomputers, technical problems like these will probably be alleviated rapidly.

However, although these systems are the result of large research projects involving dozens of people for several years they are not fully satisfactory. The major shortcomings as identified by Sleeman in ref. 11 p. 3 are:

"1- The instructional material produced in response to a student's query or mistake is often at the wrong level of detail, as the system assumes too much or too little student knowledge.

2- The system assumes a particular conceptualization of the domain, thereby coercing a student's performance into its own conceptual framework.

3- The tutoring and critiquing strategies used by these systems are excessively *ad hoc*, reflecting unprincipled intuitions about how to control their behavior.

4- User interaction is still too restrictive limiting the student's expressiveness."

These drawbacks are strongly interrelated and underline the fact that we do not know enough about how students learn. Several research projects have attempted to deduce a model of the learner from his observable behaviour as revealed by his interaction with the system. But this is a formidable task even for a very simple procedural skill on the part of the learner. For example, Brown and Burton (ref. 16) analysed student errors in performing subtraction as indicators of a "bug" in the learner skill. They identified 60 subskills leading to 110 different primitive bugs; if we consider possible combined bugs we can imagine how complex the student model would have to be in order to define an individual profile. Moreover, this is only a very crude model which does not take many other factors into consideration; among them are the student's interest and motivation, his preferences for abstract formalism or concrete examples, his reactions to humour, encouragement, etc. Last but not least, as pointed out by kayser (ref. 17) "the resulting learner model is noisy. When the computer interprets a given answer as evidence that the student knows/ignores some fact, it can merely originate in boredom, inattention, inadvertance... The construction of very sophisticated student models is thus a waste of time."

Thus the tremendous difficulties involved in the setting-up of a teaching expert system and the modelling of the learner behaviour and knowledge will remain for years. As a consequence, it seems to us that this approach will not lead to many operational applications for chemical education in the short term. A more practical approach for the present time is to be found in using computers as tools for learning rather than using them as tutors.

## COMPUTERS AS TOOLS FOR LEARNERS

It is not only a matter of feasibility that will lead to using computers more as tools rather than tutors, but more fundamentally the consideration of an alternative to the original paradigm of using computers to deliver instructions by tutoring systems. By contrast to traditional approaches in which the student cannot deviate from the learning strategy incorporated into the software, this alternative approach emphasizes learner control over the teaching material, and learning by doing. In this respect two main alternatives can be identified : to supply the students with application programs designed for specific tasks, or to provide them suitable software development environments and encourage them to engage in programming activities.

Student use of computers as tools covers many different possible applications; among them are: numerical calculation, interactive simulation, data base management, graphic processing, problem solving, etc. One possible way in which artificial intelligence techniques can facilitate any of these applications is by allowing flexible input from the student. It is possible to design semi-natural language interfaces to improve communication with the user. Usually, these interfaces include a sentence parser which accepts queries or commands expressed by the user, and converts them into a form suitable for the core application program. For example, a data base management program may accept queries expressed in the following way :

"Do you know a water soluble amine which is non-toxic ?"

The parser would:
1- recognize this sentence as a question (both by the question mark and by the phrase "do you know").
2- identify the "water soluble amine" as the target of the query (by the structure of the sentence).
3- split this target into two parts : "amine" which will be recognized as a class of compound, and "water soluble" which will be recognized as a property.
4- identify the "non-toxic" as a property (by the structure of the sentence and by the "which is".
The result of this parsing will be a set of relations expressed in an internal conventional form suitable for the data base management program. In the present case, a possible (Note a) form could be :

```
quest(x) --> compound(x)
             class(x,amine)
             solubility(x,water,yes)
             property(x,non-toxic);
```

Most sentence parsers rely on grammar rules to obtain the syntatic structure of the input sentence. A semantic analysis to obtain the meaning of the sentence must be carried out next. However, it has been shown, that direct semantic parsing is possible leading to much more flexible interfaces (ref. 18). On current microcomputers, large interfaces may require a significant amount of the machine resources and may not leave enough memory, or may take too much time to run. But, the way is open for a better man-machine interaction and this way will widen and be easier in the coming years.

The most promising possibilities created by artificial intelligence techniques for education lie in the development of problem solvers, or expert systems to be used directly by students. Almost any field of chemistry in which students have to combine several pieces of information and follow various multi-step reasoning paths could benefit from the building of problem solvers or expert systems for students. Contrary to conventional computer–assisted instruction programs, these systems do not present problems or asks questions of the user. Just the opposite, they can be used by students to solve their own problems. Not only should these systems provide the answer to students, but more importantly, they should be able to explain how the solution was reached. It is possible to conceive these systems as mixed initiative systems, a part of the job being done by the user and another part by the system, with the importance of this latter part decreasing as the user expertise increases. This is a major difference from what is normally expected of an expert system. Such a mixed initiative system could be called a problem-solver partner.

---

Note a: this form constitutes a clause in the PROLOG language.

The selection of a particular reaction for presentation to the user is achieved by using a set of rules following the conditional format :

**IF 'set of conditions'**
**THEN 'actions' ELSE 'actions'.**

Among other possibilities, the 'actions' can be the description of structural modifications applied to the target molecule to deduce the precursor. The student can use simple commands to modify existing rules, or to add new reactions to the list in order to customize the knowledge data base. Although Micro-synthese is based on a very simple algorithm, its efficiency relies on the completeness of the file of organic synthetic steps and on the accuracy of associated rules. The program has been used beneficially by students (ref. 22) and its educational value is enhanced by the fact that the file of organic synthetic steps can be enriched by the user without conventional programming. Thus it is possible to provide students with the program and a minimum set of reactions as a starting point, and to ask them to add new synthetic reactions as they learn them in courses. The performance of the resulting system would then be the responsibility of the student.

The second example is a problem—solver partner in elementary quantitative chemistry (ref. 23). This program GEORGE (Note b) uses heuristics rules to find the solution to most problems dealing with fundamental quantities : mass, volume, number of moles or molecules and the derived quantities : molarity, mass concentration, density, molar mass, etc. The major responsibility of the student using GEORGE is to properly define the problem to be solved. In the problem statements, he must correctly identify the quantity he wants to compute and the data available. These are important first steps in solving problems with or without a computer. Communication with the program is achieved by filling two types of screens : the data pages which hold the question and the available data, and the relation pages which hold the relations between quantities, specific to the particular problem in hand.

For example consider the following problem :

> A solution of 4.2 g/l NaOH is being used to standardize a solution of HCl. An aliquot of 20.0 ml of the HCl solution is used, and 18.6 ml of the NaOH solution is used in the titration.
> What is the concentration of HCl ?

In order to submit this problem to GEORGE, the user may fill the following pages (fig. 3.a and b):

| Desired Quantity | | Units |
|---|---|---|
| X: molarity of HCl in SOL.A | | mol/L |
| **Available Data** | **Value** | **Units** |
| A: mass conc. of NaOH in SOL.B | 4.2 | g/L |
| B: volume of SOL.A | 20.0 | mL |
| C: volume of SOL.B | 18.6 | mL |
| D: molar mass of NaOH | 40.0 | g/mol |

Press the ESC key to display the menu.

Fig. 3.a Data Page

Input of Relation 1

| Coef. | Quantity |
|---|---|
| | no. of moles of HCl |
| = | no. of moles of NaOH |

CTRL-L makes the next letter lower case

Fig. 3.b Relation Page

Because the problem to be solved is stated by the student, it is not possible to encode the algorithms of typical solutions for each anticipated problem. The program must be able to work out the solution to any problem in this field provided that necessary pieces of information are available. Thus, the search for a solution relies mainly on dimensional analysis and makes use of simple heuristic rules. In spite of the narrowness of the domain, the number of problems that GEORGE can solve is enormous; they range from simple unit conversion questions to complex back titration problems.

Note b: program written in colaboration with R. Cornelius and C. Cachet.

Moreover, to be beneficial from a pedagogic point of view, it is not enough for the program to find and give the answer, it is important that it "shows its work" and that the solution it has established can be understood by the student, and is appropriate for applications to similar problems. This rules out a pure combinatory search.

During the search, the program explains in plain English each step of the reasoning followed in discovering the solution. The whole solution process is also summarized by a graphic network which shows how data and intermediates are related to the answer. For the titration problem stated above, the network obtained (fig. 4.a and b ) is split into two parts linked by the application of the stoichiometry relation. By a single key stroke, the student can ask for details about each node of this network.

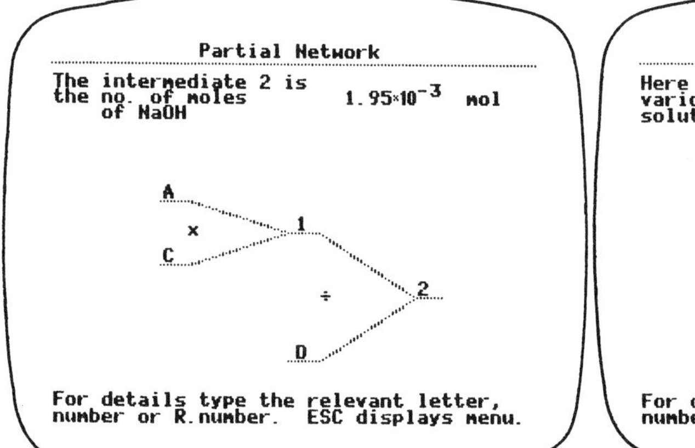


Fig. 4.a

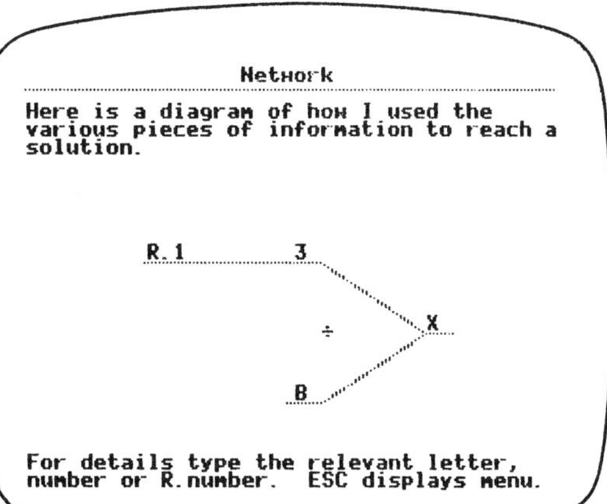


Fig. 4.b

We have observed that most failures in problem solving in this field do not originate from lack of knowledge in the subject matter, but from difficulties in the formalization of problem statements and, for problems involving solutions in several steps, in the setting up of an efficient solving strategy. Using GEORGE as a tool for problem solving forces the student to use a rigorous formalization of the problem statement; the program helps him by identifying incoherent pieces of information and/or missing and extraneous data. It also takes in charge the search for a solution. Thus the student learns by doing and following examples. Further experiments with students are needed to make more detailed observations on the educational effectiveness of this program. It remains to be seen whether students, once accustomed to using GEORGE become dependent on it or develop their problem-solving skills so that they no longer need its assistance.

In spite of their limitations, mainly imposed by the hardware on which they have been implemented, these first problem-solving partners foreshadow a new class of software for students. Hand calculators have deeply changed some aspects of science teaching. The availability of powerful portable microcomputers and of this new class of "intelligent" software raises interesting questions about the possible changes they may induce in education. If we consider that the user will have the facility to increase the capacity of these programs by interacting with them, adding new facts and rules to the data base, saving collections of problems, etc., then the capabilities of the program become the result of the student's activity. Since the capabilities of the program reflect the work and ability of the student, should we allow these programs to answer questions or solve problems for the student in examination situations?

## PROGRAMMING AS A LEARNING ACTIVITY

Another important challenge we are faced with is computer literacy for chemistry students. When the necessity of introducing computer literacy in science curricula became widely accepted, not enough attention was paid to the methodology of building programs nor to the choice of a computer language for teaching. Because of its apparent simplicity and ease of implementation on microcomputers, BASIC was the language most commonly taught to students. It is now widely recognised that this was not a good choice because BASIC does not encourage good programming style, and that the use of a structured language, such as Pascal, would be better. Nevertheless, BASIC, Pascal and similar languages fall in the same category of procedural languages. Procedural languages are well suited for describing algorithms by a series of instructions to be executed sequentially. This kind of programming is called imperative programming. In these languages the data to be processed and the instructions are completely separate, thus the program is static.

Clearly, many important applications of computers in chemistry are algorithmic, but the limits of procedural languages for representing and manipulating complex information, which constitute the bulk of chemical knowledge, are severe. Artificial intelligence achieved significant results by using special languages which do not maintain the separation between data and instructions. The ancestor of these languages for artificial intelligence is LISP, and is still widely used. Numerous other languages also have been used in this field (ref. 24); some are available on current microcomputers, among them the most promising for education are LOGO, SMALLTALK and PROLOG. Very short comments will be given on the two first languages, more details on the last one.

The LOGO language is well known for its famous turtle and for its use by children (ref. 25). In fact, the language itself is not restricted to its graphic aspects, but has greater potential, for example in list processing, which allows the representation and manipulation of non-numerical information.However, LOGO is still strongly procedural, in fact it can be considered as the archetype of a procedural language, thus it is not well suited for representing other kinds of knowledge.

SMALLTALK is classified as an object-oriented language (ref. 26). It is based upon the idea of communicating between objects. Every element in the system is represented by an object and every object communicates with other objects by sending messages. Objects belong to a given class and inherit the properties of the class. SMALLTALK is well suited for designing simulations of complex systems; it enables the programmer to describe both the individual parts as well as their interaction. SMALLTALK is much more than a language, it is a programming environment which makes extensive use of graphics to facilitate communication with the user. Many ideas concerning a man-machine interaction, such as the use of icons, scrolling menus, multiple windowing, pointing devices like the 'mouse', originate from the SMALLTALK project and have been incorporated in the design of the Apple's Macintosh. Whether SMALLTALK can be used efficiently by students to explore problem-solving situations or complex simulations has not been extensively studied. Some interesting attempts have been reported (ref. 27, 28), in which valuable results have been achieved by providing interactive environments for designing, manipulating, exploring, and simulating complex situations. Several aspects underlined by these studies, such as the importance of constraints, are strongly related to artificial intelligence techniques. However, SMALLTALK requires considerable computer resources; a limited version called TinyTalk runs on 64K microcomputers (ref. 29).

The PROLOG language represents a major advance toward the objective of logic and descriptive programming. The language was first defined and implemented by A. Colmerauer and co-workers in Marseille (ref. 30), and is now available on a wide variety of machines including microcomputers. A good tutorial introduction to PROLOG can be found in ref. 31 and 32. The language was chosen by the Japanese Fith Generation Project as the starting point of a new type of programming language . The goal of logic programming is to describe some assertions (facts and rules) to be true, and then ask for others to be deduced as a consequence, without having to consider how these deductions are made. Obviously, this is very different from imperative programming.

In PROLOG,the programmer may state facts and rules by using clauses, for example,

```
anion(sulfate)-->;          cation(sodium)-->;
anion(chloride)-->;         cation(potassium)-->;
anion(bromide)-->;          cation(ammonium)-->;
```

can be used to state that sulfate, chloride and bromide are anions, and that sodium, potassium and ammonium are cations. A rule is written by setting on the right side of the arrow, the conditions which must be satisfied for the term on left side to be true, e.g.

```
binary-salt(c,a)--> cation(c) anion(a);
```

states that the pair represented by the variables c,a is a binary salt if c is a cation and a an anion. Such a rule can be viewed from two aspects : it is a description of what a binary salt is, and a procedure of how to answer questions about binary salts. Asking questions of PROLOG is very easy; to obtain the list of binary salts, as it can be deduced from this small data base, one would simply type: binary salt(c,a); . This would result in an output of all possible pairs derived from the anions and cations that have been defined in the data base.

Another example will illustrate how easily PROLOG can handle recursivity. The radioactive family of radium 226 can be described by the two sets of clauses :

```
alpha(Ra226,Em222)-->;        beta(Po218,At218)-->;
alpha(Em222,Po218)-->;        beta(Pb214,Bi214)-->;
alpha(Po218,Pb214)-->;        beta(Bi214,Po214)-->;
alpha(At218,Bi214)-->;        beta(Ti210,Pb210)-->;
alpha(Bi214,Ti210)-->;        beta(Pb210,Bi210)-->;
alpha(Po214,Pb210)-->;        beta(Bi210,Po210)-->;
alpha(Po210,Pb206)-->;
```

We can now define the following rules :

```
give(x,y)--> alpha(x,y);
give(x,y)--> beta(x,y);
```

Which state that a nuclidyx gives a nucleide y if y is formed from x by an alpha or by a beta disintegration.

```
ancestor(x,y) --> give(x,y);

ancestor(x,y) --> give(x,z)
                  ancestor(z,y);
```

The first ancestor clause specifies that x is an ancestor of y if x is a direct parent of y.
The second states that x is an ancestor of y if x is the direct parent of a nucleide z, and if z is an ancestor of y.

With these simple rules, it is possible to reconstruct all of the radioctive family described in the data base. For example, the query ancestor(x,Pb210); would retrieve all the ancestors of Pb210, while the query ancestor(Em222,y); would retrieve all the descendants of Em222. Any fact that may be derived from known facts in the data base may be obtained by writing the appropriate PROLOG clauses. Other examples of the use of PROLOG in chemistry, such as : describing atomic electronic structures, finding paths for functional group conversions, analysing of infra-red spectra by pattern matching, can be found in ref. 33.

PROLOG is well suited for building data bases of facts and rules, natural language interfaces, and medium sized expert systems. Experiments in the U.K. (ref. 34) and France have shown that it can be learned and used directly by pupils in secondary schools. The declarative use of logic provides a number of advantages in the learning of programming. Problem specifications are often written in logic, and they can sometimes be translated almost directly into PROLOG assertions without worrying about how efficiently they will be handled by the computer. By its very nature, the PROLOG-language encourages top-down decomposition of problems. For example the core part of the problem solving program GEORGE mentioned above, which was originally written in BASIC, has been recently rewritten in a few days. The compression factor, as measured by the ratio of the number of BASIC statements to the number of PROLOG clauses, is about twelve. But the resulting program is slower. Obviously, this kind of language is somewhat ahead of today's microcomputer technology; only problems of medium size in which speed is not critical can be supported. However, these limits are not too restrictive for educational applications. Current research efforts are expected to provide the advances, such as parallel-processing or some other non-y Von-Newman architecture, that are required for very large scale applications.

## CONCLUSION

Representing chemical knowledge in a computer needs both procedural, descriptive and logical approaches. Thus for the present time no unique language is suitable for all these needs. However, there is no doubt that computer use will increasingly involve the description of our knowledge about the world rather than the defining of computer operations.

As judiciously recalled by Pr. Sasaki (ref. 35), chemistry is an experimental science. Most of our knowledge comes from observation and experiments. To be genuinely significant, results of theoretical quantum and physical chemistry must be confronted with known facts. Computers as used for performing calculations, for data processing and retrieval, for laboratory control and automation are readily accepted as useful tools for the chemist. The result of artificial intelligence research provide a new area of application of computers in which knowledge plays a central role: it is at once the information to be processed and the driving force of the processing. Thus, the techniques of knowledge engineering developed by artificial intelligence can help us to organize our chemical knowledge and to test the reliability of derived models and rules. We must be ready and prepare our students to use computers as tools for expanding our thoughts.

Artificial intelligence is a fast moving field and there is no general agreement on the issues, even for short-range predictions, but one must be careful about over-optimistic claims. However, efficient techniques from artificial intelligence are already available on micro computers and may be used to create innovative learning environments, preparing ourselves and our students for the future computer age.

## REFERENCES

1. T. O'Shea, Learning and Teaching with computers: Artificial Intelligence in Education , p. 4, Harvester Press, London (1983).

2. M. Stefik, J. Aikins, R. Balzer, J. Benoit, L. Birnbaum, F. Hayes-Roth and E. Sacerdoti, Artificial Intelligence, 18, 135-73, (1982).

3. E.A. Feigenbaum, B.G. Buchanan and J. Lederberg, In Machine Intelligence , B. Meltzer and D. Mitchie, eds., vol. 6, pp. 165-90, Edinburgh University Press, Edinburgh, (1971).

4. R.K. Lindsay, B.G. Buchanam, E.A. Feigenbaum and J. Lederberg, Applications of Artificial Intelligence for organic chemistry : The DENDRAL project. McGraw-Hill, New-York, (1980).

5. E.H. Shortlife, Computer-based Medical Consultations :MYCIN , Elsevier, New York, (1976).

6. R.O. Duda, J.G. Gaschnig, and P.E. Hart, in Expert Systems in the microelectronic age. D. Michie, ed., pp.153-67, Edinburgh University Press, Edinburgh, (1979).

7. G. Kearsley, B. Hunter and J. Seidel, T.H.E. Journal , N°1,90-4, (1983).

8. G. Kearsley, B. Hunter and J. Seidel, T.H.E. Journal , N°2, 88-96, (1983).

9. J. Carbonell, IEEE Trans. on Man-Machine Systems ,11,190-202, (1970).

10. A. Bonnet, in Intelligence Artificielle , pp. 59-80, Publications du GR 22 CNRS, Caen, (1980).

11. D. Sleman and J.S. Brown, Intelligent tutoring systems, Academic Press, London, (1982).

12. J.S. Brown, R.R. Burton and A.G. Bell. Int. J. Man-Mach. Stud. ,7,675-96, (1975).

13. J.S. Brown, R.R. Burton and J. De Kleer, in Intelligent Tutoring Systems, Sleeman and Brown, eds, p. 227-79, Academic Press, London, (1982).

14. D.H. Sleeman, Int. J. Man-Mach Stud. , 7,183-211, (1975).

15. W.J. Clancey, Int. J. Man-Mach. Stud. , 11,25-49, (1979).

16. J.S. Brown and R.R. Burton, Cognitive Science , 2,155-92, (1978).

17. D. Kayser, Euromicro Journal , 6,209-14, (1980).

18. R.R. Burton, ICAI Report 3 , Bolt Beranek and Newman, Inc., Cambridge Mass,(1976).

19. C.L. Forgy, The OPS5 user's manual. Technical Rept. CMU-CS-81-135, Comput. Sc. Dept., Carnegie-Mellon Univ., Pittsburgh, (1981)

20. N. Heite, D. Crabb and H.F. Beechhold, Infoword , 28,43-8,(1985).

21. R. Barone, M. Chanon, P. Cadiot and J.M. Cense, Bull. Soc. Chim. Belg. , 9,333-6, (1982).

22. J.M. Cense, L'Actualité Chimique , N°3,76-8,(1985).

23. D. Cabrol, C. Cachet and R. Cornelius, Comput. & Educ. , in press.

24. H. Boley, T.S.I. , 2,145-66, (1983).

25. S. Papert, Mindstorms: Children, Computers and Powerful Ideas , Basic Books, New-York, (1980).

26. A. Goldberg, Comput. & Educ , 3,247-66, (1979).

27. A. Borning, ACM Trans. on Prog. Lang & Systems , 3,353-87, (1981).

28. L. Gould and W. Finzer, "A study of TRIP: a computer system for animating time-rate-distance problems , in computers in Education , R. Lewis and E.D. Tagg, eds, North-Holland, Amsterdam, (1981).

29. K. McCall and L. Tesler, ACM Sigsmall Newsletter , 6,197-8, (1980).

30. A. Colmerauer, H. Kanoui and M. Van Caneghem, T.S.I. ,2,271-311,(1983).

31. W.F. Clocksin and C.S. Mellish, Programming in Prolog , Springer-Verlag, Berlin, (1981).

32. F. Giannesini, H. Kanoui, R. Pasero and M. Van Caneghem, Prolog , InterEditions, Paris, (1985).

33. D. Cabrol and T. Forrest, to be published.

34. R. Ennals, Artificial Intelligence: Applications to Logical Reasonning and Historical Research" , Ellis Horwood, Chichester, (1985).

35. S. Sasaki S., 8th ICCE , Tokyo (1985).

Acknowledgement: The author wishes to express his gratitude to Prof. T.P. Forrest for many valuable comments and suggestions as well as for careful reading of the manuscript.

# Chemical education for fostering future chemists of excellence

Geoffrey N. Malcolm

Department of Chemistry and Biochemistry, Massey University, Palmerston North, New Zealand

## INTRODUCTION

There is for all of us a fascination with the future as a topic for speculation. According to St Augustine of old, time is a three-fold present; the present as we experience it, the past as a present memory and the future as a present expectation. By this criterion the world of the 21st Century has already arrived in our consciousness, and we do well to make preparation for it. It is said that those who will not learn from history are doomed to repeat it, and equally those who will not plan for their future must suffer it.

The goal to be considered by us today is the provision of outstanding chemists for the future, with a view we trust to the better understanding of our world and the better provision of human needs. I say that we must trust that these are the purposes which people have in mind and not only the increase in gross national product by chemical means. No doubt some aspects of all three purposes are present in our minds.

A recent editorial in "Nature" with the title "Schools for another century" opened with the evocative sentence,"How should young people be prepared for life in the brave new world...?" (ref.1). Aldous Huxley's embryological solution to the person-power needs of the Brave New World may be nearer now to realization than it was when he produced his satirical work in 1931 (ref.2). But his solution, feasible or not, is still one from which we recoil because of its intrinsic problem of deciding who will control the planners. It is not so much a matter of scientific ability as of human fallibility. I turn away,therefore,from any completely technological solution to the supply of alpha-plus chemists in the future.

There is in all of our societies a pool of able young people who will be influenced by a variety of factors in the careers which they will choose. What we must do is to provide the most attractive context that we can for their learning of chemistry in the hope that some of the most able students will choose our subject. It is for this reason that we have the phrase "fostering future chemists of excellence" rather than "producing" them in the title to this lecture. This fostering is to be done by means of high quality chemical education and freedom of choice.

## THE STATE OF HEALTH OF CHEMICAL EDUCATION

I said at the beginning that the recent past is present in our memory and will influence our thinking about the future. It is appropriate therefore to review what has been attempted and accomplished in chemical education in the last few decades internationally. This is not an easy task and there are those who would cynically refer to the path which has been traversed since the 1950's as nothing more than a random walk, exhibiting the major characteristics of such walks of never ending up very far from the starting point.

I have decided therefore to change the metaphor and to regard chemical education not as a ship proceeding sedately across the seas of knowledge, but rather as a patient who in the minds of many is chronically ill.

Certainly ill-health seemed to be the case in the early years of the 1950's. Prior to that time it had been accepted that the learning of isolated facts, such as the vocabulary of a foreign language, was easier in the years up to the age of 15 or 16 than it was in later life. Therefore the learning of the basic facts of science had been allocated to the secondary schools. Partly as a result of public awareness of scientific achievements during the War there arose in the minds of both teachers and pupils a revolt against the dullness of teaching and learning science as a collection of facts. This feeling was experienced in varying degrees in many of the developed countries, but it came to a head in the United States following the launching of the Russian Sputnik in 1957. That event together with several other factors led to an enquiry into the state of science education in that country. The patient was found to be sick, and the specialists were called in. These of course were not the teachers, the mere General Practitioners as one might say, but the real science experts from the Universities and the research establishments. These were surely the people who knew what science was about.

These people brought their project approach to the problem of science education. They knew from experience and intellect what science was, its content and its structure, and it was axiomatic to them that that was what should be taught. Whether it would be possible to teach what they prescribed to school children who were still in the process of intellectual development was not seriously considered. After all the specialist's task is to prescribe the treatment; not for them the mother's difficult role of persuading the child to swallow the nasty medicine. Structure of science, method of science and content of science were the things they emphasised. In structure they saw science as a set of distinct disciplines: physics, chemistry, biology. For content they saw science as a set of unifying principles by means of which it was possible to make intellectual sense of a wide range of experiences in the physical world. Remembering the Project Approach to scientific problems which had been successful during the recent war, they set up various working parties of experts and produced ready-made packages for science education. You will know the names of many of them.

| | | |
|---|---|---|
| PSSC | Physical Science Study Committee | 1960 |
| CHEM | Chemical Education Materials Study | 1963 |
| CBA | Chemical Bond Approach | |
| BSCS | Biological Sciences Curriculum Study | 1959 |
| HPP | Harvard Project Physics | 1964 |

These all had the same sort of flavour, which was a strong emphasis on the principles of the subject. For example in CHEM Study the four main sections were entitled:

I Some major concepts of chemistry
II Principles of chemical reactions
III Principles of chemical bonding
IV Applications of chemical principles to descriptive chemistry

That is exactly the order of presentation of material in most of the American College Chemistry texts published in the following two decades. Their titles alone are sufficient to indicate their approach, illustrated by the following list.

| | | |
|---|---|---|
| 1966 | Chemistry: Principles and Properties | Sienko and Plane |
| 1966 | Chemistry Principles | Masterton and Slowinski |
| 1967 | Basic Principles of Chemistry | Gray and Haight |
| 1967 | Principles of Chemistry | Patterson |
| 1969 | Chemical Principles | Dickenson, Gray and Haight |

Before these books the titles were usually "General Chemistry", for example the books by Pauling, and earlier still was Holmyard's "Higher School Certificate Chemistry". Such was the treatment for Science Education offered in North America. Through textbook sales, aspects of these schemes were imported into many other countries.

In England also during this period curriculum changes were made, but not under the guidance of university or research scientists. Science teachers in the English Grammar Schools were on the whole very well qualified, more so than their counterparts in some other countries, and they guarded carefully their professional freedom as teachers. Accordingly the Nuffield Science Teaching Project was produced by able teachers seconded for this purpose. So it was good General Practitioners who both prescribed and administered the treatment here. Instead of presenting a complete curriculum package defining aims, objectives and content, complete with student textbook, films and laboratory guide, the approach adopted was "to develop materials that will help teachers to present science in a lively, exciting and intelligible way". The Teacher's Guide rather than the student's text is the heart of the Nuffield programme. Unfortunately straight adoption of such a programme into another country is not easy if the average quality of the teachers is not the same, as so much in the Nuffield course is left for the teachers to decide. But the emphasis in the Nuffield courses was very much on the presentation of the principles of science, those grand ideas through which we can make sense of our experience in the natural world.

What now has been the result of these projects with their strong emphasis on the principles of science? Broadly speaking the patient didn't like the medicine, and failed to respond in terms of increased student popularity.

Accordingly another group of specialists moved in. These were the professional educationalists. Like the professional scientists they did not actually do any secondary school teaching, but they claimed to know about the educational process. They introduced a flurry of discussion about aims and objectives, about methods of assessment, and about the psychology of learning and child development. What is wrong, they said, is not the content - after all the professional scientists had decided what that should be - but the method of presenting it was wrong. Analyse more carefully what you are doing! Define your aims, and particularly your behavioural objectives! Furthermore you must learn your Piaget!

Piaget was a natural scientist, who came to study the learning behaviour of some children. He has aroused great interest with his suggestion that in the development of our intellectual ability we all pass through a series of distinguishable stages. Only the last two of these stages are relevant to our present discussion, and have been termed

(a) <u>the concrete operational stage</u> - in which we have the ability to carry out mental operations on observed data such as classifying, ordering, noting simple relationships ; and

(b) <u>the formal operational stage</u> - in which we have the ability to carry out mental operations on abstract entities or concepts (as we do in algebra).
(Note that we are concerned with <u>mental</u> operations in both cases).

A lot of this must be received with caution. It is one thing to define an orderly set of stages. It is quite another thing to demonstrate (a) that these stages are really distinct rather than strongly overlapping in the population as a whole; (b) that it is possible to decide by tests whether a child is clearly in one stage or another; (c) that if these stages actually exist, the transition from the concrete to the formal stage can be facilitated. Nevertheless there is much food for thought here, and some of it does ring true to experience. At least we all know that children of a given age vary considerably in their ability to think about abstract ideas.

Piaget himself did not show in detail how his ideas could be made the basis of educational strategy. But a number of people have tried to do so. An example from the work of Ingle and Shayer is as follows(ref.3).

(i) Suppose firstly that you could plot the Piagetian mental stage against age for children, as shown by the three curved broken lines in Fig.1. The authors distinguished three different groups of children according to their Intelligence Quotients, and assumed that they would acquire the different Piagetian stages at different ages.

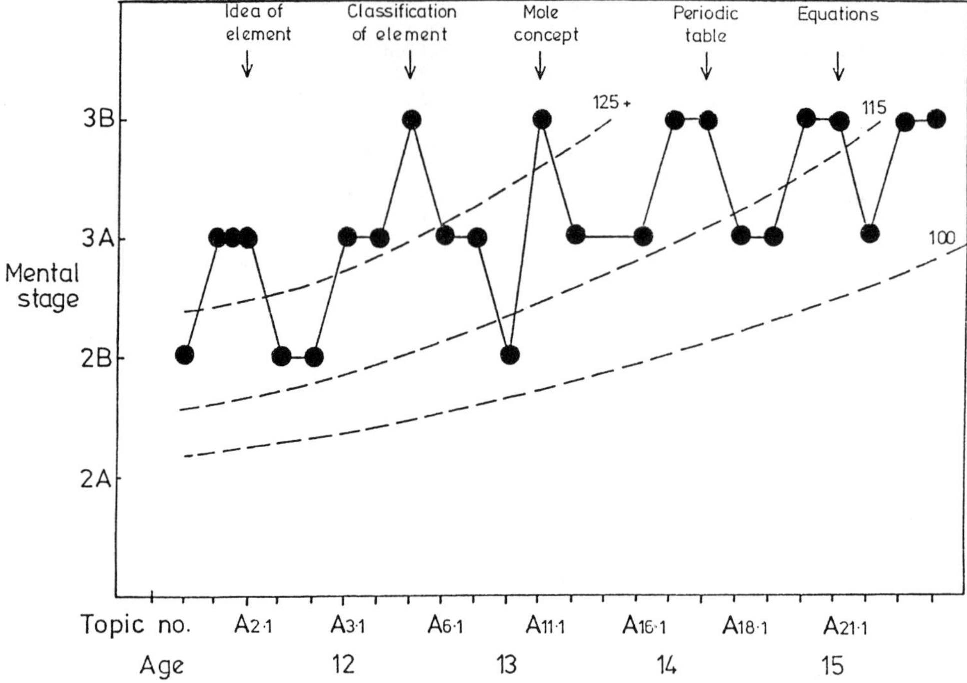


Fig.1. Classification of concepts in Nuffield O-level Chemistry course according to Piagetian mental stage, ( ref.3).

(ii) Secondly they analysed all of the concepts in the Nuffield O-level course for their Piagetian conceptual level. How one does this is still a very arbitrary matter. J.D. Herron, for example, has suggested an analysis of concepts based on whether they have perceptible examples and perceptible attributes, as set out in Fig.2 (ref.4).

There are many other such schemes for providing what some people delight to call a taxonomy of concepts. In whatever way one does it one can eventually allocate all of the syllabus concepts to their supposed Piagetian level.

| Concepts / Characteristics | Class I Concrete | Class II Formal | Class III Formal |
|---|---|---|---|
| Perceptible examples | yes | yes | no |
| Perceptible attributes | yes | yes | no |
| Instances | solid, liquid | element, compound | atom, molecule |

Fig.2. Some characteristics of concepts.

(iii) Very fortunately the Nuffield scheme has an order of treatment related to the age of the child. You will see that there are two variables plotted on the horizontal axis of Fig.1, Topic Number and Age. For each age the concepts to be introduced are plotted at their appropriate Piagetian level which can then be compared directly with the Piagetian mental stage of the child. It is suggested that only concepts underneath a given I.Q.line will be understood by that kind of child at that age.

This is very speculative, and we need to be properly skeptical about it. Nevertheless this approach has a lesson for us. I suspect that very often we are so familiar with our subject that we forget how difficult to grasp, or even how subtle, many of our science concepts are.

This then has been the prescription provided by the educational specialists. There are dangers of course in leaving these very speculative medicines in the hands of the teaching general practitioners. Concerning the matter of aims and objectives, Alex Johnstone (ref.5) has a cautionary tale to tell of the members of a working party in England who seriously prepared a list of no less than 450 objectives for the last 2 years of an O-level chemistry course! J.D. Heron (ref.4) writes of teachers who have suggested to him that as formal concepts may be impossible to grasp for children at the wrong Piagetian stage, those concepts should be left out of the course. It is as though we were trying to teach pupils tennis, and because some pupils find overhead serving difficult we should say to them : "Never mind, leave serving out of your game"! I think tennis without overhead serving is badminton (which I will admit some people enjoy). Science without abstract concepts is incapable of definition.

Despite all this specialist advice from firstly research scientists and then educationalists and psychologists, science education still did not seem to glow with health. Even some of the very bright pupils, who should have had least difficulty with formal operational thinking were found to reject science courses in favour of studies in the social sciences. Accordingly a third group of specialists came onto the scene. These we can designate as the humanists. What was needed they said,was to put science in a social context, to develop in children self-awareness, empathy with the environment, and a sense of social responsibility. Chemistry had been particularly hard-hit by the "science-is-evil" syndrome, with the cutting accusation that it produced only the effluent society. There was a move therefore to develop science courses "with a human face". A new Physics text once came onto my desk entitled "Physics, with Human Applications". The mind boggles at what the alternative applications might be. In Canada there was the CAUSS programme, which stands for "Chemistry: An Approach to Understanding Science in Society", designed to "lead students to an appreciation of the spirit of science and its role in civilisation". From this point of view there is less concern about the structure of science in disciplines or about its content of major explanatory ideas. Rather we must think of its social relevance. The emphasis is what is called by some the "affective domain" of education, i.e. changing a child's attitudes, rather than on the cognitive and psychomotor domains - i.e. on <u>knowing</u> and <u>doing</u>. Who is to decide what are the desirable attitudes and on what basis is of course never revealed! To change attitudes it apparently doesn't matter what content you use, and so a wide choice of teaching modules is offered.

But the story of the treatment of the science education patient is not yet up to date. A more recent phenomenon is that the first group of specialists, the university and research scientists, have had a change of heart. There were murmurings of this in 1975, in the David Mellor Lecture in Australia by Professor Adams (ref.6):

> "We must consider", he said, "how best we can strike a balance between the increasingly well-understood logic content and the factual-descriptive content of a course. ...........I think that the basic error has been to build theoretical frameworks at the secondary school level which are far beyond the present need".

Stronger sentiments were expressed in 1976 by Professor Gillespie in Canada (ref.7) in the following words:

> "My contention is that there is far too much emphasis on theories, and particularly on fictional theories, and far too little on facts. The uncomprehending learning of facts, which has been much deplored by the innovators of new curricula in the last 15 to 20 years, has been largely replaced by the uncomprehending learning of theories...... We do not encourage students to appreciate that theories are invented to explain facts; hence theories tend to assume the role of the 10 commandments, to be adopted without question by the believers. This can have some unfortunate consequences."

Thus we went full circle, from complaints about more descriptive teaching in science by the research scientists to calls for more descriptive teaching in science by the same group of specialists. Do you wonder that I remarked at the beginning of my lecture that progress in science education has been like a random walk, in which for all one's walking one never gets very far from the starting point.

## THE TRENDS IN THE 1980s

Despite a temptation on such an occasion as this to adopt an attitude of criticism to what has been done, there is no doubt that in the sixties and seventies much that was good for chemical education was accomplished in many parts of the world. I don't think it matters very much that we have oscillated back and forth between extreme ideas. What is more important is that there has been a spirit of self-assessment, some professional educational and scientific input and a willingness to try new methods. What concerns me more is that there seems to have been much less of this kind of activity in the 1980s. What has been happening in several countries in the eighties is a steady shift in the distribution of the school population with a greater proportion of the students continuing into the senior forms. In a sense this is good but it has led to demands for a change in the curriculum, as no longer are the majority of students in the senior forms preparing for the tertiary level education leading on to the professions. But what is the requirement (if not the demand) of the majority in science education? Is it for courses which are thought to be relevant (like Chemistry and Cosmetics) or which are full of environmental and social concern, or which will enable citizens to cope with the increasingly sophisticated products of technology? I don't object to any of these goals in themselves, but I believe that we are missing the mark in science education if we accept that that is all that it is for. It betrays in us an instrumental view of science and a materialistic view of life. Life and culture must be guided and stimulated by a spiritual quality, and science properly understood is an activity of the human spirit. "Science as the understanding of Nature" has an important place in the education of all of our citizens just as much as "Science as the control of Nature".

There is a further danger in a too facile acceptance of curriculum changes to suit the majority. The courses for the average become the courses for everybody, and the very bright students, those who may become the future chemists of excellence, are discriminated against.

## TEACHING FOR EXCELLENCE

What then can be done to produce and maintain a climate of chemical education in our schools which will be conducive to the fostering of future chemists of excellence?

### Those who teach

We must look first of all to our teachers. Interviews with famous chemists about their careers frequently show that the influence of a good teacher in arousing the interest in chemistry of an intellectually superior student is greater than that of many other factors such as the overall rating of the school or university. We are short of qualified chemistry teachers for two reasons. Too few graduates are being attracted into the profession and too many of those who do enter it as chemists are transferring to the teaching of other subjects. This is not a new phenomenon but it is one which chronically we have been unable to solve. The reasons for it are various: partly financial, but more importantly lack of teaching facilities and lack of recognition for dedicated and successful work. All of us like some status in society; some acquire it through money and others through peer recognition. Teachers by and large receive neither. The only reward for being a good chemistry teacher is to become a headmaster or headmistress after which one has little further teaching contact with chemistry. A suggestion made some years ago was that occasional honorary degrees might be given to teachers by the universities which undoubtedly profit by their work. Chemical Societies also might consider giving appropriate recognition by prestigious awards. Some already do, but the practice could become more widespread. In New Zealand, the Department of Education grants a certain number of teachers each year leave with pay to spend a year working in a university department in their teaching subject. That programme was negotiated first of all by chemistry professors, and has since spread to other subjects.

I referred earlier to the discouragement of promising science teachers through lack of facilities and lack of time for preparation which leads some to transfer to the teaching of other subjects like mathematics. I call on those who have influence in educational affairs in their respective countries to press hard the claim that laboratory science teaching is more time consuming and more expensive in staff and equipment than the teaching of some other subjects and for this reason deserves special recognition. The claim will be criticised from within the profession but it happens to be true. For the fostering of future chemists of excellence good teachers are essential, and we will retain them only if they are afforded better professional recognition and better facilities than at present.

## A philosophy of science

Once we have good teachers we can consider the content of their science teaching. This must not be simply factual or excessively theoretical, neither the "uncomprehending learning of facts or the equally uncomprehending learning of theories" (ref.7). Rather, our teaching must be guided and stimulated by a well-developed philosophy of science. Science is one of our most notable achievements and is part of our common culture. It is characterised not by mere description of observed phenomena nor by ways of producing commodities but by ideas and explanations. The great explanatory theories are the essence of science, and it is the ideas of the famous people which have led to these theories which should be woven into our teaching at school. Such things can be discussed at several levels of sophistication so that they do not all need to be reserved for senior classes or for an elite intellectual group. By discussing with all students the topics which constitute the peaks of the history of chemistry we will educate them, that is, we will inform them of intrinsic elements in our culture and will prepare the minds of some of them to follow in these ways themselves. I don't go along altogether with the recent suggestion that we can actually "educate for the serendipitous discovery" (ref.8), but I endorse a comment made several years ago by Melvin Calvin (ref.9):

> "It is no trick to get the right answer when you have all the data. The real creative trick is to get the right answer when you have only half of the data in hand. When you get the right answer under those circumstances you are doing something creative.....Creative science involves taking a step beyond the information you have at hand. If you don't do that you might just as well be a computer."

Humans are better than computers and "Science is human". Without sacrificing its value as an instrument for material progress we must give it a proper place in education as a major achievement of the human spirit (ref.10).

After one has an idea in science there is an obligation to test it. But the separateness of the creative and the critical components of scientific thinking is far from obvious in practice because "the two work in a rapid reciprocation of guesswork and checkwork, proposal and disposal, Conjecture and Refutation" (ref.11). Imaginative thought and criticism by experiment are equally necessary in all of science, whether it be in the "basic" or the "applied" sector, and some of those who are educated to see this will be attracted to devote themselves to it.

I doubt it these views of the nature of science are well understood by many teachers. Yet how can we teach without a guiding philosophy? A colleague with whom I shared these thoughts remarked sadly, "No one ever calls conferences to consider such basic questions." He is right, but why don't we? Here I make my second challenge to the powers that be. Can we have a conference in the field of chemical education on the basic philosophy of science?

## Curriculum and textbooks

After the quality of the teachers and a philosophy to enlighten their teaching we come to the topics of curriculum and textbooks. Here we face the problem of the vast size of our subject which can lead, if we are not careful about it, to what A.H. Johnstone has rightly identified as the phenomenon of "learning overload" in our students (ref.12). Obviously selection of material is required but the critical question is on what basis is that selection to be made. At school the acquisition of facts is a concrete achievement with its own satisfaction but the "main objective should not be the filling of the memory but the guiding of the mind" (ref.10). Such teaching is an art and requires constant decisions about what to leave out rather than what to include. "The tactics and the strategy of science can be understood best from a few well-chosen and fully developed examples" (ref.13). At the university level choices must also be made. Here, depth of treatment of what are judged to be significant topics for probable developments in the immediate future should take precedence over a broad coverage. Along with this should be presented what has been called by my own university teacher H.N. Parton, and what he himself excelled at, " the perspective view". There are some who are saying, and they exist in my own university, that the length of time for a science degree must be extended. The case for such extension follows naturally from a belief that scientific education consists of the assimilation of specialised knowledge. But as P.B. Medawar has remarked (ref.11), "the length of university schooling is far more important to those whose education ends with graduation than to those for whom education is an indefinitely continued process." And from another source,"Education should produce certain qualities of mind. So it is not what people have learned which is important so much as how they have learned it; not how much knowledge they have but how well equipped they are to acquire knowledge as they need it" (ref.10).

I believe it is from our textbook writers that we should receive help in coping with the selection of material for teaching General Chemistry. Many of those we have today are magnificent visually (with their colour plates), quite good scientifically and absolutely ridiculous in selectivity! You simply can't cover all that is in them in the time available. I myself do not write textbooks, so I must be careful about what I say, but because we offer our courses extramurally I frequently write Study Guides to accompany textbooks. Much of my time is taken up explaining what can be left out! In our Department we have consistently avoided chasing after the biggest and the glossiest books as though they were the best. Some of them you can hardly carry. I am happy to find that I am supported in this criticism by no less an authority than Linus Pauling. It simply proves the truth of the quip that not only great minds think alike! In a recent address to science teachers he is reported as saying (in summary) that today's General Chemistry textbooks and probably the courses based on them are too heavy, too long and too advanced, and serve to turn interested students away from chemistry instead of attracting them into the field (ref.14). Advanced textbooks have suffered from the same faults. I recall with interest the celebrated review by E.A. Guggenheim of the monumental book on Physical Chemistry by Samuel Glasstone (ref.15).

> In a certain city a councillor with grandiose ideas conceived the curious desire of having a picture of the whole city painted on a single canvas. The city indeed contained many fine modern buildings as well as older buildings of historical interest, of which good pictures already existed. But these did not satisfy the councillor who insisted that he should have one complete canvas of the whole city; every building large or small, ancient or modern, beautiful or ugly had to be included. Eventually, however, a certain travelling artist, who happened to be paying a visit to the town councillor, undertook the task of painting the whole city on a single canvas. He used a very large piece of canvas and he worked hard for two whole years, painting so many buildings each day and even working over weekends. At last his task was completed and true to his word every building, large or small, ancient or modern, beautiful or hideous,useful or useless was included on his canvas, but the product of his toil was not a picture because he had omitted to pay any regard to perspective. The name of the city was Physical Chemistry and the name of the painter Samuel Glasstone.

Brevity in science writing is possible. Sir Cyril Hinshelwood wrote in the preface to his book entitled The Structure of Physical Chemistry, "I thought I would like to write a book of moderate compass which should lay emphasis on the structure and continuity of the whole subject and try to show the relation of its various parts to one another."

We need good textbook writers and I marvel at what some of them have accomplished. But I would like to challenge them from this conference to give us something in General Chemistry that will restore a student's confidence that we have a course of learning which is of manageable size. And let it be a book which is bold enough to restrict itself to simple explanations where these are helpful to understanding at a certain level. It will be in keeping with the nature of science for some of these simple explanations to be modified or even replaced as the student proceeds to higher level courses.

## CONCLUSION

It has been said that there are certain subjects on which wise people keep silent. I am beginning to wonder if chemical education for the future is one of them! I must leave that for you to judge. My thesis has been that there is no magic formula for producing excellent chemists to order, but they must be attracted to our subject out of the general population of able students. The most significant attractive force for this purpose will be ourselves. It was at Harvard University, so I read in Nature some time ago, that there was a debate concerning a proposal to introduce a core curriculum for all undergraduates. What courses should be included and how could they best be taught? I am proud to report that it was a professor of Chemistry who made this response: "The best courses taught will be those which someone is eager to teach." May that attitude characterise us all.

## REFERENCES

1. Nature, Lond. 314, 391 (1985).
2. A.Huxley, Brave New World, Chatto and Windus, London (1932).
3. R.B. Ingle and M. Shayer, Educ. in Chem. 8, 182-183 (1971).
4. J.D. Herron, J.Chem.Educ. 55, 165-170 (1978).
5. A.H. Johnstone, in New Movements in the study and teaching of chemistry, D.J. Daniels (ed.), p.218, Temple Smith, London (1975).
6. D.M. Adams, Proceedings Royal Australian Chem.Inst., 163-171 (1976).
7. R.J. Gillespie, IUPAC Intl.Newsletter on Chem.Educ. No.6., (1977).
8. R.S. Lennox, J.Chem.Educ. 62, 282-285 (1985).
9. M. Calvin, Chemistry in Britain 9, 564-569 (1973).
10. H.N. Parton, Science is Human, University of Otago Press, Dunedin, (1972).
11. P.B. Medawar, The Art of the Soluble, Methuen (1967).
12. A.H. Johnstone, J.Chem.Educ. 61, 847-849 (1984).
13. J.B. Conant, The Growth of the Experimental Sciences, Cambridge, Mass. (1949).
14. L.Pauling, The Science Teacher, 25-29, September, (1983).
15. E.A. Guggenheim, Trans.Farraday Soc. 38, 120 (1942).

# Lessons from the past and prospects for the future of chemical education in Brazil

Reiko Isuyama
Institute de Quimica, Universidade de Sao Paulo
USP Caixa Postal 20780, cep 01498, Sao Paulo, BRAZIL

In the society we live in- technologically based, developing, with a hundred and thirty million people whose medium age is fifteen, and where half of the pupils who enter primary school leave it in the first year - we need to produce professionals capable of participating in the construction of our country.

We must solve many problems - so they need to know how to solve problems; we need to develop urgently - so, they must be updated, prepared to attend to our future needs; we need to develop chemistry, a crucial point of our technological development - so they must be excellent chemists.

The apprehension about the level of excellence of the chemists we need led me to a question: -does an excellent student become an excellent professional? -Not necessarily - is the answer I have found by my experience.

Our society does not need just an excellent student but an excellent professional. It needs clever chemists who have enough knowledge of chemistry so as to permit them to think critically, to find solutions for our problems; but it needs especially those who show action that reflects such qualities.

In general, one thinks of a variety of qualities in order to consider a chemist as being of excellence: he/she must know facts; understand the nature of chemical principles and theories in depth; it is important for he or she to understand that chemistry is not merely a collection of isolated facts but a consistent body of knowledge, and needs to be able to think logically using arguments derived from that knowledge; he or she needs to have the ability to distinguish between scientific evidence and personal opinion by enquiring and questioning.

Pondering about all the necessary and important qualities, it seems to us that they all refer to one central quality, and that is free-thinking or a flexible mind.

Logical thinking is the first skill a chemist needs and the development of the various stages involved in the scientific thinking process - observing and describing, raising hypotheses, searching for regularities for generalization and explanation - needs freedom of thought, thinking without prejudice.

## WHAT IS PREJUDICE?

Prejudice, according to the dictionary, is a "preconceived opinion or judgement, formed before one has adequate knowledge or experience, leaning towards one side of a question from other considerations than those belonging to it, a concept conceived beforehand or independently of experience or thought". People tend to acquire prejudices towards everything they experience, be it individuals, groups, political and religious institutions, moral and philosophical systems.

Prejudice negatively influences the basic steps of scientific thought:
- observation - the ability to see critically. One must be able to see things that others do not see. We must be free of prejudice, since prejudice disfigures the vision, making one see things that do not exist and not letting one see those that do exist.

- Raising of hypotheses - to propose hypotheses one must evaluate critically the information obtained through observation. If there is no predisposition to certain beliefs and opinions it is possible to evaluate ideas objectively.

- Generalization - to analyse critically the possibility of extending the concept to all cases to which it could be applied, one must be able, again, to think without preconceived ideas.

A person without prejudice can tackle and solve problems considered insoluble. We can all remember some problem we could not solve merely because we did not have the ability to change our focus. Changing focus means removing prejudice.

A free-thinking or flexible mind is not a quality required only by chemists or scientists. It is fundamental for any professional and for everyone. Our society needs men and women with flexible minds.

As an intellectual discipline, chemistry is an excellent means for developing a variety of important mental skills. It can serve as the basis for helping a student understand that preconceived ideas may lead to wrong conclusions. There are many examples in chemistry that can be used to create an impact on the students demonstrating to them that concepts are not absolute. For example, we can use the stoichiometry of one simple acid-base reaction to show that one plus one can equal one; or that by mixing colourless substances we can obtain a beautifully coloured compound, or that can impress the student and show him or her that conclusions based on previously accumulated knowledge can be erroneous.

It is important for the student to admit that there could be other factors that must be taken into account when we analyse some conclusions. He or she must be aware that the concepts accumulated by him or her were based on experiments that occurred under a certain set of conditions and if conditions are changed the same experiment can lead to different results. Chemistry allows the students to understand that external conditions determine the kind of reaction; so, if the conditions are modified we can obtain new products with completely different properties than those obtained under the former conditions. The essential statement of chemistry - a set of external conditions determines the behaviour of compounds and everything that derives from it - can be useful as an educational instrument to develop the flexibility of the student's mind.

To educate the students in such a way, firstly we chemical educators must be able to think freely. Thinking now about the current problems of chemical education in Brazil, it seems to us that our teachers' preconceived ideas about curriculum, methodology and evaluation have some harmful aspects. Although new and innovative approaches to chemistry teaching, which demand flexible and creative minds, are needed, presently we are not able to design any solution.

A number of approaches to chemistry teaching can foster good chemists. There is of course not only one route for the solution of the problems in chemical education.

Considering the complex structure and the dynamic nature of the problems in education, it is very difficult to submit these problems to a logical analysis, i.e. searching for all factors that are influencing the problem, studying each of their effects, and evaluating the results in order to find a solution. With the current level of knowledge about the development of mental skills, learning processes, human behaviour, etc., it looks to me to be difficult to submit the problems of chemical education to a simple analysis based on a cause and effect relationship. The search for a solution may be based on the future. In these projections we must have in mind that our impression of the future is often biased by predictions based on a linear projection from past experiences,i.e. we cannot look to the future carrying forward preconceived ideas.

The beginning of the chemistry course at the University of Sao Paulo, fifty years ago, can be reported here as an example of experience in chemical education fostering chemists of excellence. This first academic chemistry course was conceived and organized by two renowned German professors. The course structured by them followed a German model for 34 years. Many of the students who obtained their degree during this period were considered in our country as chemists of excellence.

The teachers of that period had a clear educational philosophy based on their perceptions of chemistry. According to this philosophy they stressed to their students the development of the so-called chemical thinking process, the ability of reasoning through phenomena.
This is obviously a statement of the belief that chemistry is an experimental science and that much is to be gained, intellectually and physically, by introducing students to the laboratory.
Students who were exposed constantly to experimental work, learned the vital knowledge in depth. Chemists from that school feel chemistry. They acquired not only the ability of reasoning in chemistry, but also an intuition in chemistry. This is the result of a hard educational process, both theoretical and experimental. This is our past.

Our present begins seventeen years ago when the University of Sao Paulo went through a total reform, in which a system of departments was substituted for the structure based on one full professor with many assistant teachers. At this time the massification of education at university level also occurred. The demographic explosion of students certainly is the main reason that dictated a big change in the mentality of our institution.

The current structure is different and the scale is also different, but it is usual to find professors who want to teach as they were taught at the old university. The preconceived idea that teaching was good is ingrained in our minds and does not allow us to realise that only at that time, for about fifty very well selected students, with ten professors involved in the teaching and for the chemical knowledge of that time, i.e. under those conditions, such a course ran very well and was successful in fostering good chemists. Today, it is a different situation: now, we have two hundred chemistry students and about two thousand students from other institutions attending the chemistry course; a hundred and forty teachers and the chemical knowledge duplicating every fifteen years.

The current structure of departments contributes to the spread of the content of chemistry across many short-term disciplines, reduces the contact among teachers and as a consequence, the level of our students declines.

Added to the various kinds of problems resulting from the reform of the university, there is another one due to our condition as a developing country. We must make an enormous effort to reach the development that is going on in the world, because the gap is continuously increasing.

It is widely known that scientific and technological knowledge doubles every fifteen years. Given this, one infers that any young scientist starting his or her career today will find, after 35 years of work, that about 80% of the scientific progress had occurred in front of his or her eyes. With this in mind, it is important to change the objectives and aims of education. Once it was sufficient for the students to learn and to understand what was taught, but nowadays the student needs much more. One has to make an effort to build one's own edifice of knowledge mostly by one's self. The current rate of expansion of knowledge requires a greater effort on the part of teachers and students to remain up-to-date.

At present, students may have different learning capabilities and thinking abilities, but they must have creative skills. If it is difficult for us to predict the nature of future relevance, it would be useful to develop creativity in our students in order to allow them to solve future problems regardless of what they think will be relevant. The development of mental skills takes time and such time cannot be economized.

On the other hand, the time necessary to learn the accumulated chemical knowledge would be shortened by the use of computers.

The astounding volume of chemical knowledge makes a re-examination, not only of the methodology but also of the curriculum, indispensable. Since the contents to be covered increase every day and the duration of the course is fixed, it is necessary to re-select the course contents.

This selection of content has to be made objectively in order to teach up-to-date chemistry. One criterion for the selection of essential contents of an undergraduate course could be the analysis of the knowledge in chemistry we must achieve to be able to use meaningfully the current instruments used by local chemists. Analyzing the current chemistry content needed for this purpose would give us an idea of the maximum level we must reach in an undergraduate chemistry course. Then, we could look for the concepts necessary to understand this content level and identify the basic ones.
Following this line of reasoning, we can select only those concepts essential to prepare a good chemist. I am not advocating a teaching method starting from high concepts and going to the basic ones - the proposed idea is valid only for selecting the essential subjects.

The time available for students and teachers during a course is limited and thus extremely precious. It is not acceptable to waste time with superfluous subjects, such as those which became obsolete or which are too advanced or of little importance:these simply being obstacles standing in the way of the best use of limited time. We can take as an example, a garden labyrinth used in olden times for amusement. Starting at the beginning and going on, choosing the way to reach the end, one faces many obstacles because if one takes a wrong way, one must return to take a new one, until facing another obstacle and so on.

Everyone knows that it is easier if you take the opposite route, starting at the end and going back to the beginning. Following this direction you can find the right route easier than by using the former way, because there is no chance of entering a wrong path. The final objective is reached certainly, because one can go directly to it.

In reality one cannot, of course, begin at the end but must start at the beginning; however, if one has clearly in mind the final objective one can reach it choosing the shortest route.

The society we live in, which is in a transition stage, undergoing many changes, requires not only free-thinking chemists or chemical educators but such men and women in general, more than in times of stable societies. Those men and women have to think freely, without prejudice, not only in the scientific field but in all aspects of human life.

# The teacher as the key to excellence in chemistry

Peter E. Childs

Chemistry Department, Thomond College of Education, Plassey, Limerick, Republic of Ireland

Abstract - Teachers are the key to excellence in chemistry, as they work at the interface between professional chemists and the citizen. The second-level teacher in particular is responsible for presenting 'chemistry for the chemist' and 'chemistry for the citizen', and must strive for excellence for all students, whatever their ability or sex. Knowledge and love both of chemistry and of students are identified as the important qualities of good teaching. These qualities must be looked for in selecting teachers for training, they must be developed during appropriate preservice courses and their growth encouraged throughout the career of the teacher. More emphasis needs to be put on inservice education for all teachers at regular intervals until retirement. The cost-effectiveness of various methods of inservice education is evaluated, and IUPAC and UNESCO are encouraged to put more funds into national and regional ventures. The chemical community needs to exert its influence to improve the pay and prospects of second-level teachers, if we are to attract and keep the best chemistry teachers working at the most important level of the educational pyramid.

## INTRODUCTION

The thesis of this paper is that there can be no excellence in chemistry, at any level of education, without good chemistry teaching. In particular I want to focus on the crucial role of chemistry teaching at secondary level in providing the foundation for the edifice of chemistry. There seems to be a crisis of confidence in chemistry at the moment, which is brought into focus and magnified by disasters like Bhopal, Love Canal, Seveso, Flixborough, Minamata, etc. If it's good news in science then physics and biology get the credit; if it's bad news then usually chemistry takes the blame. There is widespread misunderstanding of and alienation from chemistry in the general public and anything 'chemical' is suspect. In some countries there is also a crisis in chemical education with grave shortages of qualified chemistry teachers at secondary level - they have been described as "blue whales" in their rarity (ref. 1). Is there perhaps a connection between the backlash against chemistry and the inadequacies in chemical education at the level where most citizens receive their first and last exposure to chemistry? Any major change in attitudes towards, or the status of, a profession must start with education at the lowest levels of the educational pyramid.

Over the past twenty years or so many remedies have been proposed to cure the ills of chemistry and chemical education. Some of these are noted below:

- new curricula have been designed and implemented (and redesigned) in many countries;
- shortage of materials and resources has been identified as a major weakness;
- new teaching methods have been tried;
- new teaching aids have been loudly acclaimed and then quickly dropped, with microcomputers being the latest in the line;

- o relevance has been a rallying cry in trying to interest pupils in chemistry;
- o improving pay and promotional prospects for teachers;

  etc.

We could all add our own examples to this list. Vast amounts of effort, money and time have been expended and the results are largely disappointing. None of the changes seem to have made a major or long-lasting effect on the system and the crisis deepens year by year - perhaps we have missed the real key to the problem? None of these improvements alone, or indeed all of these together, can ensure excellence if the teachers are not able to teach well. Teachers are the key to excellence in chemistry. The quality of schools differs widely, despite similar provision and curricula - the difference usually lies in the teachers. Many great chemists have acknowledged the positive influence of a teacher early in their careers, (e.g. ref. 2) and many of us can remember the negative influence of a bad teacher. The 'good' teacher will succeed in passing on her/his love of chemistry and the necessary skills to her/his students despite the curriculum followed, the lack or abundance of resources or the latest teaching aids, and whatever the teaching method used. The good teacher manages to overcome all obstacles and succeeds - the poor teacher often fails with all the advantages. The American Chemical Society appeals to its advertisers: "Capture a teacher and you capture a generation". In any subject, in any country, the teacher is the main medium by which the message is communicated to the students and through which excellence is achieved. This means that our main effort and resources must be directed towards improving the quality of teachers as a first priority, both during their initial preservice training but more important during their whole inservice career. We need to know who we are trying to teach chemistry to, what it is that makes a good chemistry teacher, how we educate and train such teachers initially and throughout their careers and how we can attract teachers into the profession. I want to discuss each of these aspects further.

## EXCELLENCE FOR WHOM?

When we talk about achieving excellence in chemistry through education it is important for us to define our goals. Excellence must be a major goal for a committed chemistry teacher but we must ask the question, "excellence for whom?" In the past chemical education has been concerned to produce excellence for a few, who would become the successful researchers and prize-winners of the future. This is important but it only represents the tip of the chemical pyramid, which requires a substantial and firm base for stability. Our neglect of applied in favour of pure chemistry may result in economic decline, but the neglect of chemical education in favour of pure chemistry, typified by high-level research, may result in a complete collapse of the structure. The chemistry teacher, particularly at second level and to a lesser degree at third level, works at the interface between professional chemists and the citizen (see Fig. 1).

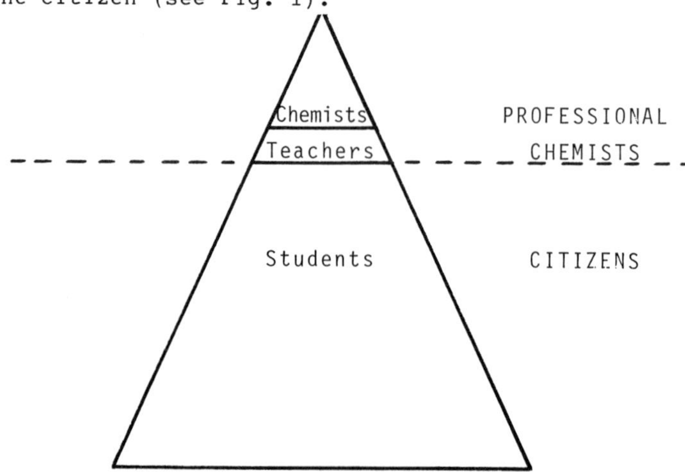


Fig. 1 The chemical pyramid with teachers at the interface between chemists and citizens

Most students of science and chemistry at secondary level will never take advanced courses in chemistry, and only a few will become professional chemists. The teacher's key and crucial role is to transmit the nature of chemistry across the interface and to communicate it afresh to each new generation. As chemists we know how important interfaces are in many areas of our subject - the teacher acts as an interfacial catalyst with a two-fold function. Firstly, the teacher has to identify, inspire, encourage and prepare those few students who will go on to be professional chemists - in academic, industrial or public spheres, and who will pick up and carry the torch of chemical knowledge for the future. To fail to do this would be to betray our subject. But secondly the teacher is also charged to educate and care for the 90-95% of her/his students who will not become chemists, and must try to inform and develop an awareness and appreciation of chemistry in them also. Excellence is required in both these jobs, not just in the first, and that is what makes good teaching so demanding. All our pupils are important whatever their sex or ability. For the chemistry teacher to concentrate only on the elite few who will become professional chemists is narrow and short-sighted. Chemistry as a discipline would survive but it would be cut-off from the majority of society and the inevitable result of ignorance is fear. We need to be as concerned with excellence in 'chemistry for the citizen' as for excellence in 'chemistry for the chemist'. We do have a real choice of being elitist or populist - we must try and achieve excellence for all our students, whatever their eventual careers. Sir George Porter stressed general education in science for all in his Presidential address to the Association For Science Education (ref.2), in which he said:

> "The ignorance of natural science among the great majority of our fellow citizens is no longer just deplorable, it is close to being a national disaster."

Working out how to achieve these two goals, often in the same laboratory/classroom, is one of the challenges for the 1980's. We must not forget though the other dimension to 'science for all', based not on ability but on gender. The reluctance of girls to study the physical sciences is a major problem in most countries and means that we are failing to develop 50% of our potential chemists properly, as well as 50% of our citizens. Many studies have shown that teacher attitudes are of major importance in determining subject choice for girls - once again we are back to the teacher as the key to the problem. So what is a 'good' chemistry teacher?

## WHAT MAKES A GOOD CHEMISTRY TEACHER?

If good teaching is so important for the future of chemistry in schools (and at higher levels also) then we need to know what characterises a good teacher. Although there is no single profile applicable to all good teachers, there are some general features common to good teaching that have been identified by various authors and research surveys. In training teachers we should be looking for these characteristics and trying to develop and improve them. A report from H.M. Inspectors in 1982 (ref.3) identified misguided selection as a major factor in the deficiencies found in new teachers. A stress only on academic achievement at school or university rather than aptitude for teaching in selecting teachers was identified as a major flaw. A good teacher, they said, is marked by the qualities of "enthusiasm, common sense, tenacity, sensitivity and friendliness". This has an important lesson for those selecting and training potential chemistry teachers: at least equal weight must be given to personal qualities as to academic excellence, because effective teaching is more than just knowing one's subject - one must be able to communicate this to one's students. As practising chemistry teachers we should take note of the qualities associated with good teaching and should be seeking to improve our own teaching to reach higher goals. Becoming a master teacher of chemistry is a life-long process and should be our long-term goal. There seems to be broad agreement between different descriptions of the good teacher, of any subject (refs. 3,4,5,6,7) and I have reduced these to two main characteristics which are applied to two objects. Good chemistry teaching requires <u>love</u> and <u>understanding</u>, of both <u>chemistry</u> and the <u>students</u>. You cannot teach a subject well that you have not mastered yourself, and that to a level higher than it will be taught. But you must not only <u>know</u> your subject you must <u>love</u> it as well in order to get it over to your, <u>often</u> unwilling, students. Enthusiasm, zeal, passion, commitment, inspiration are words that are frequently used to describe the good teacher. But many top-notch researchers at third-level have both these characteristics and are abysmal lecturers. The teacher is also a go-between, a communicator between chemistry and the student (see Fig.2). We might say that the chemistry teacher is a bond-maker. If there is no real communication or bond-forming there is no learning and thus no teaching.

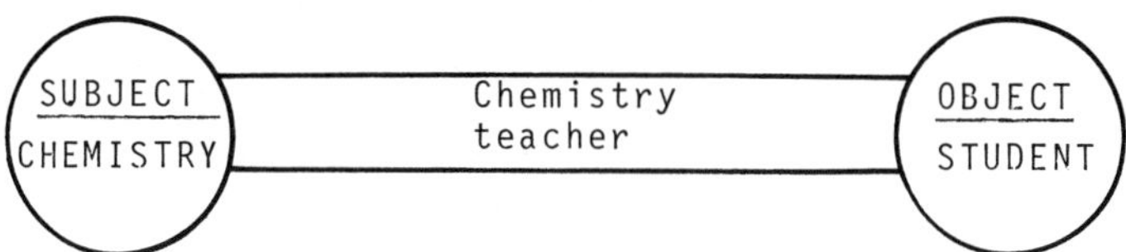


Fig. 2 The chemistry teacher as a go-between

In order to teach well the teacher must therefore also understand the natures and abilities of the students, and how children develop and learn ideas, and the teacher must also love and care for those whom she/he teaches as individuals. When students are asked what makes a good teacher, caring comes out top of the list (ref.5). I think this is interesting and significant. Teaching is a humane occupation and involves relationships, and togetherness in a shared endeavour. It is a people-centred occupation and thus personal qualities are important. Our failure in the past has been to assume that excellence in chemistry teaching is to be equated solely with excellence in chemical knowledge or understanding. The good chemistry teacher understands and cares for both chemistry and for each student individually. Students are not like atoms, best treated in bulk. In selecting prospective teachers our main emphasis should be on these vital personal qualities, and only secondly on academic excellence, though ideally we would like both. We cannot expect to find deep knowledge and understanding of chemistry or of people in an 18 to 22 year old, but we should be looking for a love and enthusiasm for both. These are the sparks that will kindle the corresponding fire in the student which will lead them to excellence. As one student said, "If I could be as good as my chemistry teacher, I'd love it" (ref.8 p.499). A teacher with the knowledge but not the personality will make a poor teacher; a poorer knowledge but the right personality will be a better teacher; a teacher who combines a good knowledge of chemistry and the right personality will be an outstanding teacher. This is what we should be aiming at in selecting and training chemistry teachers.

## HOW DO WE EDUCATE AND TRAIN GOOD CHEMISTRY TEACHERS?

A recent publication (ref.9) has surveyed the preservice and inservice education of science teachers. I want to suggest that our goals in producing good chemistry teachers are twofold: firstly, we must develop knowledge of chemistry together with acquisition of practical laboratory skills; secondly, we must also develop knowledge of children and of educational theory, together with the acquisition of teaching and communication skills. The programme must involve both chemistry and education; it must involve the development of the intellect as well as the acquisition of practical skills; it will involve both education and training. The good teacher must be more than just a skilled technician but she/he must also be more than a good theoretician. However, even a proper blending of theory and practice, knowledge and skill, chemistry and education, content and process, still misses out the vital ingredient to make this mixture palatable to its recipients. As Paul, himself an effective teacher, said a long time ago (ref.10):

> "I may have all knowledge and understand all secrets
> - but if I have no love, I am nothing"

We must also be concerned to inspire, encourage and develop a love for chemistry and for their students in trainee teachers, without which the correct mixture is insipid and tasteless. It is such intangible qualities that transform technical competence into inspired excellence in teaching. The same is true in chemical research where there are many productive technicians but only a few inspired researchers whose devotion and originality lead to major advances.

The challenge is how to include this vital ingredient in our courses, where so often it is lacking. In a few students this love will be innate and only needs watering and feeding, but in most it will have to be seeded into prepared soil, by providing appropriate role-models and exemplars of good practice. The people who teach teachers must also be good teachers if more than 'head knowledge' is to be transferred. This applies to courses in chemistry and education as well as in teaching methods. "Love is caught rather than taught" and those of us who teach teachers must be living examples of what we preach. Sadly this is often not the case.

"We teach not how we were taught to teach (theory) but how we were taught (example)" is another aphorism containing a vital truth. I have already stressed the importance of good teachers in turning students towards careers in chemistry - the same is true at higher levels also. Just as teaching chemistry at any level is both an art and a science, so too is the teaching of potential chemistry teachers. The selection and training of tutors on such courses must be such as to achieve the desired goals, and we should be concerned that trainee teachers are exposed to good teaching. I have heard too many complaints from teachers about the quality of their formal training and the poor teaching they received in college to be complacent about it.

But I want now to turn briefly to look at the ways chemistry teachers are produced. Marjorie Gardner has pointed out (ref.11) that the education of a chemistry teacher starts at primary level and should finish at retirement. The formal training part of a teacher's career takes up only a small part of the total span (see Fig.3).

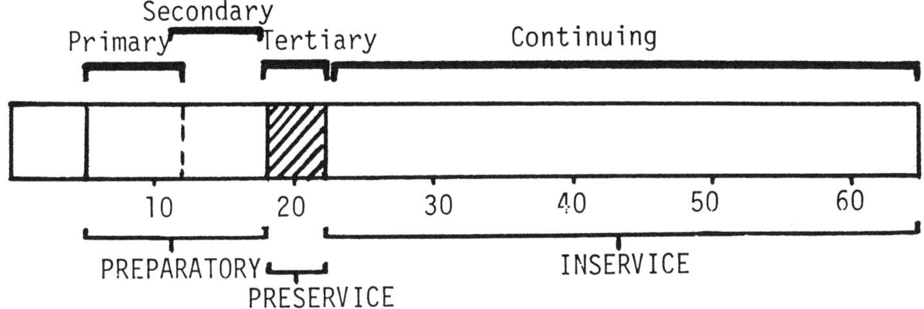


Fig. 3 The chemistry teacher's career

Preservice training is influenced by the quality of the primary and secondary science experiences that precede it,and will be modified and developed by the inservice experiences that follow it. Unless second level students see examples of good chemistry teaching and perceive teaching as a worthwhile and satisfying career, there will be no-one to fill the chemistry and science education departments at third level. There would seem to be two main routes worldwide by which chemistry teachers are produced, although there will be many variations in detail from country to country. Fig. 4 illustrates these two routes - the ends-on (or topping-up) route and the integrated route.

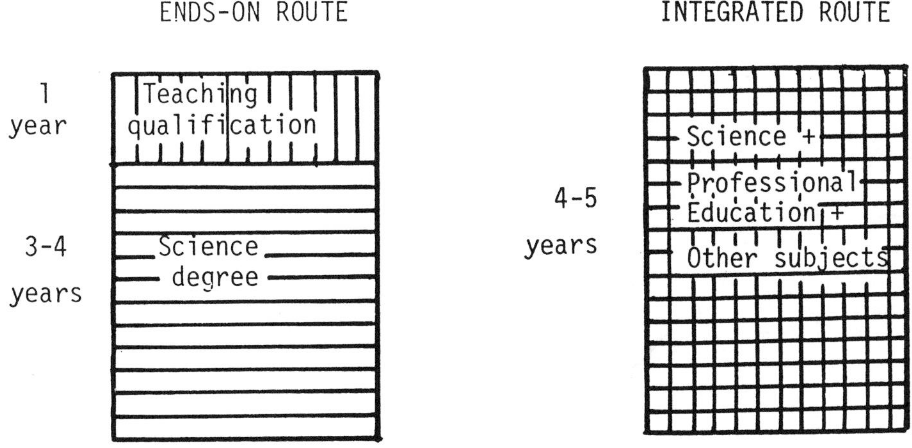


Fig. 4 Two routes for producing chemistry teachers

The ends-on approach involves a conventional undergraduate science course followed by a one year post-graduate course for a teaching qualification. In the U.K. and Ireland this is the predominant route for producing science teachers, largely for historical reasons. The one year course resembles the icing on a fruit cake: a layer of educational theory, science education, teaching methods and school practice is applied thinly on top of a conventional science degree. The science and the education/teaching components are firmly separated into different (and immiscible?) layers, taught by separate staffs with different goals. Many students (in Ireland at any rate) find such a course an inadequate preparation for the realities of teaching at secondary level. Both the chemistry they did earlier and the education theory are seen as largely irrelevant to the job in hand. The teaching practice is seen as the most valuable part of the course but rarely lasts for more than ten school weeks in total, and usually less. I myself would have severe reservations about this route for producing good chemistry teachers, though it does succeed for some. It is a subject-orientated approach to teaching giving most weight to competency in the subject to be taught, as evidenced by a degree, and downplaying the professional aspects of science teaching. The decision to teach is taken at the end of the degree course and for many it is a last option - unfortunately most of the better students go into research or into industry. Shaw's cruel aphorism "He who can does, he who cannot teaches" has too much truth in it to be ignored. We should not recruit teachers from those who can't do anything else if we want to maintain the quality of the profession. Do we believe that good doctors would be produced by a general degree in the life sciences, followed by a one year conversion course? I think we would all question the competency and professional skills of such a doctor - I wouldn't go to that doctor myself! Is our acceptance of the traditional ends-on approach for producing science teachers in many countries a failure to face up to the need for fully professional science teachers? Lunetta and Yager (ref.12) describe an ideal teacher-education programme and interestingly describe the experiences in medical terms - training is an internship and teaching practice is clinical experience. If we want to raise the status and standard of teachers we must raise both our standards of intake and the professionalism of our courses.

The alternative route to producing science teachers is the integrated approach, where students decide from the start that they are going to be teachers - as a first choice, rather than a last option. The integrated route (Fig.4) usually takes the same time to achieve the same nominal goal - a qualified teacher - as the ends-on approach, but in a radically different way. It is a more professionally-oriented course, more akin to the way in which other professionals are trained. The integrated course usually has three main elements: science/chemistry, professional education and general courses/liberal arts. Fig. 5 shows an example of an American course (ref.8, p.487).

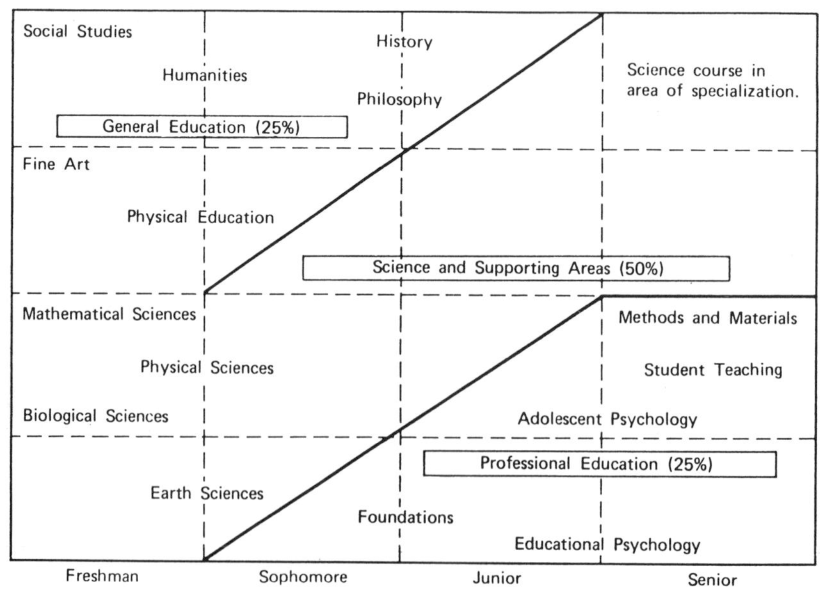


Fig. 5 An American model of the integrated route (reproduced by permission from "Science, Students and Schools", R.D.Simpson and N.D.Anderson, Wiley, New York, 1981 p.487)

The integrated route (with many variations) is the main route followed in the U.S.A. (In some countries the two routes operate in parallel - the integrated route is used to produce teachers for lower secondary years, whereas the ends-on route is used to produce more specialised teachers for the upper secondary years. Zambia and Uganda are two countries that come to mind.) Our own degree course in Thomond follows an integrated model. It differs from the American model above in introducing professional education courses, including science education and teaching practice, from year one. The main teaching area, subsidiary subjects and professional education are carried in parallel through all four years. Students need time to assimilate and integrate ideas into their own educational framework. We would all accept this premise in respect of chemistry, why not in favour of education/science education/teaching as well?

Too often though the integrated course is integrated only in theory, with a rigorous division between science and education courses and their staff. Students find it difficult to apply knowledge in different contexts. Chemistry and education must be integrated, synthesised and applied at the interface with teaching practice. This is why a component of science/chemical education must be a major component of any teacher training course for chemistry teachers. Both the subject (chemistry) and the educational theory must be made relevant to classroom practice and this is another major challenge in training good teachers. Too often the chemists who teach the chemistry know nothing of education and teaching (and don't want to); while the educationalists usually have an arts-bias with no science background. Herron and Brooks (ref.1) have sounded a clarion call to chemists to wake up and realise the need for us to take chemical education seriously, for the preservation and conservation of a vanishing species - the chemistry teacher. Each chemistry department should have a small chemical education section, just as each education department training science teachers should have specialists in science/chemical education. They might well be the same people in a given institution with joint appointments. In Thomond all the science education courses are taught by science staff. In order to produce good chemistry teachers we must have good staff members, competent in chemical education, to teach them. I want to throw out another challenge to chemists to take chemical education seriously and to approach it in a professional, rather than an amateur manner. It is chemists who must equip themselves with the knowledge and skills of education, science education and chemical education; who must get involved in research and scholarship in chemical education, so that we are able to apply the findings and insights of 'cognitive science' (ref.13) to the teaching of chemistry. I need to do this more myself. This applies not only to training secondary level teachers but also to teaching chemistry at the third level. Before leaving this topic I must also add that we must involve more practising teachers in the training of science teachers. The idea of using Master Teachers (see refs. 9 and 11) is important to give student teachers practical advice on the day-to-day problems of science teaching, and to give them appropriate role-models. Too often the people who teach teachers are out of touch with classroom realities and lack any, or at any rate current, classroom experience. Practising teachers must be involved to some extent in the training of new teachers, both in college and school-based activities.

I want to turn briefly to consider the training of third-level teachers, something conspicuous by its absence in most chemistry departments. No formal training or qualification in teaching or lecturing is needed to teach at the third level - be it in chemistry or education. Unfortunately the natural communicator is a rare species and expertise in research and scholarship does not always make a good teacher. We ought to be concerned about the quality of teaching at the third-level as well as at secondary level, because the one influences the other. If you didn't understand chemistry at university because of bad teaching you are going to find it difficult to teach it yourself. Teaching ability is not weighted highly (if at all) in university chemistry departments - research ability is what counts. Teaching large first year classes is often regarded as a chore and is given to the newest, most inexperienced staff to teach. This can have disastrous and unwanted results if you want students to continue with chemistry. In many universities students take several science subjects in their first year before deciding what to specialise in. If we want the best students to study chemistry we must put the best lecturers in the introductory courses, not the worst. Poor teaching puts students off the subject. We need to value and reward good teaching more than we do at present and people's work loads should be related to their relative abilities and interests in teaching and research. The take-it-or-leave-it, elitist attitude of some third-level chemists is dangerous for the future of chemistry. Downgrading the importance of teaching *vis-a-vis* research as an occupation, as of applied versus pure, is an insidious poison which has filtered down through the education system. One day, and maybe it's already here, we may find that there aren't enough good chemistry teachers around to ensure the continuance of the subject. With the foundation eaten away by neglect and contempt, the chemical pyramid will sooner or later start to collapse. Teachers are the key to excellence in chemistry - at any level, whether our aim is to produce chemists or chemistry teachers, and we neglect this at our peril.

## HOW CAN WE ENSURE THE CAREER-LONG DEVELOPMENT OF TEACHERS?

"Rome wasn't built in a day", so the maxim goes, and neither was a good teacher made by a three or four year preservice course, however well-designed and taught that course may be. Teaching, as most of us will agree, is largely learnt on the job and after about four year's preparation the chemistry teacher may spend upwards of forty years in service. Continuing education after the initial (or preservice) course is essential both to maintain and to improve the quality of teaching throughout a person's career.

> "Life-long learning should be seen as the master principle for the future renewal of the teaching profession ..... It renders obsolete the notion that learning to teach is ever complete, much less than that it can be totally accomplished in any initial training course, of whatever duration or excellence."

Report of the Committee on Inservice Education,
Dublin, 1984. (ref.14)

Inservice education is probably the most neglected area in chemical education at the moment, although because of the 10:1 ratio in duration of the inservice:preservice phases of a teacher's career it perhaps requires ten times as much emphasis (see Fig.3).

The importance of life-long learning for teachers to ensure continued and developing excellence in teaching is now being realised in many countries. The U.K. has recently announced increased spending on inservice education, with an emphasis on science and technology, to raise the general standards of teaching.

A recent A.C.S. report on "Priorities, Partnerships and Plans: Chemistry Education in the Schools" (ref.15) stresses the importance of inservice education for chemistry teaching - involving school districts, local chemists, third—level colleges, business and industry, and professional organisations. This report amplifies an earlier A.C.S. report (ref.16) which emphasised:

> "the necessity of continuing education for secondary school chemistry teachers"

and the need for the chemistry teacher, given the importance of the task to:

> "strive assiduously to keep up-to-date with the developments in chemistry itself and with the means of teaching them to the students."

One response to these recommendations is the Institute for Chemical Education (ICE), set up in 1983 at the University of Wisconsin at Madison, funded by grants from industry, the National Science Foundation and the American Chemical Society.

> "From kindergarten through college, ICE seeks to improve and maintain the vitality of teaching in ever-changing field of chemistry." (ref.17)

In the Republic of Ireland the major report on Inservice Education, referred to above, recommends establishing a National Council For Inservice Education (N.C.I.E.) as a central coordinating and planning body, with local councils to coordinate provision at a local level. An important recommendation is that all teachers should be released from duty for up to twelve weeks every five years to take part in inservice activities, except just prior to retirement. Unfortunately this report has not yet been implemented, probably due to the costs involved, although it makes a clear and cogent case for regular, career-long inservice education. This is an important reminder that excellence in education requires substantial investment and the rate-determining step for major change is usually determined by finances.

Proposals for radical changes in science education up to 16 in the U.K. (ref.18) recognise that:

> "Inservice training has a major part to play, because the great majority of those who will be teaching in 1990 (when the proposals are implemented) are already in schools.....It is in the nature of science, with the rapid development of knowledge and its applications, that teachers need an opportunity to keep up-to-date. But in addition, the objectives (of this report) make considerable demands upon the skills and personal qualities of teachers........The effective implementation of those objectives will be dependent as much on the teachers' ability to adopt a flexible and variable approach to their teaching methods as on a sound and up-to-date grasp of their subject."

These examples, from three countries familiar to me, make the point loud and clear. Science is changing; education is changing; society is changing - teachers must be equipped and re-equipped throughout their career to deal with these changes, in both their subject (chemistry) and in their teaching methods. The necessary financial resources have to be found to maintain excellence in teaching through adequate and appropriate inservice activities. Opportunities must be provided and made available to teachers, either on a voluntary basis or made mandatory for continued certification/registration. This should ensure that good teachers get better and that poor teachers improve, as well as preparing all teachers for changes in chemistry and in education. There is no single cure for all the ills of chemical education and no one method will achieve all the desired goals of inservice education. Some of the most important types of inservice activities are discussed below, in order of <u>decreasing</u> cost-effectiveness. The actual inservice programme in a particular country will depend on national circumstances, needs and finances.

<u>Inservice activities in chemical education</u>

Figure 6 shows the cost-effectiveness of a chemistry teachers' magazine (3 issues per year), local short inservice courses or workshops and international conferences like the current 8th I.C.C.E.: the ratio is 1000:30:2. For an expenditure of three thousand pounds I can circulate 1000 teachers with a termly magazine to help them teach more effectively by improving their knowledge of chemistry and of ways to teach it. The same sum of money would enable 30 teachers to attend a week-long inservice course/workshop/conference in their own country. Only 2-5 teachers would be able to attend a big international conference like the 8th I.C.C.E. (the number would depend on the distance). (A similar cost would be involved in a higher degree.)

<u>Publications</u> aimed at the practising chemistry teacher are the most cost-effective way of providing continuous, career-long inservice education for <u>all</u> teachers in a country. I have been involved in producing such magazines in two countries - "A modern approach to Chemistry" in Uganda (1972-1976) and "Chemistry in Action!" in Ireland (1980 - to date). "Chemistry in Action!" is a low budget production sent free to all identified chemistry teachers, and supported financially by donations from chemical industries in Ireland, with an annual budget of about three thousand pounds. I cite this merely as an example, although most countries now have publications produced for the science/chemistry teacher (ref.19). Such publications are usually produced by a national chemistry society or science teachers' organisation for their members. I believe such publications to be very important in attaining our goal of excellence in chemistry, for the following reasons:

- a regular publication provides continual help and stimulation to a teacher in her/his own school;
- over the career of a teacher such a publication builds up to a valuable teaching resource in its own right;

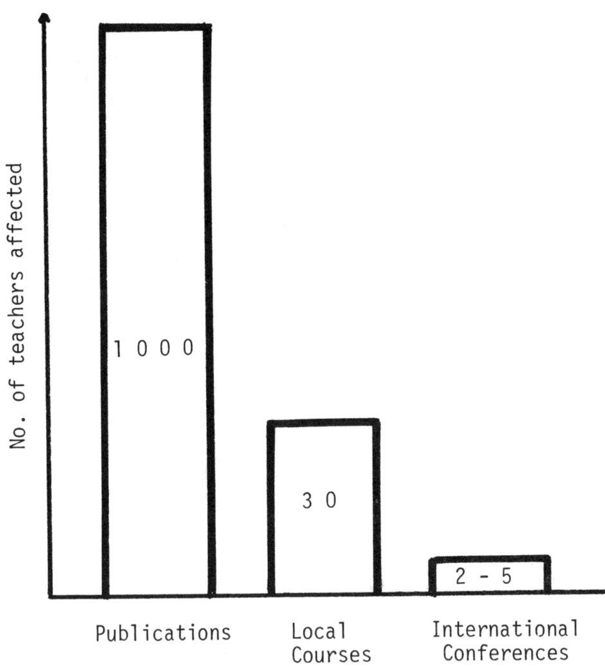


Fig. 6 Cost-effectiveness of inservice activities

- o all the teachers in a country can be reached for a fairly small cost.

Getting teachers to buy such publications is a problem in all countries - only the keen and committed teacher tends to invest in them. Thus the rich get richer and the poor get poorer. That is why I have attempted to reach all teachers with a subsidised publication, in both Uganda and Ireland, so that all may benefit. Initiatives like the Schools Publication Service of the Royal Society of Chemistry (ref.20), which started in 1985, are to be welcomed in providing a package of publications at reduced cost to a school, and other countries might consider this. The A.C.S. is also very active in producing publications aimed at secondary level school teachers and their pupils (ref.21). I would hope that IUPAC/UNESCO would provide funds to ensure that chemistry teachers in developing countries have access to publications in chemical education by supporting local publications or encouraging their initiation, and by importing suitable external publications in bulk.

Inservice courses on a local or national basis. These may be short courses/workshops - evening, day, weekend, or week-long - or longer courses leading to higher qualification. Many types of courses are offered by various institutions, e.g. school authorities, third—level colleges or national bodies. Such courses are more expensive per capita than publications, as they involve costs for tuition, travel, accommodation and materials. However, it is only through such courses that interaction can occur between teachers and other chemists, e.g. from industry or third-level institutions, and that practical skills can be developed. It is one thing to read about an experiment or demonstration, but it is another thing entirely to try it out and to get it to work. Our experience in running short inservice courses for chemistry teachers at Thomond College over the past six years has shown the value of hands-on experience. Many teachers are afraid to try something new and often lack crucial practical skills, which leads to a lack of confidence in the school laboratory. Appropriate hands-on experience in a sympathetic learning environment can produce substantial improvements in a teacher's confidence and competence, leading to better chemistry teaching. The social interaction with other teachers and the sharing of experiences and mutual problems, is also an invaluable feature of inservice courses. They are thus invaluable in the long-term development of a teacher and regular opportunities must be given for all teachers to participate in such courses during their careers. The incentive of a higher qualification leading to better pay and status is an important factor in designing longer inservice courses leading to some form of award, e.g. certificate, diploma, master's degree or doctorate. However, such programmes must be relevant to a teacher's real needs or they are just paper qualifications. A longer course makes it possible to give a more thorough up-grading of the teacher's professional skills in both chemistry and teaching, both of which are important. It should also equip the teacher for self-directed study in chemistry and teaching. Many factors make it difficult for a practising teacher to avail of such courses on a full-time basis, and part-time courses done during the teacher's free time are often a compromise solution. We are in the process of planning a part-time Master's Degree in the Teaching of Science by distance learning, in an attempt to make the course available to teachers anywhere in the country. This is a model that would be particularly attractive in developing countries as distance-learning can be a very cost-effective method of tuition, as the U.K.'s Open University has shown. Inservice courses for chemistry teachers may involve both theory and practice, in chemistry and in education, involving both mental and manipulative skills. However, I must stress that most teachers don't want more theory, in chemistry or in education, unless it can be related to their teaching situation. Teachers are pragmatic people and want courses of direct usefulness in the classroom or laboratory, taught by people who know what they're talking about.

Conferences at local, national or international levels represent the least effective means of inservice education, both economically and in improving a teacher's skills. In a conference the participant is usually fairly passive and there is rarely opportunity to practise or develop skills. Talking and sharing experiences does have a part to play in the improvement of chemistry teaching as it can change attitudes. The most important function of conferences is to inspire and motivate teachers, to give them new insights and higher goals. Unfortunately these are volatile substances and easily evaporate after a conference, sometimes even on the way home! The more teachers who can participate in a conference the better, as a bigger fraction of the teaching population is being reached. That is why I would favour local or national conferences rather than big international jamborees. I am sure that many Japanese chemistry teachers will benefit greatly from this conference, and Japan will have more positive returns than any other country as a result. I would like to see UNESCO/IUPAC encouraging more national or regional conferences in chemical education, perhaps every two years, which would feed their findings into less frequent international conferences. I think a two year interval for the I.C.C.E.'s is too short and I would like to see a four year interval, with a worldwide proliferation of smaller conferences in the intervening years encouraged and funded by UNESCO and IUPAC. Even in areas of chemistry where there is vigorous research going on conferences are often held at four year intervals: if we want to make substantial progress in chemical education by sharing ideas between countries, as at the I.C.C.E.'s, we must allow time for people to do research and complete projects between major conferences. I think we would see more actual progress in chemical education if the money channelled into a meeting like this were to be redistributed, particularly to the countries in greater need. We do need the inspiration and stimulus that a conference like this provides, but do we need it or can we afford it so often?

## HOW DO WE ATTRACT AND KEEP GOOD TEACHERS?

Even if we select and train new chemistry teachers in the best way and provide adequate career-long inservice opportunities, are we going to attract and keep good teachers? Teaching is not the attractive profession it once was in the eyes of our most able students. Many teachers are leaving the profession early for various reasons, but one important one for science and mathematics teachers is the poor pay and prospects. Most well-qualified science or mathematics teachers can get better paid and less demanding jobs in industry. Teaching is hard work in today's schools and in many countries it is not rewarded sufficiently to keep the most able and well-qualified teachers. Both in the U.K. and the U.S.A. there are moves to increase the amount of science and technology taught at school but there aren't enough teachers coming forward to fill existing vacancies, and experienced teachers are leaving for greener fields in industry. The Encyclopedia Britannica (quoted in ref.22) in 1810 reported:

> "It must be the duty, therefore, of every state to take care that proper encouragement be given to those who undertake this office (of instruction). There ought to be a salary as would render it an object of ambition to men of abilities and learning, or at least as would keep the teacher respectable."

In those days a teacher was paid as much as a ploughman and the profession was falling into disrepute. "All things change and all things stay the same." Teaching is still badly paid and has a low status in many countries, and it is not an attractive profession like medicine or other high status and highly paid careers. Merit payments for good teachers and special incentives for science teachers are possible solutions that have been tried, but tend to be divisive within the teaching profession. The rewards of teaching as a profession need to be raised for all teachers to make it attractive to young and mature people of "abilities and learning". Education appears to be a non-profit making 'industry' which only takes from the Gross National Product and doesn't contribute to it. As professional chemists and educators we need to get the message out to society and to governments that a good educational system staffed by well-qualified teachers is the foundation of our wealth and prosperity and the hidden foundation of our GNP's. Better pay in itself will not make good teachers but it gives teaching a fairer chance in the career stakes. If teachers are the key to excellence in chemistry we need to make sure that the pay and prospects are attractive enough to attract and to keep the best.

## CONCLUSION

In summary, if we want to achieve excellence in chemical education at all levels and for all students:

- we must know what qualities make a good chemistry teacher;
- we must know how to identify, produce and develop such qualities in our preservice courses for teachers;
- we must know how to encourage and develop these qualities throughout a teacher's career through appropriate inservice provision;
- we must press for governments to improve the conditions of service and pay of teachers to make it a more attractive career.

This is going to mean a major commitment of funds by the chemical community and from national education budgets to achieve success. But first, and most important, we need to change our attitudes towards education and teachers so that we will spend the money needed. I want to see the status and value of the secondary level chemistry teacher raised in the eyes of society, and in the eyes of chemists at third-level and in industry, with a realisation of their key role in the future of chemistry as a subject. I also want to raise the aspirations of practising teachers and restore their self-confidence so that they will want to strive for excellence for all their pupils. Too many good teachers are leaving the profession daily. Success in chemical education, as in other branches of chemistry, requires costly commitment. Some see teaching just as a 9-4 job, with poor pay but good holidays. Others see teaching as a worthwhile profession whose rewards outweigh its material disadvantages. But the best teachers, I believe, have always considered teaching as a vocation, a personal calling to excellence in their subject for the sake of their pupils. We all need to raise our sights higher in the pursuit of excellence and to realise that it is the teacher who _is_ the key to excellence in chemistry, whether at second or third-level.

"Behind every good chemist and every informed citizen lies a successful chemistry teacher."

## REFERENCES

1. J.D.Herron, D.W.Brooks, _J.Chem.Educ._,_61_,1088-1089 (1984)
2. G.Porter, _S.S.R._,_66_,617-627 (1985)
3. _The new teacher in school_, HMI Matters for Discussion no. 15, HMSO, London (1982)
4. G. Highet, _The art of teaching_, Methuen, London (1963)
5. M.B.Rowe, _The Science Teacher_, _44_(5),37 (1977)
6. R.M.Sutton, _The Science Teacher_, _24_(8),379 (1957)
7. P.May, _Which way to teach?_, IVP, Leicester (1981)
8. R.D.Simpson, R.D.Anderson, _Science, Students and Schools_, Wiley, New York (1981)

9. Ed. by P.Tamir, A.Hofstein, M.Ben-Peretz, Preservice and Inservice Education of Teachers, Balaban International Science Services, Rehevot (1983)
10. 1 Corinthians 13:2, Good News Bible, Today's English Version, American Bible Society (1976)
11. M.Gardner, Eur.J.Sci.Educ.,4, 137-147 (1982)
12. V.N.Lunetta and R.E.Yager in ref.9, 253-263 (1983)
13. L.B.Resnick, Science, 220, 477-478 (1983)
14. Report of the Committee on Inservice Education, Dublin (1984)
15. Priorities, Partnerships and Plans: Chemistry Education in the Schools, American Chemical Society, Washington (1984)
16. Guidelines and Recommendations for the Preparation and Continuing Education of Secondary School Teachers of Chemistry, American Chemical Society, Washington (1977)
17. R.Packard, ChemUnity, 6, 4,12 (1985)
18. Science 5-16: A statement of policy, HMSO, London (1985)
19. J.N.Lazonby, D.J.Waddington, Bibliography of Chemical Education Journals, IUPAC/UNESCO (1981)
20. Schools Publications Service, Royal Society of Chemistry, London (1985)
21. 1984 American Chemical Society Annual Report, C&EN, April 22, 41-43 (1985)
22. Times Ed. Suppl., 26th April (1985)

# The role of chemistry in a biological education

P. J. Kelly

Department of Education, The University, Southampton, SO9 5NH, England

Abstract - Chemistry is seen by biologists as a crucial, supporting, component of the study of living matter and processes. The key issue in determining the role of chemistry in biological education is the elucidation of the relationships between chemical and biological topics. Here four categories of relationships are outlined. (i) integrated, (ii) subsumed and consecutive, (iii) instrumental, and (iv) philosophical. Such categories can form a framework for course design. Certain tenets of learning theory indicate that incorporating these categories into courses requires (i) an emphasis on topics rather than whole subjects, (ii) co-ordinated teaching between biology and chemistry teachers (iii) the use of modules, and (iv) a spirally developed curriculum. Some ideas for curriculum development are outlined.

Chemistry is an integral part of biology. At the upper secondary school and lower university levels of education I would estimate that at least fifty per cent of the topics taught in biology require a fundamental understanding of some aspect of chemistry. In the earlier years of schooling this is less so but in more advanced work some biological topics appear as if they have slipped out of a chemistry course by mistake, such is the high load of chemical understanding required to comprehend them. The role of chemistry in biological education is thus a matter of considerable concern for biology as well as chemistry teachers.

In broad terms I do not think there is much of a problem in determining what aspects of chemistry are relevant to biology. We might argue about detail but such outlines as that conceived by David Samuel (1) would surely be generally acceptable. He stresses the importance of the basic laws and concepts of physical chemistry and 'that the basic essentials of chemistry - bonding and structure, synthesis and reactions - must be taught again and again'. He then points to a number of topics which he considers could be left out of a chemistry syllabus, especially from organic chemistry, and suggests that, in their place, the following should be introduced because of their greater relevance to biological studies: phosphorus chemistry, non-covalent bonds, lipids and lipid membranes, structure and properties of macromolecules and the chemistry of glycoproteins. I would merely add my personal predilection for ensuring that students are well grounded in the chemistry of the major life elements, especially carbon, hydrogen, oxygen and nitrogen. They make up 87% of the dry mass of a human and some 96% of a corn plant, so they have to be important!

Whilst the question of what chemical topics should be included in chemistry and biology syllabuses is an important one, of greater importance - and of greater difficulty - are questions to do with the relationships between topics and the way we teach them. Usually, for example, we consider the question of relating the teaching of one science to that of another in terms of how the understanding of concepts in one supports an understanding of concepts in the other. An example of this is the proposition that an understanding of the chemistry of oxygen, carbon dioxide, water and glucose is necessary before we can understand the biological phenomenon of respiration and, thus, respiration should be taught in biology classes after the appropriate chemistry has been dealt with. This, to my mind, is an approach which is both over-simplified and, potentially, educationally unsound. I suggest that, certainly, at an elementary level, respiration can be understood biologically without much chemical knowledge and, indeed, it is possible to develop a chemical understanding of the process and, at the same time, increase the motivation of students, by starting with non-chemical, essentially biological, ideas.

How sciences *per se* relate to each other is a complex and many-sided issue. How they should be related in education is even more so.

If we are to elucidate the role of chemistry in a biological education, we have first to define, or at least indicate, the perspectives of biological education. The starting point for this, I suggest, is the recognition that biology is not a singular science but a field of study served by many sciences. Its subject is the study of life, that is the nature of living things (including ourselves) and the relationships between their form, functioning and environment, both in the present and over time. It includes biochemical, physiological, and anatomical aspects on the one hand and, on the other, behavioural, ecological, evolutionary and, even, cultural dimensions. An understanding of life comes from a synthesis of all these perspectives. Whilst it is true that some of the methods of the biological sciences are reductionistic, the purpose of biological study is essentially holistic. It is also its purpose to understand _living_ things. We are dealing with biology not necrology.

Biological education has to mirror this and thus provide a holistic understanding of living phenomena whilst, at the same time, portraying their variety and that of the methods by which we study them. When it comes to an understanding of life as a synthesis, chemical concepts tend to become subsumed within broad biological principles. Thus, for example, whilst knowledge of the chemical composition of DNA is important for an understanding of the central biological principle of genetics, it is only of value in that it illuminates the procedures of heredity. To a biologist _qua_ biologist it has little significance by itself. To a chemist, of course, this is certainly not the case. For biological understanding, knowledge of the matter of life is only of relevance in that it offers understanding of the nature of life.

## RELATIONS OF UNDERSTANDING

In order to examine the ways in which chemical and biological understanding are related we can consider four areas of biological study. Each is selected to illustrate a particular type of relationship although, I should stress, they are not necessarily exclusive.

_Internal respiration_ is a biological topic in which, conceptually, biological and chemical understanding are, more or less, the same thing. The biological purpose of internal respiration is to produce energy for the metabolic processes of the cells. Beyond that, understanding is concerned essentially with glycolysis and the key role of adenosine triphosphate (ATP) in metabolism. If you look at any biology text book you will see that this comes out as chemistry. The issue for the teacher is not what chemistry is needed to teach the biological topic, but merely how the topic should be taught at different educational levels. The considerations of biology and chemistry teachers are the same. The topic exemplifies an _integrated_ relationship.

_Genetics_ is an example of a topic in which chemical understanding is essential but has a more clearly defined subsumed relationship with biological principles than, say, internal respiration. The basic principles of genetics - those of Mendelian inheritance, chromosome behaviour and the gene concept - were elucidated without resource to much chemical knowledge. It is only subsequently that advances in our knowledge of DNA and gene-enzyme relations - which require chemical understanding - have provided more detailed ideas and the means of experimentally controlling genetical processes which now feature so strongly because of their key role in genetical engineering. It follows that there can be a _consecutive_ relationship in the teaching of genetics and similar topics; biological concepts being taught before those of chemistry.

_Ecology_ is an area of study in which biological concepts are paramount. Basic ecological ideas such as food web and energy flow depend on some understanding of chemistry, as they do of physics, mathematics and other disciplines. However, as with genetics, these are subsumed within the biological ideas rather than being independent or integrated concepts.

The particular dependence of ecology on chemistry comes from the key role that analyses of the physical environment and organic material play in the experimental study of ecology. The detection of inorganic and organic chemicals - sometimes in very minute quantities - , the measurement of pH and other indications of conditions in the environment and organisms, and other similar procedures require an understanding of chemical principles if their validity is to be appreciated by the student. Chemical ideas are necessary, not so much for understanding biological principles, as for understanding investigational techniques and procedures. It is an example of an _instrumental_ relationship between biology and chemistry.

For the teacher this relationship presents problems. Ideally the chemical knowledge required for understanding techniques should arise in parallel with the development of biological knowledge to which they are related. However, the difference in difficulty of understanding between the two aspects is often so great that this is impossible and one has to rely on a 'black box' approach which requires students to take the validity of procedures as given truth.

In parenthesis it is worth pointing out that the three topics we have so far considered vary in other ways in their relation to chemistry. Internal respiration and genetics link predominantly to aspects of organic chemistry whilst ecology requires a much greater association with inorganic chemistry. All three rely on physical chemistry.

Behaviour is a topic which, again, is essentially biological in character. Indeed, some would see it as central to a distinctive definition of life. Chemical knowledge plays a part in the understanding of behaviour: the role of endocrine systems, pheromones, chemotaxes in invertebrates and the chemistry of brain function are clear illustrations of this. Its relation to behaviour is, in fact, similar to that of chemical concepts in genetics. However, this relationship with behaviour raises important issues to do with the limits of chemistry in defining life. It is a philosophical relationship and, in part, one of conflict as well as a conceptual one.

At one level life can be defined in terms of a particular combination of physico-chemical materials and processes. This is the perspective we obtain from, say, a study of internal respiration and from much of the life of plants. However, the behaviour of animals including, and especially of, man is a distinctive phenomenon revealing life as something which apparently (and I use the word carefully) cannot be defined solely in such terms. Considerations such as levels of material and psychological organization, the relation between mind and matter, autonomy of organisms and even the nature of consciousness can, and should, be taken into account.

For teachers, behaviour is not a topic which can be dealt with solely through transmitting knowledge and doing experiments. It requires also debate about unresolved philosophical issues and, in a very real sense, about the limits of chemical explanations of biological phenomena. It can, of course, also question the validity of biological explanations of life.

A similar relationship occurs in dealing with what has been termed issue studies in biology (2). These are concerned with social, economic and other issues which arise from the application of scientific and technological developments. Inevitably the approach has to be interdisciplinary and chemistry features a great deal. Invariably the issues are the subject of debate rather than resolution.

## THE LEARNER'S PERSPECTIVE

So far we have considered the relations between the subject matter of biology and chemistry. I have attempted to show that these vary with the biological topic being considered. Four categories of relationship have been identified which, to return to the title of this paper, can be said to reflect the four key roles of chemistry in biological education, namely as (i) integral components of some concepts, (ii) subsumed, consecutive components of other concepts, (iii) providing knowledge on which techniques and procedures are based and (iv) providing a stimulus for debate and problem solving.

Essentially we have examined the problem from the teacher's point of view. Now let us look at it from that of the student.

Whilst the process of human learning is still as much a matter of speculation as clear understanding, there are aspects which, I suggest, can be accepted with some confidence. These include -

(i) that learning of new concepts depends on their close links with ones already learned and established in the student's mind,

(ii) that the level of difficulty of a new topic should not be excessively greater than that of the student's current understanding,

(iii) that it is necessary to have a clear purpose for learning and a clear logic in the subject matter being learned,

(iv) that concepts have to be returned to several times and possibly in different ways to ensure that understanding will be retained,

(v) that students must have the opportunity to work creatively with new concepts in different contexts and to enquire about them within their own minds.

Put more concisely, we need to ensure that there is a close and relevant relationship between what is known and what is being learned [(i) - (iii)] and that there is adequate time in which to learn; the syllabus must not be overloaded [(iv) and (v)]. The implications of these requirements for relating chemistry to biology are that merely providing parallel chemistry and biology courses which are not tightly co-ordinated will not be very beneficial, and that we cannot expect an adequate coverage of chemistry within biology courses unless the time given to them is greatly increased - and that, I suspect, is unlikely to happen!

## GUIDELINES FOR CURRICULUM DEVELOPMENT

There is no simple answer to the problem of dealing with the role of chemistry in a biological education. During a young person's education different solutions will be appropriate for different levels of courses, different topics and different educational and scientific objectives. It is necessary for biology and chemistry teachers to be aware of the psychological and educational issues involved as well as those of a scientific nature. Real solutions will come not so much from the production of new syllabuses and text books but from biology and chemistry teachers working closely together on the details of their courses. We need to establish guidelines which will help them to do this effectively. Here are a few tentative suggestions based on the analysis I have just outlined.

The first step, I suggest, is to orientate our general attitudes towards curriculum construction in the following ways.

(i) To think less of subjects as exclusive entities. There is, of course, value in understanding the discipline of a subject. Indeed, I would argue that biologists should have some idea of what chemistry is from a chemist's point of view and vice versa. However, it is equally valid to argue that, for example, chemists should know what chemistry means from a biological standpoint. Subjects need to be treated as the concerns of many groups, not just of the discipline on which they are founded.

(ii) To think more of co-ordinated teaching within and between courses.

(iii) To think more of teaching topics than subjects and to determine carefully the purpose of teaching each topic.

(iv) To develop flexibility in courses through careful organization of the structure of their components.

(v) To follow the principle of a spiral curriculum in developing courses based on topics recurring at successive stages in order to build up understanding incrementally rather than dealing with a topic completely as a large unit at one stage in a course.

The second step is to establish relevant purposes for foundation studies in the elementary and lower secondary schools. These would have to fit in with the general educational objectives and pattern of the schools and, I suggest, should not be primarily concerned with subject disciplines. However, some should be on topics which link biology and chemistry and (a) provide central, organizing ideas on which future cross-subject studies can be built, (b) develop an integrative attitude towards the two subjects and (c) motivate the pupils towards future study. It will, no doubt, be necessary to provide teachers with curriculum modules giving ideas and materials based on these principles.

The third step is to conceive of a modular approach to course construction in upper secondary and higher education which, whilst not inhibiting the establishment of the principles of either the biological or chemical disciplines, will (a) enable biology students to take relevant chemistry modules in close association with topics in their biological studies, (b) provide for each topic, or area of study, a gradation of modules of increasing depth and difficulty which both biology and chemistry students can take to meet their particular needs (including advanced biology students taking an elementary chemistry course with elementary chemistry students) and (c) provide common modules for biology and chemistry students on topics such as macromolecules and metabolism, in which chemistry and biology have an integrated relation; more philosophical issues such as the origin and definition of life; and issues concerned with the social and other implications of developments linking the two subjects.

Such an approach is not totally new. It requires a high level of academic counselling of students and inevitably produces logistical problems. Something like it, however, is required if we are to foster effective relations between the teaching of biology and chemistry.

## REFERENCES

1. David Samuel, Chemistry and the Life Sciences, Chemistry in Britain (June, 1984).
2. P. J. Kelly, The structure of biological education in P. J. Kelly and G. Schaefer (eds.), Biological Education for Community Development, Taylor and Francis (1980).

# Education of chemists for future breakthroughs in medical sciences

Kui Wang

Faculty of Pharmaceutical Sciences, Medical Sciences University of Beijing, Beijing 100083, People's Republic of China

Abstract - To meet the increasing requirements of specially prepared chemists for the development of medical sciences, a program is worked out for the training of medically backgrounded, generalized chemists to conduct medicine related chemical research. The conceptual education is emphasized to enable the students to assimilate and transform the *in vitro* chemical concepts to *in vivo* medicochemical ones.

## MEDICAL PROGRESS AND CHEMICAL EDUCATION

The place of chemistry in medical education is related with its role in medical progress. For medical sciences, the first half of this century has been a glorious age. The physicians have conquered a number of infectious diseases by means of two new weapons: the microscope and chemotherapy. However, the predominating philosophy of the medical workers of that time was following a biomedical model. They were not aware of the importance of chemistry in medical progress. By their views, the contributions of chemistry are merely supplying various chemical substances for clinical uses and basic chemical knowledge for the study of medicine. The conventional medical education system has been worked out under such a condition. In this system, the chemistry courses are offered in the beginning of the curriculum as the basis of the understanding of biochemistry, and biochemistry for the understanding of medicine. This course-after-course presentation reflects also a traditional pedagogic thought. Just as Kelly pointed out: "They considered the question of relating the teaching of one science to that of another in terms of how the understanding of concepts in one supports an understanding of concepts in the other." (ref. 1) This presentation neglects the intercrossing of various disciplines and underestimates the intrinsic value of chemistry and its direct contribution to medical sciences.

In recent years, medical scientists have attached great importance to the studies at molecular level. Then the biochemistry course in the medical education program has been greatly strengthened and some preclinical courses revised. But the curriculum remained without any substantial alteration.

Now, the status is changing significantly. The chemists of all domains, inorganic, organic, physical and structural chemists, stepped into the medical field and took part in the medical research. Many chemistry-originated concepts and ideas have been injected into the biomedical studies and mediated various creative works. In virtue of the fitness of chemical ideas, theories and methods with the trends of contemporary medical sciences, chemistry has manifested its strong potentiality. It is now predicted that a series of breakthroughs are brewing in medical sciences and most of them are related with the medically oriented chemistry studies.

As a result of this situation, the importance of chemical education for the medical sciences should be fully considered and the educational system should be reformed. Now, we are faced with the problem of how to educate the qualified personnel to conduct the interdisciplinary research of the chemistry-medicine interface, by the reinforcement of the chemical part of the conventional medical education or by the modification of the traditional chemical education to meet the requirements for medical progress? In other words, which is the better way, give the medical scientists a sound chemical basis or give the chemists concerned an adequate background of basic biomedical sciences?

To answer this question, a general survey of the features of modern medical sciences is necessary, because the pattern we take for the preparation of the qualified personnel has to be determined by the structure of the knowledge and capabilities. And the latter, in turn, is determined by social and academic needs.

## FEATURES OF CONTEMPORARY MEDICAL SCIENCES

As a medicine-oriented chemist sees, the main aspects of the progress of modern medical sciences may be generalized tentatively as follows:

1. the deepening of the understanding of a disease through more microscopic studies(structural studies)

2. the extension of the understanding through more macroscopic studies (studies of systems).

Both structural and systematic studies of a pathological process are chemical in nature. The chemical reasoning of a disease may turn into the foundation of the insight of its nature, because this reasoning covers all the processes, from microscopic to macroscopic. That is just where the potentiality of chemistry lies. But the potentiality does not originate from the superiority of chemical methods only. Actually, the achievements of modern medical sciences are related with conceptual revolutions. In the course of the development of medical sciences, the transplantation of concepts from other disciplines played a vital rôle. These concepts, being enriched and modified with the evidence collected in biological and medical studies, are transformed into new medical concepts, which are creative and challenging. Now, the relation between medicine and chemistry is much different from that in the early twentieth century. The medicine becomes more chemistry-dependent and the dependence more conceptual. To adapt this situation, we have to pay more attention to conceptual education, especially those of the interdisciplinary concepts. Some of them are outlined below.

### Chemicopathological system

To give a theoretical answer to the question--what is the nature of disease, is of utmost importance and a long discussed problem. The chemical concepts have refreshed present medical ideas on this aspect and give a series of hints on the understanding of the general characters of various pathological processes. From a chemical viewpoint, the human body is but a system working under specific conditions. Normally, a static state is established through the coordination of the reactions involved, as shown in Fig. 1:

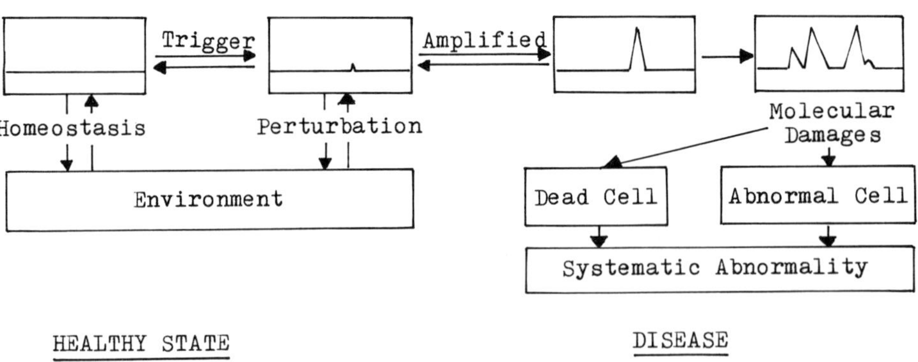


Fig. 1. Chemicopathological System

The abnormal process starts with the formation of a pathogenic factor, which acts as a trigger and induces perturbations. If there is something wrong in a certain link, the trigger or perturbation reactions may be amplified and cause more or less profound damage. For example, the trigger of the formation of gallstone pigment may be traced up to the inflammation process, which produces free radicals, enhances the secretion of mucin and promotes the hydrolysis of bilirubin glucuronides. These perturbations, if amplified, will initiate calcium bilirubinate deposition and stone formation. This is an example illustrating the effect of conceptual consideration, especially conceptual transformation. These transformations, from chemical system to pathological system, from in vitro to in vivo, may become problem-solving approaches as well as the motivational force for medical progress. To be a good medicochemist, the ability to assimilate, transform and utilize concepts are very important.

### Chemical correlation of diseases

The introduction of chemical concepts is altering the present medical cognition on the relation of diseases. From a chemical viewpoint, an inherent relationship, a missing link, may be found for a group of diseases, which may be of little relevance a in clinical sense. For instance, the free radical related diseases constitute a family of diseases, with active oxygen species as the common triggers.

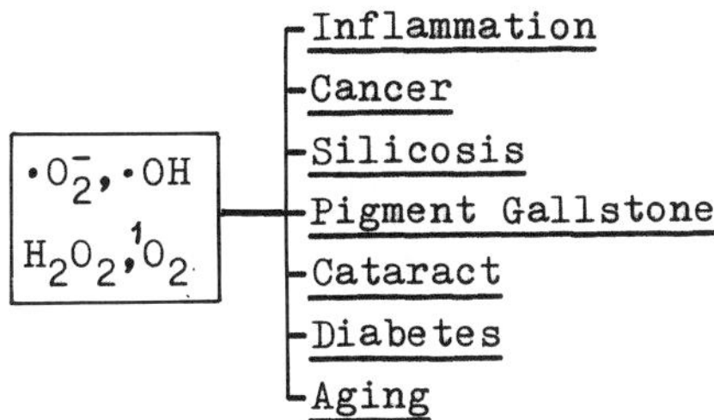


According to this consideration, a certain pathogenic factor acting on different target molecules will induce different chemical and biological effects and initiate different diseases. The concept of disease group may become a valuable theoretical basis guiding further studies of the nature of diseases and their treatments.

### The structure-property-activity relationship (SPAR)

The structural studies on the pathogenic and target molecules in relation to a certain disease is now a notable trend in medical studies. In this aspect, the chemists correlate the structure of molecules, their properties and their biological activities or functions. The SPAR has long been known as the fundamental relation determining the relation between microscopic characteristics and macroscopic behaviours of any chemical substance. Probably, it is proper to say that the whole of chemistry is devoted to the studies and applications of these relations. However, if a chemist is working on the microscopic interpretation of a disease, he is facing some biomacromolecules. He has to study the SPAR of these macromolecules, such as their activities in relation to the set up of these molecules, their conformations and conformational changes and their dynamic states, etc. Therefore, a chemist must know more about macromolecules in addition to the common chemical principles, information and methods.

Although these discussions are by no means comprehensive, we may arrive at an idea--the persons who are able to do creative medicochemical work should be qualified with a broad and profound chemical foundation and an adequate medical background. In brief, they are hybrid personnel desired from two orbitals, chemical and medical, with the former as the main body.

## COMPONENTS AND STRUCTURE OF KNOWLEDGE AND CAPABILITIES

To get a deep insight into the nature of a disease at the molecular level, one should be able to:

1. extract the chemical problems from a clinical one,

2. establish a proper model for it, and

3. solve the problems by means of suitable chemical methods.

For the study of the formation of gallstone pigment, a quaternary mode system (calcium-bilirubin-glycoproteins(polyelectrolytes)-bile salts(surfactants)) has been studied with the deposition and aggregation of calcium bilirubinate as the key step. It is obviously improper to treat this problem simply as a precipitation equilibrium system. The chemists should be able to approach this problem with consideration of the characteristics of this system--the micellar and polyelectrolyte aggregate medium. In such a background, the precipitation and aggregation processes are very different from those in simple aqueous solution. Therefore, the chemists ought to be endowed with the ability to carry out comprehensive studies in the overlapping areas of different domains of chemistry and solve a problem by integration of the methods from these domains. In other words, they should be <u>generalized chemists</u>. Moreover, the complexity of this problem lies in <u>the medical</u> considerations too. The anatomical and physiological characteristics of the hepatobiliary tract have to be considered and the relevant reactions should be brought back to the <u>in vivo</u> condition to be studied. Thus it can be seen that a medical <u>background is required</u>. Then, our formula for the preparation of medicochemists may be outlined as:

generalized chemical education + medical background education ⟶ Medicochemists

We may summarize the structure of knowledge of medicochemists as shown in Fig. 2.

<table>
<tr><td>Medicine</td><td colspan="3">Medical Chemistry</td></tr>
<tr><td>Basic Medical Sciences</td><td colspan="3">Biology-oriented Chemistry</td></tr>
<tr><td rowspan="2">Biology</td><td colspan="3">Domains of Pure Chemistry</td></tr>
<tr><td>General Chemistry</td><td>Physics</td><td>Mathematics & Computer</td></tr>
</table>

Fig. 2. Structure of knowledge of medicochemists

We have developed an interdisciplinary program "<u>Medical Chemistry Program</u>" with a guideline-- a comprehensive chemical education as the core, with biological and medical background education penetrated into it. It is an undergraduate-graduate diphasic scheme. The curriculum is mainly composed from two lines of courses, chemical (Fig. 3) and biomedical (Fig. 4).

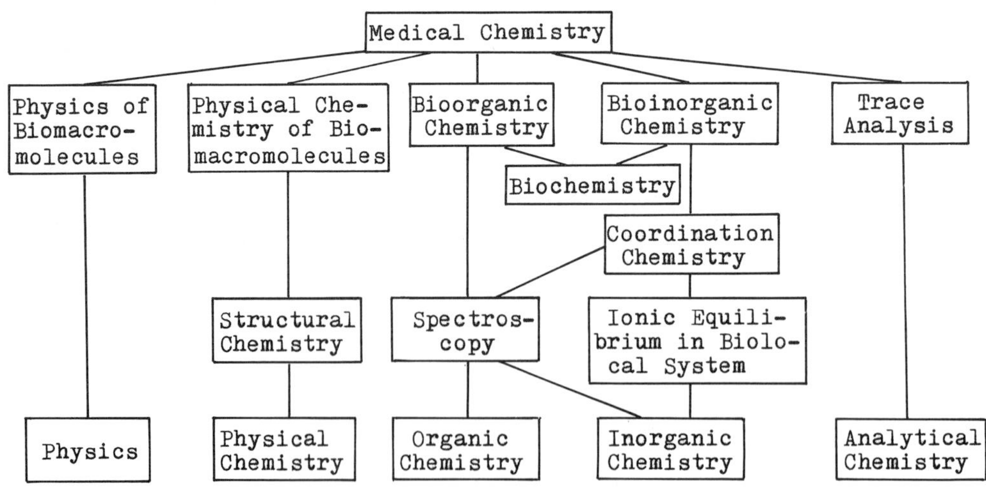


Fig. 31. The courses of the chemical line

As shown in Fig. 5, the first stage of this program comprises all the chemistry courses as usual. Toward the end of this stage, following the gradual introduction of biological ideas and concepts, the students are arriving at a <u>turning point</u>, which implies that they have already been well prepared in

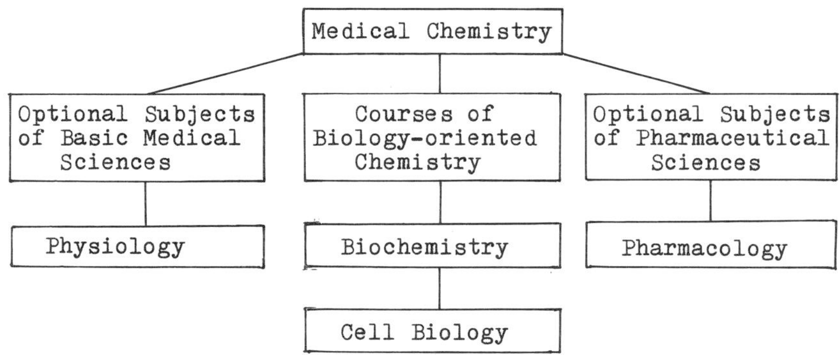


Fig. 4. Courses of biomedical line

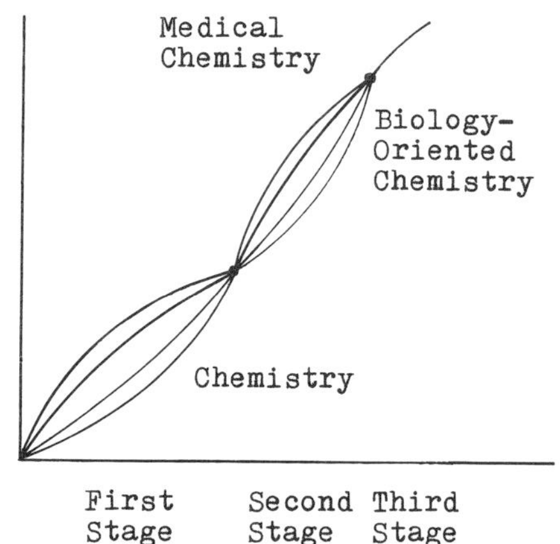


Fig. 5. Construction of "Medical Chemistry Program"

basic chemistry and become familiar with the intermingling of chemical and biological knowledge. They are ready to turn to biology-oriented chemical studies. In the second stage, begining from the first turning point, some biology-oriented chemistry courses are offered, namely, bioinorganic chemistry, bioorganic chemistry, physics and physical chemistry of biomacromolecules. The biology-oriented training enables the students to understand the relations and differences between _in vitro_ and _in vivo_ situations and transform the _in vitro_ concepts to _in vivo_ ones. Thus the second stage study may be called _transformative_ education. At the end of the second stage, an integrated training is given, in which the students learn to solve the problems using many-sided approaches. When they reached the second turning points, they are already acquainted with the biological considerations, but not yet the medical. As the final stage, a course "Medical Chemistry" is offered in coordination with thesis work. The students learn to do chemical work in relation to medicine.

## MEDICALLY BACKGROUNDED, GENERALIZED CHEMICAL EDUCATION

To achieve the medically backgrounded, generalized chemical education, we have taken some measures.

### The medical background education

To bring about this background education, the biological and basic medical concepts and knowledge are given by penetration into the chemical courses by the following means.

The contents of basic chemistry courses are resorted and rearranged in order to accept the penetration of biological and medical sciences. For example, the chemistry of oxygen is discussed in the following way (Fig. 6).

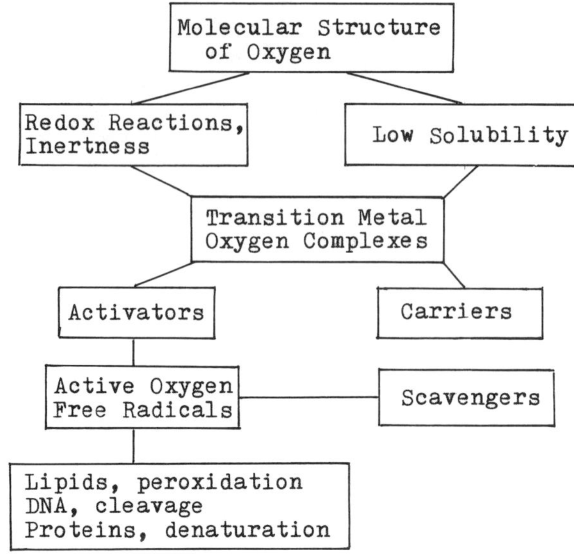


Fig. 6. The chemistry of oxygen

This scheme is somewhat different from the common ones, but the resorting and rearrangements are not pragmatical and are beneficial for the mastering of fundamental chemical principles.

We have developed some courses by the reformation or reconstruction of some original courses. They are designed by following some new guidelines, rather than the simple revisions on the original basis. For example, we have offered a course "Ionic Equilibrium in Biological Systems" as the reconstructed form of the theoretical part of traditional qualitative analysis. The contents are organized in order to establish a general working program for the treatment of biological ionic systems (Fig. 7).

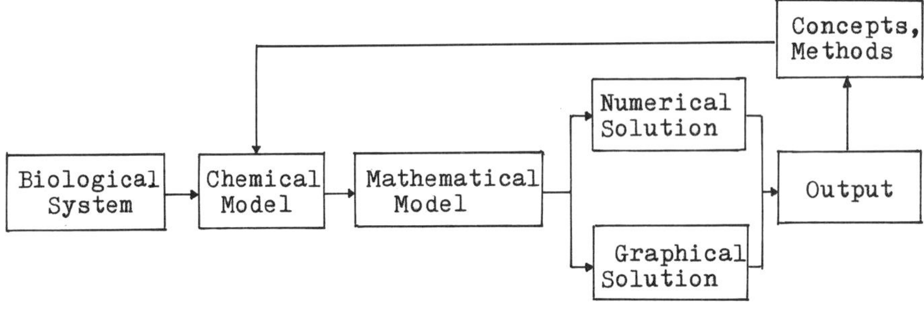


Fig. 7. The theme of "Ionic Equilibria in Biological Systems"

The students are asked to answer some chemical questions concerning a certain clinical case. For urinary stones, the questions raised are:

1. How to predict the risk of stone formation for a certain person? (given the composition of urine).

2. What chemical substance shall crystallize out at first from this urine?

3. What chemical substances shall remain at equilibrium state?

The students learn to establish a simple model and try to answer these questions by appropriate methods. Perhaps the students will be somewhat puzzled by the deviation of their chemistry-originated answers from the clinical reality, but a further discussion will stimulate the students to persue this problem on a new level. In this way, the in vivo concepts are gradually intensified and the in vitro-in vivo differences are gradually understood.

Of course, it is unnecessary to explain how the biology-oriented chemistry courses are designed with concern for the interpenetration of chemistry and biology. But a comment should be made on the general characteristics of these courses, because they are the essence of backgrounded chemistry education. The characteristics are that the discussions of all these courses are focusing on two subjects, biomacromolecules and biological systems. In relation to urinary stones, we have a subject "biomineralization" given in bioinorganic chemistry. The deposition and growth of crystals are discussed with the considerations of the open system of the urinary tract, emphasizing three steps of stone formation — nucleation, retention and crystal growth (Fig. 8).

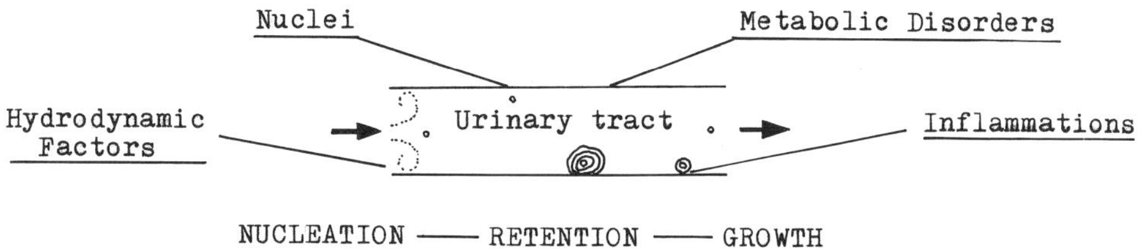


Fig. 8. The dynamic model of the formation of urinary tract culculi

Through these discussions, the students may get the idea of how to transform a chemical consideration, crystal formation and growth, to a biologically oriented consideration.

In the course "Medical Chemistry", all the medical and biological background components are summed up and sublimed to the chemistry-based discussions on the initiation, development, prevention and treatment of some group of diseases. For instance, following the simplified studies of urinary stones in the former courses, the discussion is extended to contact with the real cases and the scope is enlarged to include all the related diseases. Under the topic "Diseases Related to Abnormal Mineralization and Demineralization", the etiology, pathology, diagnosis and treatment are discussed by means of chemical principles. This course is of conceptual and methodological importance, instead of informative. We do not attempt to give our students the systematic knowledge or practical skills of medicine. Instead, they are trained as the _chemists_.

_From in vitro to in vivo_

Normally educated chemists may find it difficult to work with medical workers in the beginning. The chemists usually raise some oversimplified interpretations or put forward some suggestions, which are reasonable chemically but impractical medically. We may generally attribute these underpreparations to the departure between the _in vitro_ chemical education and the _in vivo_ medical realities. Thus, to convert the scope of concern from _in vitro_ to _in vivo_ is the most important part of our program. To develop the _in vitro-in vivo_ transformative ability, the contents of courses are revised with concern for the following aspects.

_Systems from in vitro to in vivo_. In traditional chemical education, learning to treat a chemical system is a major topic all along. Generally, some examples are discussed in the aim of the elucidation of some principles. Therefore, the examples are selected on a tutorial basis and are typical, simplified, and ideal. In contrast, the real life systems are far more complex. They are all multicomponent, multicompartment, open systems. The states are all dynamic and the reactions highly concerted.

_Media from in vitro to in vivo_. The medium effects are known as the prevailing factors either _in vitro_ or _in vivo_. However, this topic is usually discussed in chemistry courses as the empirical basis for the introduction of some concepts of solution chemistry, such as Debye-Huckel theory. But the biological media are much different from the simple aqueous systems. For instance, most of the biological media are _organized_, such as micelle, liquid crystal, gel and molecular sieve.

Reactions from in vitro to in vivo. The different behaviours between in vitro and in vivo systems often stem from the differences in reactivity, and the latter from the differences in media of the systems. Seeing that, we have revised and rearranged the contents of the reaction chemistry in order to adapt the in vivo considerations. For instance, the students have learned redox potential and are acquainted with the inertness of metallic copper to water and non-oxidative acids. But when we turn to an in vivo system containing copper, the discussion cannot exclude the possibility of its dissolution in some body fluids. In the presence of oxygen, chloride and amino acids, copper may dissolve according to the reactions given below:

$$Cu \xrightarrow{O_2} Cu_2O \xrightarrow{Cl^-} CuCl \xrightarrow{Cl^-} CuCl_2^-$$

$$CuCl_2^- \xrightarrow{O_2} Cu^{2+} + \cdot R \qquad R = \cdot OH \text{ or } \cdot O_2^-$$

$$Cu^{+} + O_2 \xleftarrow{H_2O} Cu^{2+}$$

$$Cu^{2+} \xrightarrow{AA} Cu(AA)_2$$

This is an enlightening supplement to chemical principles. The students may learn how to predict or interpret the in vivo reactions with the theories or knowledge they acquired from the in vitro chemistry.

Generalized chemical education

As I have mentioned, due to the complexity of life systems, both in chemistry and in biology, any medical topic involves many different aspects or chemistry. The chemists have to put inorganic, organic and physical chemistry all together to work out an approach or to get an idea. Thus we prefer a generalized chemical education. In fact, the studies of different domains of chemistry are discrete in the first stage and most of the second stage for better understanding. These discrete, specialized components are integrated at the end of second stage by a course "Integrated Laboratory Works". This course is given in the form of lab work, because there is no better way to put the students into a real case study and research atmosphere, which is essential for the development of the important qualities of the generalized chemists (generalized brains, eyes, ears and pens, as Moore discussed)(ref. 2). One of our integrated works is given below to exemplify our intension.

Lipophilic Cupric Salicylate
as Topical Anti-inflammatory Drugs

* Synthesis of cupric salicylate alcoholate
* Analysis
* Spectroscopy and structure
* Solubilities in organic solvents
* Stability constants in aqueous solution
* Superoxide dismutase(SOD)-like activity

In addition, the course "Medical Chemistry" is entirely devoted to generalized chemical education.

## FOSTERING RESEARCH ABILITY

We devote much attention to the training of research ability, because most of our students are expected to engage in research work. The training is distributed throughout the whole course of study, but the objectives and forms are different in different stages.

Case studies

In the first two or three years, our goal is merely to enable our students to know the research work. For this purpose, some cases are studied to demonstrate how the problems have been raised and solved. Case studies may be included in any course, even the freshman chemistry, but the studies become increasingly profound and comprehensive following the progress of study.

Besides the improvement of cognitive ability and mode of thinking, the vision of students becomes wider step by step. In the beginning, the discussions are mainly chemical and become medicochemical later.

Integrated laboratory work

Being a part of generalized chemical education, this course teaches the students to solve a problem by integrating different chemical knowledge and methods. They learn from this course, how research work is designed and how it is worked out. The contents of this course are essentially the repeating of some published works, but they are reconstructed as a mimetic research. It will give the students adequate preparation, both theoretical and practical, before they undertake any real work.

Thesis Work

The thesis work is required from both the undergraduates and graduates. The students will learn much from this study, because they work and discuss with the members of the research team for one (undergraduate) or two (graduate) years.

What I have discussed is a general problem concerning the interdisciplinary education. Baiulescu et al. have discussed this problem and suggested the interpenetration of different disciplines by offering some interdisciplinary lectures (ref. 3). But in my opinion, if our purpose is to educate the hybrid personnel working at the interface of two sciences A and B, we have to decide which way is the best:

undergraduate (A) + undergraduate (B),

undergraduate (A) + graduate (B),

or undergraduate + graduate (A+B).

I prefer the latter, with one of the disciplines, A or B, as the primary one.

REFERENCES

1. P.J. Kelly, Plenary Lecture, The Eighth International Conference on Chemical Education (Tokyo, 1985).
2. J.W. Moore and E.A. Moore, J. Chem. Education, 57, 17 (1980).
3. G.E. Baiulescu, C. Patroescu and R.A. Chalmers, Education and Teaching in Analytical Chemistry, P. 23, Ellis Horwood, Chichester (1982).

# An approach to chemical education from the biological point of view

Isao Nakayama

Azabu High School, Tokyo 106, Japan

Abstract - Under the present system, the treatment of chemistry and biology in Japanese high schools is widely different, so the order between these two subjects cannot be recognized. The purpose of this paper is to introduce the example of educational practice which emphasizes scientific conceptions with regard to life science education.
In particular, the contrast between the quality of bio- and synthetic-high polymers, a principle consideration about the spiral structure of bio high polymers, and the utility of the periodic table in relation to life science, are emphasized. A list of effective experiments and exercises is also drawn up. An evaluation by school graduates may support the validity of these trials. A proposal is made that an integrated approach to be used for chemistry and biology would contribute to the development of science education.

## INTRODUCTION

Modern biology has been changed in its interpretation since molecular biology was introduced. Hence, the introduction of Organic Chemistry into Biology should be emphasized. However high school Chemistry was not affected by the merging of these two fields at the research level. There are many items relating to molecular biology and/or biotechnology, on television, and in newspapers. Therefore, pupils hope to receive more precise knowledge of this new field.

However, there were many trials aimed at giving a course including the so-called integrated or interdisciplinary approaches, in high school science; but almost all of them failed. From this, it was understood that there is a possibility for successful integrated courses only at primary school or at least undergraduate level. This indicates that integrated knowledge can be taught to very young pupils or to those who have a good understanding of the background of the specific fields to be integrated.

The author would like to share his experience in giving a high school course, which is a kind of combined course of Chemistry and Biology, although there are many difficulties in trying to teach different subjects in the same course. The author, a biologist, hopes that this approach will offer one example for those who were not satisfied with the current "Course of Study".

## ANALYSIS OF LIFE SCIENCE IN HIGH SCHOOL CHEMISTRY AND BIOLOGY IN JAPAN

Firstly, the author would like to present the items relating to Life Science, which appear in the "Course of Study", and the textbooks for Japanese high school Chemistry and Biology. Since all items described in the Biology textbook are concerned with the fundamentals of Life Science, only those relating to Chemistry are mentioned. (Table 1)

In these items, the typical and transition elements are used for the explanations: for example, magnesium for photosynthesis, iron used in hemoglobin for breathing, sodium ion and potassium ion for active transport and exitation of cell membrane potential, and calcium ion for muscular contraction and/or blood coagulation.

Some items appear in both Chemistry and Biology Courses, such as Starch, Cellulose, Proteins, Oxidation-Reduction; however, the explanation treatments given to these differ, and there are no special arrangements in these courses or specific orders. Although there are different objectives for each course, it might be mentioned that there are difficulties as far as the teaching of Life Science is concerned.

Table 1. Items and contents of life science in high school chemistry and biology

(CHEMISTRY)

| Items | Contents |
|---|---|
| Organic Compounds | Sugars, Amino Acids, Aliphatic Acids, Cellulose, Polypeptides, Proteins, DNA |
| High Polymer Chemistry | Chemical Bonds : Peptide Bond, Hydrogen Bond<br>Structure : $\alpha$-helix, $\beta$-structure of Proteins<br>Role : Materials of Organisms, Transportation (Hb), Movement<br>Reaction : Hydrolysis of Starch by Acid Detection of Protein by Means of Color Reactions |
| Theoretical & Inorganic Chemistry | Solution, Colloidal Solutions, Osmotic Pressure, Acid, Base, pH, Oxidation-Reduction, Complex Compounds, Chemical Equilibrium, Chemical Bonds |

(BIOLOGY)

| Items | Contents |
|---|---|
| Chemical Compounds in the Cell | General Properties & Biological Roles of Water, Proteins, Lipids, Sugars, Nucleic acids & Inorganic Salts |
| Function of Cell Membrane | Osmotic pressure, Semipermeability, Active Transport |
| Metabolism (Exchange of Materials) | Enzymes, (Reaction Velocity, Activation Energy, Functions, Classification, Co-enzymes), ATP<br>Anabolism : Photosynthesis, Nitrogen Assimilation<br>Catabolism : Respiration, Biological Oxydation, Gas Exchange (Hb)<br>Digestion (Hydrolysis of Organic Nutrients), Excretion (Ornithine Cycle), Movement (Muscular Contraction), Bio-luminescence, Biosynthesis |
| Homeostasis | Hormones, Immunoglobulin |
| Genetics | Molecular Genetics (Structure & Function of DNA & RNA) |
| Evolution | Molecular Evolution. Chemical Evolution |
| Circulation of Materials in the Ecosystem | Oxygen, Carbon Dioxide, Nitrogen, Phosphagen, Biological Con-centration |

TRIALS TO TEACH LIFE SCIENCE

- An Approach to Emphasize the Scientific Initiative Consideration -

Because of the importance of the larger effectiveness of Life Science teaching in high school some trials using an interdisciplinary approach to teaching "Chemistry" and "Biology" have been made by the author.

The following are some examples:

1. Specifications of Bio High Polymers in Comparison with Organic Compounds

(1) Comparison of Bio and Synthetic High Polymers

The number of functional groups in bio high polymers is larger than that for synthesized high polymers. Because of this variety, the order permutation of functional groups relates to genetic information, and hence the complexity of life is explained. The number of functional groups in bio high polymers is as follows: for DNA and RNA, there are 4 bases in the nucleic acids, which are data bases (genetic code), hence this information is transferred. There are 20 amino acids for proteins, which have many activities. These complexities do not happen in synthetic high polymers, e.g. polyethylene and/or polystyrene, which have only one functional group in general. In comparing synthetic high polymers to starch and cellulose, students can understand that the roles of the latter are simply to preserve cell structure and/or to act as energy reservoirs for the body.

(2) Explanation of the Reasons why Bio High Polymers Tend to have Spiral Configurations

Simple mathematics should be used for an understanding of Life Science.

The figures show the illustrative model enabling pupils to understand the spiral configuration.

Referring to Fig. 1.(a), the king piece of "Shogi", the chess-like Japanese game, as shown. As illustrated, the piece is symmetric around the vertical center line but the thickness varies at the top and bottom. A circle is formed when the king pieces having the same figure and size are placed on their sides close to each other. Fig. 1.(b) shows the king pieces with shaved right-hand side back placed one upon another repeatedly to make up a left-handed spiral. Fig. 1.(c) shows the king pieces with shaved left-hand side, to make up a right-handed spiral.

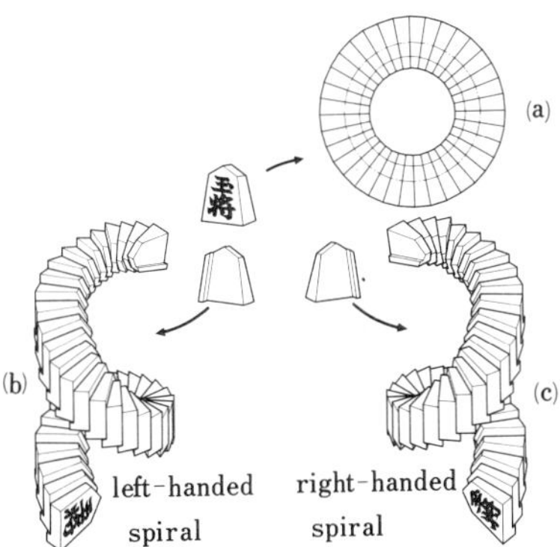


Fig. 1. The schematic explanation using models of the reasons why bio high polymers tend to have spiral configurations

By those models, pupils understand that because the structural units making up bio high polymers are asymmetric, their combination results a in helix configuration.

The teaching of the fundamentals of bio high polymer structures rather than descriptive explanation of the double-helix in DNA and the α-helix model in protein, stimulates the pupils' interest.

For pupils capable in mathematics the stereo structure of such molecules can be explained using the formula taught in the Mathematics Course.

Let the spiral under consideration be composed of some units and let the units be numbered in an appropriate way.

Now we consider a fixed point in each unit. For example, let Po be a fixed point (x, y, z) in the 0-th unit and let Pn be the corresponding point to Po in the n-th unit (Fig. 2, 3).

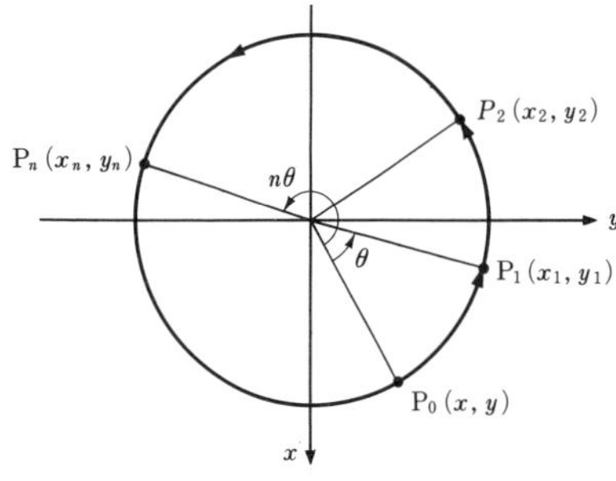


Fig. 2.

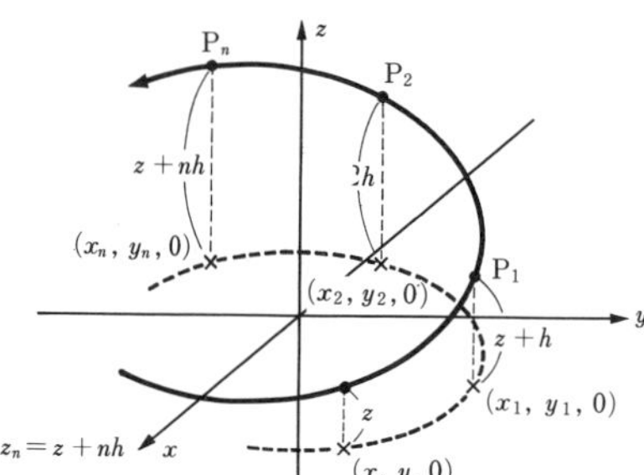


Fig. 3.

Henceforth we set Pn = (Xn, Yn, Zn). In the X-Y plane, rotating Po by nθ we can obtain Pn.

Therefore, by using the rotation in Matrix, we have the equality (1);

$$(1) \quad \begin{pmatrix} Xn \\ Yn \end{pmatrix} = \begin{pmatrix} \cos n\theta & -\sin n\theta \\ \sin n\theta & \cos n\theta \end{pmatrix} \begin{pmatrix} X \\ Y \end{pmatrix}$$

On the other hand, since Pn is the n-th point, Zn-Z is equal to n times the "height" of each unit.

Namely we have; Zn = Z + nH (Fig. 3)

Hence we can denote Pn by (2);

$$(2) \quad (Xn,\ Yn,\ Z + nh),$$

where (Xn, Yn) is given in (1).

Now (2) may be written in Matrices as follows;

$$(3) \quad \begin{pmatrix} Xn \\ Yn \\ Zn \end{pmatrix} = \begin{pmatrix} \cos n\theta & -\sin n\theta & 0 \\ \sin n\theta & \cos n\theta & 0 \\ 0 & 0 & 1 \end{pmatrix} \begin{pmatrix} X \\ Y \\ Z \end{pmatrix} + n \begin{pmatrix} 0 \\ 0 \\ h \end{pmatrix}$$

The pupils can check the equality (3) by calculating the right-hand side.

Next, let's consider (3) in a particular case in order to explain (2) intuitively. That is putting;

Po = (1, 0, 0)

They get by (3).

Then we can give (4) some explanation corresponding to Fig. 4.

$$(4) \quad \begin{cases} Xn = \cos n\theta \\ Yn = \sin n\theta \\ Zn = nh \end{cases}$$

By the definition of the trigonometric functions and Fig. 4, we have

$$(Xn,\ Yn) = (\cos n\theta,\ \sin n\theta)$$

On the other hand, as for the Z-coordinate, "we go up stairs by n-times from Po to reach Pn."

Therefore (cos nθ, sin nθ, nh) is seen to denote the point (Xn, Yn, Zn) in Fig. 4.

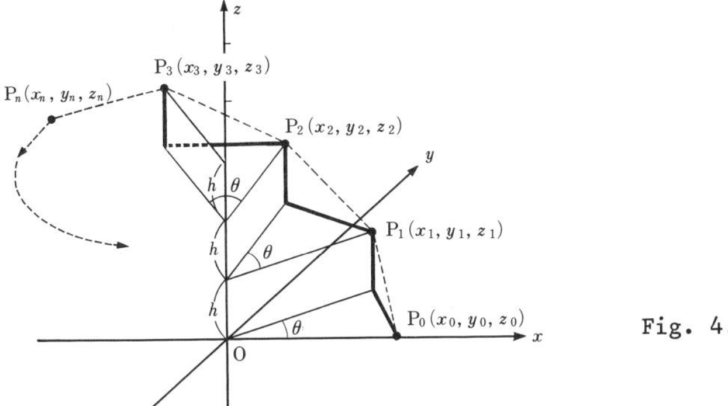


Fig. 4.

Fig. 5. shows schematically one polynucleotide chain out of the DNA double helix. The DNA molecule is 2 nm wide. The distance between neighboring nucleotide pairs is 0.34nm (shown with height h along the z-axis). The angle of rotation θ is 36°.

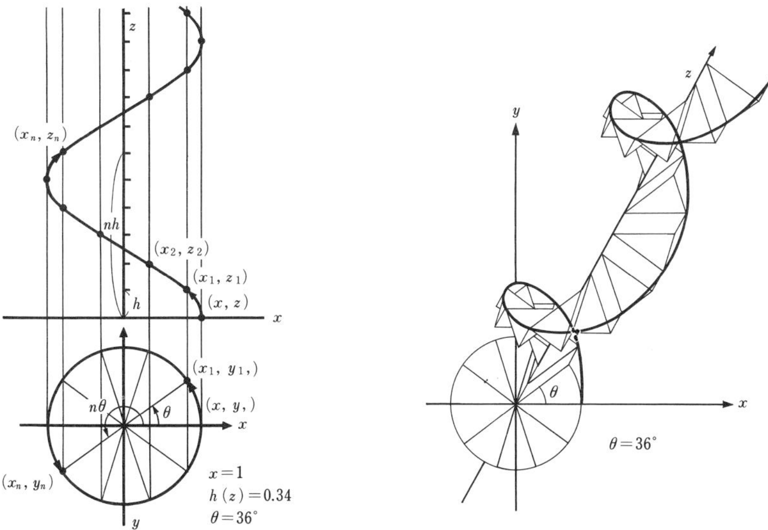


(a) vertical and horizontal (b) perspective

Fig. 5. One polynucleolide chain of the B-DNA double helix

(3) Optical Rotation

In addition the relationship between D- and L- for the amino acids may be used to illustrate the complexity of bio high polymers.

2. The Use of the Periodic Table for the Understanding of Life Science

The Periodic Table of the elements can be applied for investigation of the specialities of Life, if the Table is used from the biologist's standpoint. Most elements in biosubstances are located in periods 1 - 4 of the Table. "Typical Elements" containing s- and p-electrons are components of biosubstances.

In biosubstances, there are a few 4th period elements - transition metals. These elements are involved in the coordination bonds in cytochrome and hemoglobin, thus indicating that there is a phenomenon relating to bioactivities through coordination bonds.

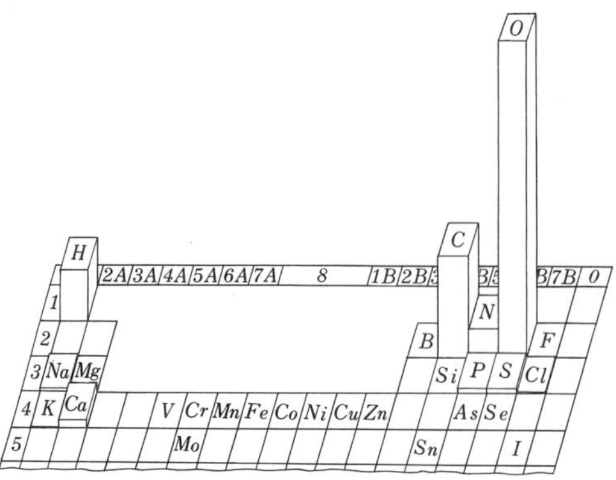


Fig. 6. Periodic table and the elements in biosubstances

However, the Periodic Table was only used to show the arrangement of the elements in Chemistry, and pupils never thought of its importance in daily Life. If the utility of the Table is emphasized in the Chemistry and Biology Course, pupils will show interest in the constitution of substances.

3. Experiments and Exercises

Experiments and exercises should be emphasized to obtain a good understanding of a science course. In the case of Life Science, Chemistry and Biology experiments can be combined even in the high school curriculum.

The following are the examples the author tried to apply:

a) Extraction of DNA from biosubstances
b) Hydrolysis and paper chromatography of DNA and RNA
c) Construction of DNA molecular model
d) Construction of protein molecular model (Myoglobin)
e) Preparation of plant protoplast and cell fusion

## ANSWERS TO THE COURSE ASSESSMENT QUESTIONNAIRE BY GRADUATES

The above-mentioned aspects were clarified via a questionnaire administered to 37 students who had graduated from the school.*

Table 2. Classification of the returns

| | Life Science Major | Non-Life Science | in 1975 | after 1976 | Total |
|---|---|---|---|---|---|
| Post doctorates | 6 (1) | | 5 (1) | 1 | 6 (1) |
| Graduate students (DC) | 11 (2) | 1 | 2 (2) | 10 | 12 (2) |
| Graduate students (MC) | 7 (1) | | 6 (1) | 1 | 7 (1) |
| Bachelor | 1 | 9 | 9 | 1 | 10 |
| Undergraduate students | 2 | | | 2 | 2 |
| Total | 27 (4) | 10 | 22 (4) | 15 | 37 (4) |

* Among the 37 students, 22 graduated from the School in 1975 (total graduates: 300). About 20 of the 37 students were admitted to Life Science Courses in University, and six of them received the Doctor's Degree.

However in the table, the numbers in parentheses mean the numbers of students who answered only questions 1) and 2).

The questionnaire aspects were: 1) To what extent the graduates were impressed by the Chemistry and Biology Course; 2) How to determine the make-up of the future courses (whether the lessons were effective or not); 3) Which subjects among about 35 items were found to be interesting. The respondents were asked to answer using numbers from 5 to 1 (best to worst assessment).

The questionnaire objective was mainly to find out the students' impressions, since it was the first time the author had tried to teach such a combined course.

What kind of impression did the pupils get from the lessons described above? Next, the author refers to it.

On subject 1) of the questionnaire: Many pupils answered by saying the lessons on molecular biology were centered by the so-called "central dogma". Other subjects which impressed them included kinetics and energy theory of enzymes, biochemistry of respiration, etc., and the non-locality of electrons in the benzene ring.

On subject 2) of the questionnaire: It was found that many pupils were stimulated by such lessons to select fields of life science for their major in universities. Those who answered the questionnaire and now have life science as their major are further classified by their major below. A student who studies in an interdisciplenary field is counted only once according to his major since many students are in this situation.

Table 3. Classification of the science majors

| Major | Numbers | Major | Numbers |
|---|---|---|---|
| Molecular Biology | 5 | Animal Physiology | 2 |
| Biophysics | 2 | Nerve Physiology | 2 |
| Immunochemistry | 2 | Plant Physiology | 1 |
| Biochemistry | 3 | Cytogenetics | 1 |
| Organic Synthetic Chemistry | 2 | Taxonomy | 1 |
| Organic Photo Chemistry | 1 | Marine Biology | 2 |
| Applied Chemistry | 1 | The others | 2 |

Actually it is almost impossible to strictly separate the fields related to life science, such as chemistry, biology, biochemistry and pharmacy, etc., because the boundaries between them become obscure.

Some major subjects they study now are as follows.

1) Development of a three dimensional reconstruction system in electron microscopy and image analysis of electron micrographs (for high-resolution study of the structures of bio high polymers which cannot be analysed by X-rays due to difficulty in crystallization).

2) Analysis of bacterial genes by cloning and of the arrangement of bases in DNA.

3) Analysis of genetic functions of insect mutants.

4) Molecular genetic study on cell differentiation and variation of the phenotypic expression of genes.

5) Development of a marking method for nucleic acids without using radio isotopes.

6) Proline carrier of *E. coli*'s proteins - in relation to products of cancer genes.

7) Study on control systems of information transmission between cells with an immunity system taken as a model.

8) Synthesis of new types of medicines related with enzyme and receptor

9) Establishing a microinjection method for cell culture.

10) Nervous system in feeding behavior of toads.

11) Ecosystem-level biophysics by physico-mathematical analysis.

12) Evaluation of coastal upwelling effects on phytoplankton growth by simulated culture experiments.

13) A 45,000-mol-wt protein-actin complex from unfertilized sea urchin egg affects assembly properties of actin.

14) An electrophysiological study of the olfactory system.

The answers of the respondent who replied to questionnaire item 3) are given below. The average points (M) for each item are obtained through the points the item received and the standard deviations ($\sigma$) are calculated through the distribution of them. The following tables indicate the items in order of decreasing points score (Table 4, 5):

Table 4. Items in order of student interest in chemistry

| Chemistry Items | Total | | Life Science Majors | | Non-Life Science Majors | |
|---|---|---|---|---|---|---|
| | M | σ | M | σ | M | σ |
| 1. Electronic Configuration of Atom, Chemical Bond | 3.8 | 0.92 | 3.9 | 0.88 | 3.4 | 0.92 |
| 2. Chemical Reaction Equation | 3.5 | 1.05 | 3.6 | 1.01 | 3.4 | 1.11 |
| 3. Specific Properties of Organic Compounds | 3.5 | 1.18 | 3.6 | 1.21 | 3.4 | 1.11 |
| 4. Battery, Electrolysis, Ionization Potential | 3.5 | 0.89 | 3.5 | 0.97 | 3.9 | 0.83 |
| 5. Periodic Law | 3.5 | 1.13 | 3.3 | 1.12 | 3.8 | 1.07 |
| 6. Atoms, Molecules, Concept of Mol | 3.4 | 1.01 | 3.5 | 0.97 | 3.2 | 1.08 |
| 7. Chain Compounds, Cyclic Compounds | 3.4 | 1.01 | 3.5 | 0.97 | 3.2 | 1.08 |
| 8. Oxidation-Reduction | 3.3 | 0.94 | 3.3 | 1.00 | 3.3 | 0.78 |
| 9. Natural & Synthetic High Polymer | 3.3 | 1.06 | 3.3 | 1.08 | 3.4 | 1.02 |
| 10. Properties of Gas, Equation of State | 3.2 | 1.21 | 3.0 | 1.16 | 3.9 | 1.04 |
| 11. Reaction Velocity & Chemical Equilibium Equilibium Constant | 3.2 | 1.10 | 3.0 | 1.04 | 3.7 | 1.10 |
| 12. Acid, Base, Salt, pH, Neutralization | 3.2 | 0.80 | 3.2 | 0.78 | 3.2 | 0.87 |
| 13. Non-metals and their Compounds | 3.2 | 1.06 | 3.0 | 0.98 | 3.6 | 1.11 |
| 14. Metals and their Compounds, Complex Compound | 3.1 | 1.10 | 2.9 | 0.97 | 3.4 | 1.28 |
| 15. Properties of Solutions (Boiling Point, etc.) | 3.0 | 0.90 | 2.8 | 0.72 | 3.4 | 1.11 |
| 16. Colloidal Solutions | 2.9 | 0.99 | 2.7 | 0.87 | 3.6 | 0.92 |
| 17. Solubility of Solids in Water | 2.8 | 0.89 | 2.7 | 0.81 | 3.1 | 0.94 |
| Average | 3.3 | 1.06 | 3.2 | 1.04 | 3.5 | 1.06 |

The results of the questionnaire 3) can be analyzed stochastically under the limitation that the respondents are not numerous. The results of this kind of investigation must be useful (like clinical reports on new medical technology) in order to discuss how future scientific education should be developed.

Table 5. Items in order of student interest in biology

| Biology Items | Total | | Life Science Majors | | Non-Life Science Majors | |
|---|---|---|---|---|---|---|
| | M | σ | M | σ | M | σ |
| 1. Molecular Genetics (DNA, RNA) | 4.4 | 0.89 | 4.6 | 0.64 | 4.0 | 1.18 |
| 2. Origin of Life, Chemical & Molecular Evolution | 4.2 | 0.88 | 4.4 | 0.71 | 3.7 | 1.01 |
| 3. Mendel's Genetics | 4.0 | 1.06 | 4.2 | 0.89 | 3.6 | 1.32 |
| 4. Reproduction & Development | 3.9 | 0.95 | 4.0 | 0.74 | 3.8 | 0.98 |
| 5. Evolution | 3.8 | 0.91 | 4.0 | 0.74 | 3.4 | 1.24 |
| 6. Enzyme, ATP | 3.8 | 0.87 | 3.9 | 0.88 | 3.6 | 0.92 |
| 7. Cell Division | 3.8 | 1.00 | 3.8 | 0.92 | 3.8 | 1.17 |
| 8. Energy Utilization (Movement, Biosynthesis) | 3.8 | 0.87 | 3.7 | 0.81 | 3.9 | 1.01 |
| 9. Structure & Function of Cell | 3.7 | 1.02 | 3.8 | 1.02 | 3.6 | 1.02 |
| 10. Respiration | 3.7 | 0.97 | 3.9 | 1.08 | 3.3 | 1.10 |
| 11. Homeostasis | 3.7 | 1.15 | 4.0 | 1.30 | 3.4 | 1.24 |
| 12. Sensation, Response & Behavior | 3.7 | 0.96 | 3.6 | 0.87 | 3.8 | 0.98 |
| 13. Photosynthesis, Nitrogen Assimilation | 3.5 | 0.93 | 3.5 | 1.10 | 3.6 | 1.02 |
| 14. Ecology | 3.4 | 0.93 | 3.5 | 1.35 | 3.3 | 1.10 |
| 15. Tissue & Organs | 3.3 | 1.32 | 3.3 | 0.84 | 3.3 | 1.15 |
| 16. Excretion | 3.2 | 0.87 | 3.0 | 0.75 | 3.4 | 1.02 |
| 17. Phytohormones | 3.0 | 0.95 | 3.1 | 0.78 | 2.8 | 1.25 |
| 18. Classification & Phyletic Line | 2.8 | 1.18 | 2.7 | 1.12 | 2.8 | 1.33 |
| Average | 3.6 | 1.07 | 3.7 | 1.03 | 3.5 | 1.14 |

The results of the investigation are summarized below.

(1) Generally they tend to show that there is more interest in subjects of the biological fields than of the chemical fields. We cannot say at present whether it is a tendency peculiar to the author's school or common in Japan. There is an opinion that it is rather common. To clarify this aspect, the author would like to receive cooperation from many schools.

(2) Life science undergraduate majors are more interested in biology than non-life science undergraduate majors. However, in chemistry no such significant difference is evident.

(3) The difference in items in which they show interest appear to be attributable to the curriculum and learning method.

The students were very much interested in molecular genetics in the biology field but not so much in high polymer chemistry which corresponds in the chemistry field to the biology subject. This indicates that, in dealing with chemical substances like proteins and nucleic acids, they tend to be interested but much more so when these are related to life science than when such substances are introduced only from the viewpoint of the structures of matter and chemical reactions. This appears to present a guideline when considering how chemical education should be taught for those who learn life science.

It is interesting to note that the pupils' major in university was determined by interests developed in high school. Thus, pupils who choose their major in the Science area prefer to study systematic and logical explanations in the high school Course; and there are no difficulties in introducing higher concepts of modern research, although the "Course of Study" has no such recommended approach.

## CONCLUSION

We are living in the Computer Age which means we are exposed to a flood of information. Therefore, even the youngsters who have no strong background in science have some concept of very advanced scientific knowledge. Biotechnology is one example. Pupils know that biotechnology, developing now via Life Science based on Molecular Biology, is becoming a main part of the most advantaged technology, although these remarkable advantages are due to the talents and efforts of the specialists in that area.

These new fields must receive attention from people who are relating to Science Education. As the author has mentioned, some trials to innovate the Biology Course on the basis of integration of the subject with Chemistry and/or Mathematics were welcomed by the pupils. There are some restrictions existing in schools, such as the level of the pupils and the limitation of time available, however, it should be emphasized that curriculum development was necessary to stimulate pupils to study Life Science in order to meet their needs and interests related to their random knowledge received from various information sources.

The "Course of Study" used now was established in 1982, however, there are no descriptions in it of the integration of Chemistry and Biology, or items of modern molecular biology relating to biotechnology. The situation for Chemistry is the same, i.e. there are no descriptions of liquid crystals, semiconductors or, of course, biotechnology.

There is strong opinion that education should be based upon establishments and that it is dangerous to try to cover new items in the school curriculum. There is also a theory among the pedagoguse that integration of two different subjects can only be done when they are taught completely. Thus, integration should be applied at the university level. However the amino acids were taught both in the Chemistry and Biology Courses, fairly independently, and there are no relations between Mathematics and Science. This separation in the school curriculum seems strange to the pupils, since they know that mathematics can be used for the interpretation of nature, and that biosubstances are also chemical materials.

Chemical education is indispensable to the pupils who learn life science since the structures of matter and chemical reactions are the basis of the phenomena of life. In high school education, however, it is not always necessary and appropriate to offer detailed knowledge on chemistry.

Even if a teacher intends to teach the structures of nucleic acids, for example, in detail, pupils could not learn the subject well unless they have a knowledge of the fundamentals of chemistry, including organic bases. If the high school curriculum were similar to that of the university, the results would not be good, rather more pupils could feel an aversion to science.

In relation to nucleic acids, for example, it is possible to teach the fundamentals of molecular genetics by schematically explaining the principles of replication and transcription occurring between adenine and thymine as well as guanine and cytosine, while not detailing the structures of the purine and pyrimidine of which nucleic acids are composed.

Thus, if the pupils who have learned some first steps of chemistry look at the phenomena of life, they may have a feeling that they are able to straighten out by themselves the "entangled threads of life", which they considered rather improbable. This is an important idea to meet the aims of science education. In high school education, it is essential that pupils receive strong scientific impressions and flexible ideas so that they can become capable of doing scientific research.

Therefore, the author proposes the following to innovate the high school science course (at least for the Chemistry and Biology Course), although this is only in light of his limited experience: 1) To rearrange the items appearing in the Chemistry and Biology Courses; 2) To consider the subject order: most of the high schools teach Biology first without giving knowledge of pH, and Organic Chemistry. However, if we try to introduce Molecular Biology, the pupils should know at least the fundamentals of Organic Chemistry; 3) To create linkages between the mathematics and science courses. There are connections between Physics and Mathematics, for example, differential equations applied to the analysis of motions. But, mathematics (Matrix) can also be applied to explain the spiral DNA model, as shown earlier.

There would be many ideas for curriculum changes relating to the Life Science Course, and the author wishes to be the initiator of such new approaches through his talk at this Conference.

# What does industry expect from an education in chemistry?

W. J. Beek

Unilever Research Laboratory, Vlaardingen, Delft Institute of Technology,
The Netherlands

## INTRODUCTION

When I was invited to present a lecture on industry and chemical education, I accepted with pleasure. I felt that as one who combines two jobs, as the head of a multinational, industrial research laboratory and as a teacher in an Institute of Technology, I could perhaps contribute to this theme.

In the end, however, I did not find this to be an easy task. Not only are cultures different and nations in various states of development, but industry as such has never voiced a coherent view on education and certainly it has not sustained any clear thinking on the issue. Of course, we find examples of enlightened captains of industry commenting on education. However, speaking generally, industry is not your best adviser on teaching and on the design of a curriculum. My arguments for making this bold statement are twofold: we cannot predict social demand for the next generation and we cannot predict individual development. Therefore, industry as well as the state itself should stay at arm's length from the direct responsibility of the school and its teachers. Our inability to forecast demand and individual progress form the very rationale with which to plead for freedom of teaching, as the best gamble for a society.

Professor Oki in his opening speech, which set the scene for this symposium, asked the following question: can chemical education in school and college become directly useful for industry? My answer to this question is a blunt "no". Schools should not attempt to be <u>directly</u> useful to industries. They should be useful to students. The real service which schools and colleges deliver to industries is to enable the latter to recruit self-confident and motivated youngsters. As long as that is so, no industry will complain about having to pay taxes for contributing to this purpose.

This liberal statement has limits, of course. The history of education is full of examples where the state has intervened because schools did not meet actual demands. Sometimes, these interventions have been prompted by industrialists. Hence, a school which wishes to be up-to-date on developments in industry and its labour market for graduates should be engaged in a dialogue with people from the industries who have an open mind on educational issues.

The first thing then to clarify in such a relationship is that more often than not the questions asked differ from the questions answered. I will explain this by an example. I am often asked to give a view from the industry on the selections made for a certain curriculum. I find that a relatively uninteresting question, albeit that I always attempt to answer it politely and to my best judgement. But, it is a question outside my industrial point of view, for the following reason. As an employer of graduates I am not particularly interested in <u>what</u> students have covered as subject matter, at least within broad limits. My interest is in <u>how</u> they did it. I base my judgement on personal abilities and motivations of the graduates and I value their teachers above all as professionals in teaching. We in Unilever sponsor quite a number of schools and university chairs not because we might get good results cheaply (which is seldom the case), but because we judge them to be eminent educators in science, whose student we seek.

## AN INDUSTRIALIST ON EMINENCE IN EDUCATION

What then, in our view, is an eminent education? Here, I have to explain the different outlook on knowledge and technology between a school and an industry.

The school transfers knowledge from one generation to the next, so that students know for themselves, by inquiry and study, what may be considered to be accepted reasoning, proven fact and validated evidence. Industry, on the other hand, uses science and knowledge to improve the technology of their products and processes with the aim to effect markets for products and services, attempting to be more economic and more effective than others. Where industry engages itself in the more fundamental aspects of science, (which happens), it will do so for the same practical reasons and not out of a mere desire to foster basic science. So, the result of an education is measured by industry as the capability of its graduates to effect change, by relying on knowledge. An example of a successful education is the case of the biologist who was entrusted with building the first nuclear power station in the UK and who performed what at that time was an excellent engineering job.

Do not ask me how exactly this maturing process is promoted in a student, from absorbing science matter to novel applications or innovative combinations of known, but still separated corners of science. The poster sessions and the demonstrations in this symposium have given a reassuring number of examples where teachers challenged students to seek out their own abilities and their own limits and this, I think, has to be the basis for educating maturing people. I learned three things from these posters and demonstrations.

Firstly, too much sophistication is bad. People learn from simple truths. Bring issues back to grasping the essential, to the kernel of the matter. I remember my chemistry teacher in high school, who came into the school canteen where we were enjoying a yoghurt provided by the school. He told us he would talk that afternoon on enzymes, and then asked us to fill the drinking straws with the thick yoghurt and to measure the time for emptying the straws by gravity. Then, he asked us to spit into the yoghurt, to shake it well for a few minutes and to repeat the experiment. All he asked us to do was to prove that a dilution effect was an unlikely explanation of the measured difference. Such experiences stick in the memory of young people as did his subsequent lecture.

Secondly, I learned over again that starting from real, concrete examples taken from daily life is a better approach to learning to appreciate abstract science than the reverse. Industry, for that matter is always concerned with matters of everyday life.

And last but not least, teaching a course according to a pre-cooked syllabus is taking all life and all motivation out of an education as Kelly so rightly stated yesterday. Science has to be fun, in school and in industry. This is what we expect industrial research managers as well as teachers to bring to it. Both are most challenging and demanding jobs, which leave no room for the pursuit of individual hobbies. The only satisfaction of those jobs is seeing how others achieve their purpose.

## A JAPANESE SUMMARY SO FAR

So far for the differences and similarities between science in schools and in industry, I could summarize all this in the words of Arinori Mori, who became at the age of 25 Japan's first Ambassador in Washington and of Fukuzawa Yukichi, the founder of Keio University. They belonged to the Meirokusha, the Movement of the sixth year of the Meiji Society, which later would develop into the Risshisha, the Society of Independent Men.
They wrote: *

-"The urgent task of government is to tell people what to do on education,
 - not to do the job itself".

* From the Speeches of Fukuzawa, W.H. Oxford, Hoduseido Press, 1973, Tokyo

- "The purport of learning is not merely to read books. In society, both ignorance and the lack of common sense are pitiful. Scholarship is not to be considered the only valuable thing in life. One who pursures knowledge to become engrossed in learning per se will turn into just another playboy. In view of this, it is not necessary that all have to be profound scholars or be specialists in some kind of art. The important thing is to have such an educational background as to grasp the essentials of the physical sciences and the liberal arts".

Mori's and Fukuzawa's speeches on education were revolutionary in that they broke away from the view that science should be taught for science's sake only, the old ideal of the ruling, leisured classes. They spoke about a professional, vocational commitment, much the same as industry will look upon an education today. I am not saying that the pursuit of science for science's sake is bad. It is a cultural asset, but society, and with it industry, need more than the values of a new leisured class. It relies on schools and invests in science because it needs people who are entrepreneurial and capable of effecting change, preferably because they learned to think logically and to some purpose. As Fukuzawa wrote: "Education may be regarded as a business and should be judged upon as a power to generate national and individual wealth and well-being".

Logic with purpose, analysis and synthesis, specialization without losing the ability to generalize when experience grows, are issues which pose some eternal dilemmas to teaching. I will elaborate somewhat on two of these dilemmas: today's schools as mass institutions versus an appropriate individual education and the dilemma that young minds tend to take scientific truth naturally as the equivalent of sound judgement.

After having dealt with these, I will finish my talk by reviewing in the light of our needs for the future our skills to influence favourably our societies by science based technologies, which I call engineering.

The best way to distinguish, in my opinion, between the school as an institution and an education is to compare an examination paper with an interview for a job.

A school examination requires the correct answer and proof that the student has mastered a given course. Here, even clever mistakes score poor results. In an interview for a job, the pecking-order of school examinations is taken for granted, but the acid test is how does the graduate perceive his future, what is his or her motivation, how does he or she approach a problem, what is the reaction to criticism and mistakes, how open is the communication?

Did you ever ask graduates to write a letter of application for a job? In my industry, we write off six out of ten applicants for a vacancy, because they prove to be unable to write letters which make their points. Further, most of those who enter our service feel ashamed to admit mistakes, because you as teachers have demonstrated the value of impeccable truth, even when in doubt. It takes us years to teach them that making incorrect assumptions, holding hypotheses which can be falsified and communicating failed experiments are part of our struggle for progress, but serve nothing when they are hidden. A good teacher is aware of this and does not shy away from showing his struggles nor his mistakes. Scientific exploration is communication with a very human face, at least in an industrial establishment.

For these reasons, I favour not too many introductions into something, which all have to be examined. I dislike too much memorizing. But I favour a curriculum of a limited number of subjects, which are taught fundamentally and are examined in a challenging, vocational way. Let them prove after a basic chemistry course that the farmer did well who, after knocking down a brick-built stable, transferred his chicken shed to the spot for two years and only then integrated the area into his vegetable garden. Here again, I find Fukuzawa on my side: "Our curricula are too complete rather than inadequate and are not properly adapted to the students' needs. As a result, many students grow lazy. Why do we ask students to study everything and do we accept to turn them out in the same frame of mind, as if they had studied nothing? Let us think of the off-campus society as being another university".

The difference between knowledge and judgement poses another dilemma to the young graduate. His perception is, after school, for understandable reasons, that scientific proof equals judgement.

I have been involved in creating a new industrial research laboratory in a developing country. It took ten years before the new group had grasped two things: the relevance of certain scientific disciplines to our products and the position of our products in the market. That had nothing to do with a lack of education. They all had come from the best schools in their country, which had given them a direct entrance to a graduate school in the USA, the UK or Germany.
It had to do with understanding common products of every day use and consumer behaviour. The scientific challenge is there, once one starts to think about it, as with everything one is motivated to study, but to match science and objectives and to judge on opportunities and approaches, asks for more than scientific proof. Generally, it will take about 5 years of guidance and support before a young graduate in industry knows its products and markets so well that he can rely on his own initiatives and his own design of experiments. Some, who are introvert and poor communicators, may never reach this relatively independent situation. Social abilities prove to be a first prerequisite for a successful career.

From such experience, I draw the following conclusions. The generic vocational skills in a science education are the flair to perform experiments, the simpler the better, the proper design of these and the correct design of equipment. Qualitative observation is equally important as quantitative measurement. This, again, can better be achieved by studying in depth a limited number of subjects, than by covering many subjects superficially. Experimentation, interpretation and validation is communication, and experiments in school should serve as exercises in communication.

And last but not least, to end this chapter, experience with experiments teaches a student better than text books his ultimate potential.

## SKILLS FOR SOCIETY'S DEMANDS

We come now to the last part of my talk. In all the countries that I visit regularly, I hear that scientists lack mobility, afraid to enter into new fields of application and that the skill to implement new findings, the engineering skill to effect things, is underdeveloped. Governments and industries seem to agree on these issues. I think this gloomy talk is partly due to the world recession, but it is also true that engineering, the skill to implement on the basis of scientific insight, has to be developed further. Our culture and with it our schools, take it too much for granted that engineering follows science, almost automatically. That is a remnant of the past, in which we favoured the cultural value of science much more than that of engineering.

One may judge this situation also from the fact that engineering goals are often nuclear. An actual exmple of this is biotechnology. All governmental science policies mention it, and have made it fashionable, but the scientific challenge in these policies is much more clear than the ultimate objectives to be engineered. The state of affairs may also be judged upon by the lack of engineering teaching in all non-technical education. In a world so dependent on technology, it can hardly be believed that the interest in how we do things is so limited, be it the production of drinking water or the generation of electricity. The remarkable thing is that all our teaching touches upon it, but stops short after explaining the fundamentals, leaving it to the pupils to fill the gap, which only few do. As long as this is the case, the engineering language will not develop adequately along with the technical systems which we use and our engineering heritage will be without history. It is not difficult to imagine the effects this could have on our cultures. What will be progress for one, will be alienation for the other. The organisers of the Tsukuba Expo 1985 have sensed this, but the solution to this dilemma is not enthusiasm for technology derived from the childish amazement at a fair, but a new outlook on science education at all school levels. Industry will have to support schools and teachers to achieve this. I will give you one example.

It is no longer difficult to foresee how the telephone will develop into a device for transferring voice, data and eventually images. The service functions of this device to a household, office and factory will be many. This future can already be observed from the present development of the working place of a chemist in a modern laboratory. His personal computer annex screen, linked to a network of main frame computers as prime movers of information has replaced his telephone. The prime movers assist him in the retrieval of information or in the comparison of spectral data. His dedicated computer power is not only for individual data and message handling, but also for the design of experiments, such as generating a random input to separation columns for obtaining (with cross-correlations between output and input) much better signal to noise ratios than would be possible by a stepwize change in the feed.

The latter is novel, because it is applying radar signal theory to chemical analysis. Excellence is quite often found in new combinations made across the boundaries which our curricula create. This is not unexpected. Tinbergen was awarded the Nobel prize for applying principles of fundamental physics to economics. The new breakthroughs in biochemical engineering marry up a profound insight into the laws of selective environmental pressure with chemical engineering principles. Or to add one of my new ideas, studying kangi script bears a direct relationship with studying chemical structure and nomenclature.

This brings me to the following new conclusions.

Science and technology should be part of all education. This can be achieved by applying fundamental principles outside traditional fields. In chemistry education similar forward-looking approaches are to be fostered, be it in combination with physics, life sciences, signal theory, data handling, languages or structural imaging. Overall, the structural and qualitative side of chemistry is less developed than the analytical side and needs strengthening. That is why I defend physical chemistry against further deterioration, because it deals with structure covering the scales of the molecule, the unit cell, the domain, the grain, the dislocation, the bubble and the droplet. It is still more qualitative than quantitative, but it deals with everything industry is interested in, from metals to coal pellets, from ceramics to food stuffs and from plastics to microchips.

New combinations between chemistry and other sciences will not make life easier for the teacher of chemistry. But it is a necessity if we look at the trends for the future. These are, on the one hand: low cost manufacturing, but also a further diversification in products. A more efficient use of materials and energy. A market need for even better quality products, often of a high chemical purity.

On the other hand, jobs will be upgraded in industry, but fewer in number. Jobs for technologists in the service sector will increase. Overall, there will be a less clear distinction between professional specializms. This is both a threat to, and challenge for graduates. In industry, we observe already how a development chemist in a factory, having access with his computer to all marketing research, may develop, if this suits his ability, into a market research or a brand manager.

These trends bring me to my final conclusions.

We have to consider a more fundamental split into two-level exit examinations of our school system, that is for those who will operate in the size front of our knowledge and those, the majority, who will be skillful operators within the existing knowledge pool. The latter will mature by the experience gained in practice, because free enterprise will more and more exploit and share knowledge and experience internationally.

This inevitably means that for the majority of the graduates the final education and training will be on the job. For them knowledge will grow with experience, in a competitive environment. It could well be that this leads to joint ventures between some academia and industries, in a common pursuit of progress. Industry looks for people who know how to study science, but who need not have covered a wide range of subjects, who have been taught to be entrepreneurial by inventive teachers who on their part foster dialogue, communication and improvisation and do not overemphasize sophistication.

Or, to say this in Fukuzawa's words: "How one should study will vary to each man's abilities and environment. Each man will have to select a path appropriate to himself. The one area in which competence is most urgently needed today, is the writing and speaking of foreign languages. In short, our problem today is international intercommunication.

## CONCLUSION

I confess we had little time in this hour to digress on chemistry. Other issues were too important, as seen from my side. Perhaps this came as a disappointment to your professional pride. The message is we value chemists, but we value real, confident and entrepreneurial people above all. The essence of my talk is given in the following, well-known short story of a learned American friend.

A student was asked to measure the height of a sky scraper, given a barometer. He answered I measure its shadow and the shadow of the building. The ratio gives the height of the building in barometer units. My friend asked for another answer. The student said I walk the steps and with my pencil I mark its height in barometer units. My friend grew angry and said he could have one more try. The student said I walk to the roof and measure the time for the barometer to fall to the ground. If that does not satisfy you, I make a pendulum with the barometer and a string and measure the frequency of oscillation of the ground and on the roof. From the ratio of the two I derive the height, be it not very accurate, if you allow me to check in an encyclopedia on the radius of the earth. In anger, my friend asked him if he did not guess what he was after, of course, the student said, but why should we discuss an approach on which we both agree already.

I think the student was right, and I thank you for having tried to follow me, up to this final conclusion.

# The teaching of industrial chemistry—widening the scope of high school chemistry

Avi Hofstein

Department of Science Teaching, The Weizmann Institute of Science, Rehovot 76100, ISRAEL

Abstract - The need to change chemistry curricula to meet the goals of science education in the 80's and to meet the needs of not only those who are going to be future scientists but also future citizens is discussed. This article is in fact a call for widening the scope of school chemistry by including industrial units. These industrial units should be presented using an interdisciplinary approach, i.e. include also social, environmental, economic and technological dimensions as well as the scientific dimension. In this article, the problems of development and implementation of industrial units is also discussed and demonstrated with examples from various countries.

## DOES OUR CHEMISTRY CURRICULA REFLECT THE SPIRIT OF THE TIME? THE NEED FOR CHANGE.

In an article titled "Is school chemistry relevant?", Barbara Presst in the U.K. wrote in 1970 that (ref. 1):

"Looking at school chemistry syllabuses, it is difficult to realize that we live in a world which is increasingly dependent on the work of a chemist. School chemistry courses show almost no sign for the change which has taken place in the last few decades; for present day needs they are completely irrelevant".

It seems that since then very little has been accomplished in this area, since in his 1983 lecture at the 7th ICCE in Montpellier, Prof. Kempa called for development of an "effective chemical curricula and courses", he wrote:

"... The courses or curricula should adequately reflect the spirit of the time as well as being educationally effective." (ref. 2)

Do our chemistry curricula reflect the spirit of the times? On reviewing school chemistry curricula we find that in the 80's we continue to ignore one of the most important issues of our everyday life, namely chemical industry.

Almost all the secondary science curricula developed throughout the western world during the golden decade of curriculum reform (1960-70) were based on the conceptual structure of the discipline. They tried to show how a chemist works in a research laboratory, emphasizing laboratory work and concept formation and usually did not attempt to include any technological applications of the scientific concepts studied. The developers of the 'new' science curricula at that time considered technology to be at best irrelevant and, at worst, an interruption in the order of development of the structure of the scientific discipline being studied.

The teaching of chemistry without discussing various aspects of the chemical industry omits one of the most important features of modern life and, by implication, indicates that technological achievements are not important. At the same time, in the media, there is over-emphasis of the hazards of pollution, carcinogenesis, chemical warfare, radioactive waste disposal, etc. which in turn help to form negative attitudes in the general public towards chemistry as a whole.

There is always the danger that the public's image of chemistry is influenced by catastrophes that happen around the world from time to time. For example, this year the tragedy in Bophal involving the Methyl Isocyanate and the panic in Sweden concerning the leakage of 40 tons of Oleum ($H_2SO_4$ $SO_3$) revealed chemical industry in its most negative connotation. These issues are indeed part of the problems of daily life, but the publicity that they attract produces an anti-industry attitude and gives an unbalanced picture of the nature of science, in general, and of chemistry, in particular (ref. 3).

There is no doubt that constant exposure to such news items will eventually influence the public's image of chemistry in general and students' attitude towards the study of chemistry in particular.

We must ask ourselves if it is not our duty as chemical educators to provide the students with a more objective and balanced picture of the chemical, technological and societal role of chemical industry. Is it not our duty as educators to reverse the negative image of chemistry as a destroyer of quality of life?

This is particularly important if we assume, as I believe we all do, that chemistry at the school level should aim not only to produce future chemists but also to educate "tomorrow's citizens" so that they will have sufficient understanding of the many problems related to the industrial production of chemicals so that they will be able to make rational decisions based on knowledge and not on emotion. It is our duty to change the attitude of those who are going to be future citizens from being hostile, to being more open-minded about issues concerning the chemical industry.

## THE IMPLEMENTATION OF INDUSTRIAL CHEMISTRY LEARNING UNITS - PROBLEMS AND SUCCESSES.

During the last fifteen years many articles have been written about the need to incorporate some aspects of industrial chemistry into the regular chemistry curriculum (ref. 3, 4 and 5). Why is it that although so many have called for redefining the goals of science teaching to meet the needs of society, so little has been done?

Before examining the difficulties involved in the development and implementation of industrial learning units, the objectives of such a course should first be defined.

In the past, mainly in the 30's teaching industrial chemistry meant:

"to devote a lesson or two to such large-scale general operations as grinding and crushing; rolls and mills; grading, shifting and screening; sedimentation and filtration; centrifugal machines, vacuum dryers, lixiviation, crystallization, calcination, reverberatory furnaces, and kilns; evaporation and distillation; conveyance of solids, liquids, and gases; elevating liquids; refrigeration; the hydraulic press".

(It was also suggested that)

"... few minutes at the end of a lesson would suffice to outline a manufacturing process, illustrated if possible by pictures and sectional diagrams". (ref. 6)

Much time has elapsed since then and together with our changing views about ways of teaching science I believe that the goals of teaching a course on the chemical industry should be rephrased as follows: (ref. 3 and 7)

i. To consider the basic chemical principles as applied to the production of chemicals on an industrial scale;

ii. To demonstrate the importance of the chemical industry to society and to the economy;

iii. To develop a basic knowledge of the technological, economic and environmental factors involved in the establishment of a particular chemical industry;

iv. To investigate some of the specific problems faced by the local chemical industry (e.g. location of industrial plant, supply of raw materials , labor, etc.).

These goals, however, pose some problems:

### The problem of the teachers

The main difficulty in the implementation of a course on industrial chemistry at the high school level is no doubt, the chemistry teacher. The traditional training of a chemistry teacher at a University or Teacher Training College hardly touches upon the applications of chemistry in general, or upon industrial chemistry in particular. Even in the rare cases that teachers have an appropriate background, up to date information on industrial chemistry is not readily available. (ref. 4)

For this reason many teachers feel insecure in discussing in class, the economic technological and environmental issues of a chemical plant. Teachers do not feel competent to handle the many facets that such an interdisciplinary program touches upon. On the whole, most chemistry teachers feel reluctant to get involved with such courses, unless they have had a comprehensive and intensive training course in industrial chemistry. (ref. 8)
The second problem is that most teachers regard their major task as being in the cognitive domain and that the coherent development of scientific ideas is of the greatest importance.

These teachers should be trained to find the right balance between application issues and scientific ideas (ref. 9).

### The influence of universities and external examination boards

The science curricula of the 60's and 70's, were greatly influenced by the various science disciplines as taught in the universities. As a result, universities and external examination boards are today fairly influential when it comes to school science curricula .

Although university academics and administrators give verbal expression to the importance of programs which incorporate industrial applications, their entrance requirements and the syllabuses of the external examinations, are geared to the conceptual structure of the discipline (ref. 10). There must be more than just a verbal commitment on the part of the universities and examination boards as to the value of courses on the chemical industry. Only then will such courses have a chance to make a significant impact on the chemistry school curriculum.

### The local industry

The successful development of a course on industrial chemistry requires the active cooperation and if possible, the participation of the chemical industry itself. The support of the local industry is important both in the development stage of a course as well as during implementation in schools. This cooperation and support is important for many reasons. Informing teachers about developments in industry, the organisation of visits in industrial plants, replying to teacher's inquiries, supplying lecturers to schools and in the design and preparation of teaching aids (e.g. films, slides, booklets about industry, etc.). Information centers exist in the U.K.; The School Information center on the Chemical Industry (S.I.C.C.I.) and in South Africa (ref. 11) which provide teachers with a continuous flow of information, background reading material and lectures as well as with guidance of how to organize visits to different chemical plants. British Petroleum (B.P.) in the U.K. has recognised the importance of encouraging the two-way flow of information between the company and schools, providing information and material that can be used by the chemistry teacher in his classroom. There is no doubt that creating an active relationship between schools and industry is an important aspect of the implementation of such courses.

### The planning of visits to industry

It was found, that outdoor experiences are an important method of achieving some of the goals of a learning unit in the chemical industry (ref. 12). Through school visits to a certain industry, students can observe the applications of "classroom" chemistry to the real world firsthand. Seeing the chemical industry at work can be both stimulating and informative (ref. 13). The process of converting raw materials into useful products, provides a real lesson not only in chemistry, but also in technology and economics.

Such visits may also be used as a way of preparing our future citizens to understand the role of industry and eventually as a citizen to be able to make informed decisions in our highly technological world. Therefore, the time and effort expended by a teacher in preparing for such visits, is more than offset by the educational experiences obtained.

Visits to industrial plants are only effective, provided they are properly prepared and a detailed scheme is prepared by the teacher with the help of representatives from industry. A preliminary visit by the teacher to the plant will often insure the success of the visit by the school (ref. 12). In fact Combs *et al.* (ref. 14) in York found that there is positive correlation between students' attitude towards those industrial chemistry and industrial visits that were organized and properly planned and conducted.

In summary, programs on industrial topics demand new types of curriculum and implementation strategies. All the well-known problems of implementation of a curriculum are present, in addition to the new ones mentioned above.

## THE DEVELOPMENT OF LEARNING UNITS AND PACKAGES ON THE CHEMICAL INDUSTRY.

### Review of developments around the world

In most countries over the past 15 years, few attempts have been made to introduce material about the chemical industry into the school chemistry curricula. Several attempts were conducted, for example, in the U.K.:

Johnstone, Percival and Ried (ref. 15) have described the use of "learning packages", which include written material and audiovisual aids designed for developing, among other things, a social awareness related to chemistry and the chemical industry. For example, they developed the "Amsyn problem" which involves the student in a simulation problem where students are instructed in role-playing on an environmental problem. 'Amsyn' is an industrial chemical plant which produces aromatic amines. Its waste disposal has 'killed' the local river and the local authority is ashamed of this eye sore. To improve the ameneties of the town the authority has begun to build a sewage treatment plant and has ordered Amsyn to make its effluent suitable for treatment in the plant. A number of courses of action are possible. Students are asked to discuss and try to solve the "Amsyn problem". By doing this, they study the relevant science, apply scientific knowledge to real problems and participate in the simulated resolution of socially relevant questions.

Interactive teaching units to be used by sixth form chemistry students were also developed - by a group in Glasgow (ref. 7). Six units were written based on the idea that discussion—based methods and decision-making procedures, are effective in changing students' attitude (ref. 16). The following units were developed:

1. 'To market a drug' - an examination of the factors involved in drug safety.
2. 'Nitrogen for the nineties' - choosing a fertiliser.
3. 'Titanium' the aerospace metal.
4. 'Zinc and you' - The practical outcome of the thermodynamics of metal extraction.
5. 'Organic liquids containing oxygen'.
6. 'Vinyl chloride'.

For some years the Nuffield Project included in its A-level program (grades 11-12, age 16-18) industrial options for those majoring in chemistry. One option describes the differences between small scale laboratory research and large scale industrial production. Optional units were also developed in: food science, extraction of metals, and the production of fossil and fuels (ref. 17).

A series of learning packages on the chemical industry were developed by Waddington and co-workers at the University of York (ref. 18).
These were developed for use by A-level students and involve them in interaction with each other and with their teachers. The following were developed: ammonia and proteins, the manufacture of chlorine and sodium hydroxide, mass spectroscopy, methanol and the synthesis of a gas, acetic acid, aluminium and ethene. The Joint Matriculation Board in the U.K. (ref. 4 and 18) offers an alternative chemistry syllabus at A-level which includes aspects such as chemistry in relation to industry, society and the environment.

The syllabus is divided into a main core and a series of optional topics. The core includes a substantial section on the social and economic aspects of chemical technology as well as industrial processes and chemical principles. To teach these aspects six case studies were developed: 'sulphur and environment', 'salt based industries', 'a study of effluent disposal', 'routes to phenol', 'the spray-steel process' and 'the use and abuse of drugs'. The options include: agricultural chemistry, polymers, electrochemistry and metallurgy. Both the J.M.B. and the Nuffield projects are interesting attempts to integrate pure and applied chemistry.

Recently, a new course in chemistry for citizens concerning issues from Science-Society and Technology was developed in the U.S.A. (ref. 19). The course is called CHEMCOM. These issues oriented Chemistry curricula consist of 8 issues:

1. Supply of our water needs.
2. Conserving chemical resources.
3. Petroleum.
4. Freezing or cooling.
5. Nuclear chemistry.
6. Our health and chemistry.
7. Chemistry - air - climate.
8. The chemical industry; promise and challenge.

CHEMCOM can be considered as a course that examines chemistry though the consequences of its applications.

## The Teaching of Industrial Chemistry in Israel

I would like now to give an example of our work in Israel.

In an attempt to introduce relevant science-society-technology issues into our chemistry classes, we decided to develop an optional unit that consists of three case studies* on the chemical industry in Israel (ref. 3, 20 and 21). The objectives of this unit have been listed earlier.

The three case studies we chose were:

I. Copper Production in Timna: The study of a copper mining in Israel, which included such topics as the construction of the plant, copper production in general, cost factors, and considerations that led to the periodic reopening and shutting down of the mine.

II. Plastics: This case study was chosen to illustrate the importance of polymers and plastics, as well as the technological, economic and the ecological aspects involved in this industry. It includes the electrolytic production of $Cl_2$, aspects of petrochemical industry and the polymerization of vinyl chloride to polyvinyl chloride - PVC; the environmental problems and threat to the health of the workers in PVC plants and the associated problem of mercury poisoning (which is used as the cathode in the electrolytic production of $Cl_2$) that is polluting the sea.

III. Life from the Dead Sea: a study of the production of bromine and its compounds and their use.

I have chosen to elaborate on this topic, as an example of the case study approach in industrial chemistry.

The bromine case study begins with a discussion of the geography of the Dead Sea, its mineral content (a saturated solution of 330 mg/dm$^3$ magnesium, sodium and potassium chlorides and bromides). This is followed by a discussion of the world food shortage, the need for fertilizers, and the role played by the production of potash (potassium chloride) by the evaporation of the Dead Sea brines. How these issues relate to the Dead Sea is "brought home" by specifically considering bromine. Bromine is produced at the Dead Sea from the "end brines" of the potash plant containing about 12g of bromide ion ($Br^-$) per liter. In fact, about 350,000 tons of bromine are available for extraction each year. The students soon realize that the benefits of extracting bromine from the Dead Sea may have world wide application since bromine compounds are widely used both as pesticides (methyl bromide) in agriculture, and as anti-knock additives in gasoline - dibromethane ($(CH_2)_2Br_2$) now used as a substitute for lead ($Pb(Ethyl)_4$). Students, in the laboratory, compare the various ways of producing bromine from brine by (i) electrolysis of brines, (ii) reaction of bromide ions with chlorine and (iii) 'oxidation' of sodium bromide with potassium permanganate. Students discuss the production and uses of such compounds as dibromoethane and methylene bromide; perform simple experiments to compare the different methods. Finally, with data supplied to them, determine the availability and cost of the raw materials, the energy needed, the safe disposal of by-products, and technical and economic problems involved in scaling up to full production. Students also discuss and analyse the factors relevant to deciding on the location of a new plant (e.g. availability of raw materials, energy, waste disposal manpower, etc.) for producing bromine and bromine compounds.

Figure 1 illustrates the overall components of the interdisciplinary approach of the bromine and bromine compounds case study.

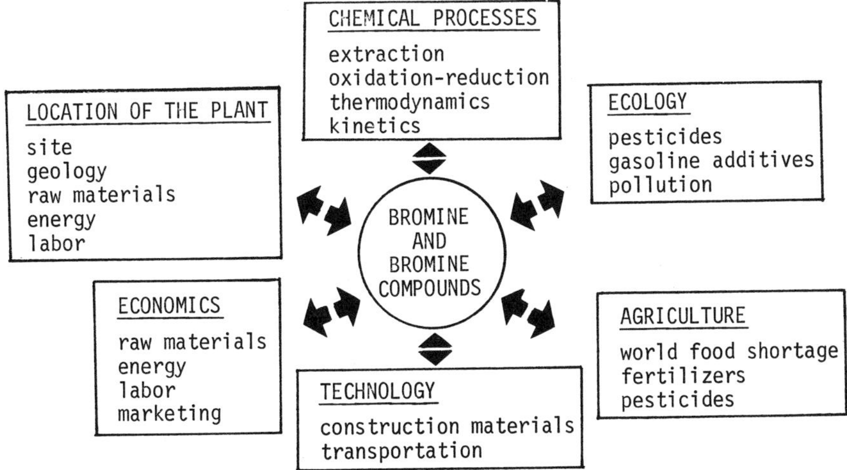


* A case study was defined by Walker (ref. 22) as a study in detail of a particular issue or problem.

Using such an interdisciplinary approach the teacher can involve his students with a variety of teaching/learning techniques, e.g. small group discussion, (involving students in decision-making processes), simulation games, laboratory experiments and visits to industrial plants. All these help in breaking the monotony of the chemistry classroom and eventually will help in motivating students who learn chemistry (ref. 23).

## The training of teachers to teach the industrial case studies

As was mentioned above, the main problem in the implementation of industrial courses is the lack of preparation of high school chemistry teachers. In order to overcome this problem, a series of in service teachers' courses were held at the Weizmann Institute of Science, for teachers intending to teach this course. These included (a) one day conference on chemical industry in general, with lectures (on the industry of fertilizers, pharmaceuticals and plastics) from experts in the fields, (b) eight lectures by chemists from some of the major chemical industries and visits to some of these industries, such as the Dead Sea potash and bromine works and to the heavy chemical industries (ammonia, nitrates, detergents) round Haifa Bay, and (c) a four day inservice training course, specifically designed for teachers who were going to teach the new unit. In this course, each case study was discussed in depth, the experiments were tested in the lab, and various administrative and didactic problems were discussed. During the preparation of this course, contacts were established with many of the local industries which helped to provide some financial support for the project and in organizing visits (ref. 8).

Another aspect which is worth mentioning is that in recent years it has been suggested that teachers themselves should become more involved in the process of curriculum development and implementation (ref. 24). In fact, the idea is that the teacher will become the developer of a substantial part of science curriculum he/she is implementing in his/her classroom. Recently we made an attempt to train teachers to develop their own case studies. We organised a longitudinal inservice course for senior chemistry teachers. The teacher came one day a week during the academic year and were trained in improvement of instruction, scientific subject matter as well as didactis of chemistry. At the end of the course teachers, in groups, were asked to prepare a final project - a case study in the chemical industry. In these case studies, teachers were asked to use a variety of instructional methods to which they were exposed during the course. Teachers made contact with different industries, developed the story board of the industrial plant, suggested learning activities and experiments to simulate the industrial process. Information about technological, economical and environ-mental issues concerning the plant was provided.

The following case studies were developed:

- The soap and detergents industries.
- The industrial production of MgO - (a raw material used in high temperature bricks for furnaces.)
- Chemical fibers for the textile and related industries.
- The batteries industry.
- The manufacture of rubber tyres.

Similarly the science education group, at the University of York has involved a group of scientists, secondary school teachers and chemists from industry and university in the development of the A'level Chemistry Curriculum which cover industrially based topics. (ref. 25 & 26).

## APPLICATION TO OTHER COUNTRIES AND OTHER CURRICULA.

In all countries there is a need to open a window to the "real world" outside the school. I am sure that the chemical industry can provide an interesting and useful opportunity for introducing relevant issues to pupils in senior high schools. The guidelines in planning such a course are the same in most countries, even though obviously the actual industries described and the details of the methods of production vary from place to place. Each case study should be chosen using the criteria described, keeping in mind the importance of that particular industry to the country in question.

In some countries it might not be possible to add a full course to the existing curriculum. In these countries there is merit to using the 'case studies' method, which allows the teacher, or the examining board, to incorporate one or more case studies as a part of the curricula, according to the time available.

The teacher could use the model suggested by Lewis (ref. 27) for the design of science curricula. Lewis suggested that in the future, learning units in science should be made up of three components: "science for the inquiring mind, science for action and science for citizens" (figure 2).

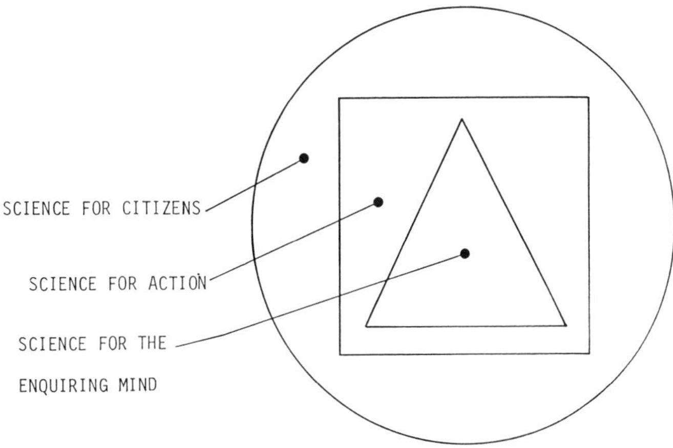


Figure 2 : Lewis' model for science teaching

An example of such a unit is the course developed on the chemistry of fertilizers (ref. 28). The scientific component (science for the inquiring mind) curriculum is the production of $NH_3$ by the Haber process:

$$N_{2(g)} + 3H_{2(g)} \rightleftharpoons 2NH_{3(g)}$$

The 'action' dimension is the technological industrial application of the process and the 'citizens' dimension is the production of nitrogen fertilizers for growing better crops for the over-populated world as well as some environmental - ecological considerations.

It is suggested that Lewis' model could help for future development of industrial learning units.

Though the chemical industry in different countries is based on different raw materials, an interdisciplinary approach can be used to enable the student to understand the overall concept of the industrial endeavor, to combine chemical principles with problems of technology and economics and to foster a positive attitude to the preservation of the environment.

In conclusion, let me mention a recent call to redefine and reformulate the goals of science education. Project synthesis, a comprehensive research project conducted in the U.S. (ref. 29) considered current needs and corrective actions in science education. Four goal clusters divided learning outcomes into categories of relevance:

(a) Personal needs: Science education should prepare individuals to utilize science for improving their own lives and for coping with an increasingly technological world.

(b) Societal issues: Science education should produce informed citizens prepared to deal responsibly with science-related societal issues.

(c) Career awareness: Science education should give all students an awareness of the nature and scope of a wide variety of science- and technology-related careers open to students of varying aptitudes and interests.

(d) Academic Preparation: Science education should allow students who are likely to pursue science academically as well as professionally to acquire the academic knowledge appropriate to their needs.

Programs on industrial chemistry topics which are based on the interdisciplinary approach described in this paper have the potential in enabling students to attain, at least partially these four goals.

## SUMMARY

I believe that given a proper training, the average chemistry teacher can develop and teach a course concerning a local industrial plant. In order to do so it is our duty as chemical educators to:

1. provide the chemistry teacher with certain flexibility of time, and of course budget so that they will be able to select the most relevant industrial issues for their students.

2. create active cooperation with local industries which will provide teachers and curriculum developers with relevant information and will help in organizing visits to chemical plants.

3. convince boards of examination as well as universities of the importance of teaching about the industrial chemistry in a broader sense.

4. We, chemical educators and teachers must convince ourselves that, as Gallagher (ref. 30) stated:

   "For future citizens in a democratic society, understanding the interrelations of science, technology and society may be as important as understanding the concepts and process of science".

## REFERENCES


1. B.M. Presst, Ed. Chem. 7, 66, 1970.
2. R.F. Kempa, Proceedings of the 7 ICCE, Montpellier, (1983).
3. N. Nae, A. Hofstein, and D. Samuel. J. Chem. Ed., 57, 366, (1980).
4. W. Hughes, Chemistry: Pure vs Applied in D.J. Daniels ed. New Movement in the Study and Teaching of Chemistry.
5. A. Hofstein and R.E. Yager, Sch. Sci. Math., 82, 539 (1982).
6. F.W. Westway, Science Teaching: What it was, What it might be. Blackie and Son: London (1929).
7. D.F. Hoare and A.H. Johnstone, Ed. Chem., July, 120, (1984).
8. N. Nae and A. Hofstein in P. Tamir, A. Hofstein and M. Ben-Peretz. Preservice and In-service Training of Science Teachers, Balaban International Science Services, Rehovot, (1983).
9. I. Holman, Proceedings of the International Conference on Science and Technology Education and Future Human Needs. Bangalore (1985).
10. A. Cowton, The London Times, 13, February,(1978).
11. P. Spargo, Personal Communication, (1985).
12. N. Nae, V. Mandler, A. Hofstein and D. Samuel, J. Chem Ed., 59, 582, (1982).
13. W.A.H. Scott, Ed. Chem, 22, 5, (1985).
14. S.D.Coombs, J.N. Lazonby and D.J. Waddington, Ed. Chem., 21, 93, (1983).
15. A.H. Johnstone, F. Percival, and N. Ried, Stn. High Ed., 6, 77, (1981).
16. N. Ried, Unpublished PhD. Dissertation, University of Glasgow, (1978).
17. D.J. Daniels, Ed. Chem., 7, 108, (1970).
18. Several Reports published by the Science Education Group at the University of York, 1985.
19. H. Heikkenen and W.T. Lippincot, Proceedings of the International Conference on Science and Technology Education and Human Needs. Bangalore, (1985).
20. N. Nae, A. Hofstein and D. Samuel, Ed. Chem., 19, 20, (1982).
21. N. Nae, Unpublished PhD. Dissertation, The Weizmann Institute of Science, Rehovot, (1980).
22. I. Walker, Ed. Chem., 11, 58, (1974).
23. W.B. Kolesnick, Motivation: Understanding and Influencing Human Behavior. Allyn Bacon, Boston, (1978).
24. N. Sabar, J. Curr. Stu., 15, 43, (1983).
25. D.J. Waddington and P.E. Nicolson, University of York, Science Education Activity for the year 1984-5. The University of York, (1985).
26. C.J. Garrat and B.J.H. Mattinson, Proceedings of the International Conference on Science and Technology Education and Human Needs, Bangalore, (1985).
27. J. Lewis in U. Ganiel. Proceedings of the 7th GIREP Conference. Balaban International Science Services, Rehovot, 1980.
28. A. Hofstein, New Trends in Integrated Science Teaching, V, 119, (1980).
29. N.C. Harms and R.E. Yager, What Research Says to the Science Teacher, Vol. 3, N.S.T.A. Washington, (1981).
30. J.J. Gallagher, Sci. Ed., 55, 329, (1971).

# Linking chemical education to industry—Korean experiences

Wha-Kuk Lee

Department of Chemical Education, Chonbuk National University,
Chonju 520, Chonbuk, Republic of Korea

## INTRODUCTION

The links between chemical education and industry have continually been weakened as more and more theoretical chemistry has been introduced in school and university chemstry courses and as sophistication of industry has been accelerated (ref. 1). School chemistry has become more "pure", getting rid of elements which are related with everyday life and industry. Teaching and researching in university chemistry has become more separated from industrial technology and research & development. Recent attempts to strengthen chemical education-industry links including the 8-ICCE have exposed not only many problems of chemical education-industry links or cooperations but also possible solutions for them (ref. 2-4).

The purpose of this paper is to introduce some Korean experiences in school & university-industry cooperations with respect to chemical or science education. Korea, as a typical developing country, has successfully developed chemical education and industry out of the dust of the Korean war in the 1950s. Though industry-school & university links were not very tight and strong, some forms of industry-school and university cooperation have evolved in Korea. In the first part of this paper aspects of school & university-industry links with Korean experiences will be presented.

The development of industry-university cooperation in Korea is the second theme of this paper. Developmental stages of Korean industry and patterns of industry-university cooperation at each stage of industrial development will be described. Sometimes establishment of special institutions might be necessary and effective to promote school & university-industry cooperations. Introduction of the Korean Traders Scholarship Foundation established to stimulate Korean industry-university cooperation is the final theme of this paper.

## ASPECTS OF SCHOOL & UNIVERSITY-INDUSTRY LINKS

Typical aspects of cooperative links between school & university and industry are shown in Fig. 1 in which main functions and possible supports of school & university and industry are presented.

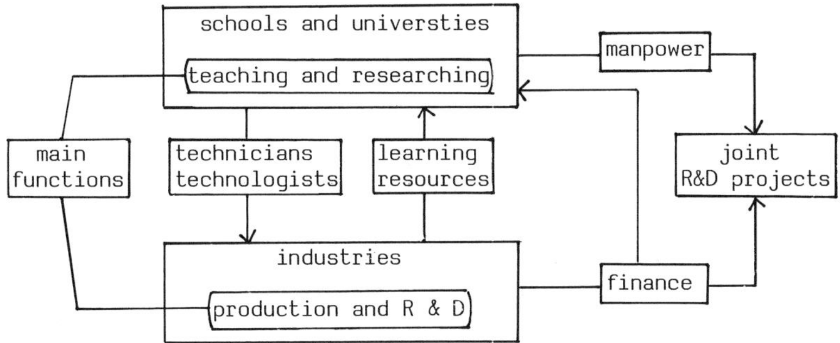


Fig. 1. Schematic representation of school & university-industry links.

In school & university-industry cooperation what school & university could provide to industry are education, and manpower for consultation or joint R & D. On the other hand, learning resources and finance might be of major help when supplied from industry to schools and universities (ref. 5-7). Some Korean experiences in this two forms of cooperation are discussed below.

Education of industrial manpower

Forms of industrial manpower training could be divided into initial, cooperative and continuing education. Initial training of technicians, technologists and industrial researchers is one of the major goals of chemical education in schools and universities. In Table 1 student enrollment in the 1984 academic year are shown at different types of Korean schools (ref. 8).

Table 1. Student enrollment in 1984 academic year

| Types of institution | Total years of study | Number of students (unit: thousand) |
|---|---|---|
| Kindergarten | 1 | 254 |
| Primary School | 6 | 5,041 |
| Middle School | 3 | 2,736 |
| General High School | 3 | 1,200 |
| Vocational High School | 3 | 892 |
| Junior Vocational College | 2 | 232 |
| College and University | 4 | 884 |
| Graduate School | 2-5(*) | 63 |

* :doctorate course.

In the Republic of Korea over ten million students, approximately a quarter of the whole population, are enrolled in various schools and universities and this clearly demonstrates the importance of education in Korea. Recently, college and university graduates are being over-produced and youth unemployment has become a new social problem. Graduate schools have been expanded rapidly but high quality researchers to carry out scientific R & D are not sufficient.

The curriculum of departments of chemical engineering and industrial chemistry is well linked with industry but in departments of chemistry and chemical education the curriculum is hardly related with industrial needs. Cooperative programs or sandwich courses operated in some countries are not common in Korea. In Ulsan Polytechnic, however, several sandwich programs have been operating since its establishment in 1972 by the Korean-British Cooperation. Placed near to the Ulsan industrial park and supported by British experiences many students have successfully completed the sandwich courses (ref. 9).

In Korea continuing education to retrain and refresh the industrial personnel is mostly carried out formally or informally by the industrial training centers rather than by the schools and universities. And it has been one of the dissatisfactory aspects of school & university-industry cooperation related with chemical education. Regarding cooperative and continuing education, a special type of college called "Open University " needs to be mentioned here. In 1983 Gyonggi Junior Technical College was reformed into a college to train technicians most of whom are high school and junior college graduates for their B.S. degree. Several more junior technical colleges were transformed into Open Universities to take part in the combined role of sandwich courses and continuing education for the technicians.

In general, education of industrial manpower as an aspect of school & university-industry links can't be considered satisfactory in Korea. More attention should be paid to enhance cooperative education as well as continuing education and furthermore chemical education should be linked to industry.

Supply of learning resources

Industry can provide many forms of learning resources to schools and universities such as work experiences, visits, finance and learning materials. As an initial work experience "vacation employment", which is called "Vacation Arbeit" in Korea provides students not only work experience but financial aid. The vacation employment which started recently in Korea, would and should be expanded since it was welcomed by both industry and students. Through cooperative education described earlier more work experience could be provided to the students. Visiting factories and other work places enables students and teachers to learn about industrial life and processes. In Korea students visit industry once or twice during school and universitiy mostly as excursions. So more careful planning and clear objectives are needed to make visits to industry more meaningful beyond sightseeing (ref. 10).

The most important role of industry in school & university-industry links is the provision of financial assistance to schools and universities. Some possible forms of financial assistance are scholarships, fellowships, endowments for buildings and equipment and research grants for pure academic or joint R & D projects. The Korean Traders Scholarship Foundation (KTSF) founded in 1974 has supplied various forms of financial assistance to schools and universities on a small scale. The development of school & university-industry cooperation in Korea and the activities of the KTSF will be described in following sections.

Learning materials prepared by industry are rarely found in Korea. Due to lack of those learning materials and the lack of techers' industrial experiences,the industrial aspects of science education aren't well integrated in secondary school chemistry courses. The Korean industry has doubted the short and long term benefits of industry-school & university cooperation. Some reasons for those industry's attitudes can be found in the developmental history of industry-university cooperation in Korea described as below.

## DEVELOPMENT OF INDUSTRY-UNIVERSITY COOPERATION IN KOREA

The pattern of industry-university cooperation has changed depending upon the stages of industrial development in Korea. Since its early development out of the dust of the Korean war, Korean industry has evolved rapidly following several developmental stages of a decade. Some Korean experiences in industry-university cooperation in various stages of industrial development will be described in turn (ref. 11).

### Stage I (1955-1965)

Light industries to produce daily commodities evolved just after the Korean war. In this stage industry-university cooperation was neither necessary nor possible because most industrial techniques were imported along with the hardware and the poor quality of the university scholars. Nevertheless the efforts to import, adopt and modify foreign techniques enabled the recognition of the importance of science and technology education for the development of industry.

### Stage II (1965-1975)

Light industry producing export goods had developed in this stage. The export industry required not only quality managers and well trained technicians but also information about foreign technology. Establishment of the Korean Institute of Science and Technology (KIST) and the Ulsan Polytechnic to train engineers and technologists were natural consequences of those needs. The Industry and university, both recognising the importance of cooperations in R & D, met frequently but failed to find common language to assure mutual benefits from industry-university cooperation. The Korean Traders Scholarship Foundation was establisbed to foster communication between industry and academic communities in 1974.

### Stage III (1975-1985)

Heavy industries were developed to open a highly technical society in this stage. But high technology for the construction of heavy industry hasn't developed in Korea but has been imported from foreign countries and it caused a retreat of the industry-university cooperative movement.

The quality of university scholars has matured greatly by this time but industry was only eager to secure high quality manpower in order to import foreign technology and to handle the problems arising from adoption of that technology. Many doctorates trained in foreign countries contributed to the quick adoption and accumulation of foreign high technology in this stage.

### Stage IV (1985 - ? )

Maturation of high technology and heavy export industry will occur during this stage based on strong industry-university cooperation. The industry and university will actively promote industry-university joint R & D projects. The Korean government will ultimately realize that new technology needed to transform a developing country into a developed country is not available without heavy investment into R & D in science and technology. The Korean government is planning to double current R & D expenditure which is about 1 percent of the GNP in near future.

Development of industry-university cooperation in Korea discussed so far, clearly demonstrates several barriers that make the cooperation difficult in a developing country.

First of all, most technology is imported from developed countries rather than developed through industry-university cooperation.

Secondly, as industry has thought that academic research is not related to industry it was unwilling to provide funds for academic research.

Thirdly, many university scholars considers that their "pure" research is intellectually superior and are not ready to seek mutual benefits through industry-university cooperation.

In a developing country, therefore, promotion of communications between industrialists and university scholars could be catalyzed by institutions similar to the Korean Traders Scholarship Foundation. A brief discussion of the Korean Traders Scholarship Foundation follows.

## THE KOREAN TRADERS SCHOLARSHIP FOUNDATION

In Korea lack of mutual understanding and communication has been identified as a main obstacle to industry-university or business-academic cooperation in Korea. Similar difficulty might be experienced in many countries, especially in some developing countries. One major goal of The Korean Traders Scholarship Foundation (KTSF) founded in 1974 by the Korean Traders Association and its member firms was to lower the barrier between industrial and academic communities. The initial endowment of the KTST, 2,600 million won, has been increased to 10,500 million won in 1983 which is approximately 13 million US dollars. Annual working capital comes from interest on the endowment, business earnings and other miscellaneous income.

Programs and services provided by the KTSF and the 1985 annual budget for each of them is shown in Table 2. Objectives, contents and retrospects of each program based on a decade's experiences will be described in turn (ref. 12).

Table 2. Programs of the KTST and annual budgets of '85.

| Program and service | Annual budget (1985) million won (US $1,000) |
|---|---|
| Funding researches | 270 (309) |
| Scholarship grants | 245 (280) |
| Aid to academic societies | 50 ( 57) |
| Subsidy for scholarly conferences | 50 ( 57) |
| Aid for international cooperation | 140 (160) |
| Assistance for other programs | 25 ( 29) |
| Total | 780 (892) |

### Funding research

Research grants have been provided to selected university faculties in order to stimulate industry-university cooperation and academic research. Also involvements of university faculties in the management of industry and industrial R & D are supported. About 5 billion won (5.7 million US dollars ) has been provided in last 10 years to about four thousand persons and two thousand R & D projects.

The research grants were not enough for academic research but they certainly were useful in stimulating research activities of the universities. Research grants have been provided for all sorts of research but KTSF's ten year experiences suggest that more projects related to industry should be supported and a matching-fund system needs to be introduced (ref. 12).

### Scholarship grants

To attract talented students to industry and to induce students to diligent study the KTSF has awarded grants to more than ten thousand students. Children of the workers attending a college or university with good academic records were preferentially selected for this program. The KTSF's scholarship grants which occupies half of the whole private scholarship grants in Korea seems to be a successful example of industry-university cooperation in Korea.

### Aids to academic societies

Publication of journals by small academic societies has been subsidized. About half a billion won was provided for this service but many subsidized academic societies are still in an infant stage. Development of a self-sufficient academic society seems to require much greater amount of aid.

### Subsidy for scholarly conferences

Various seminars, conferences and meetings are subsidized. Though emphasis was given to economic and industrial growth many sorts of conference themes including the following were included in this program.

- development of primary school education and the community.
- life and heredity
- planning of digital system using LSI.

Aid for international cooperation
The objective of this program was to assist activities involving international cooperation devoted to the development of the Korean economy and promotion of external commerce. Most of the 3 million US dollars of this program was provided to propagate Korean Studies through famous foreign institutions such as Harvard, Hawaii, Toronto and Sheffield Universities.

Assistance for other programs
Various other programs related with promotion of industry-academic cooperation were assisted. About one billion won was provided to 146 programs over 10 years.

The KTSF has contributed to the promotion of university-industry cooperation for over a decade. The objective and programs, however, should be evaluated and modified to maximize its function. Furthermore without substantial increase of endowment and working capital the role of the KTSF could but be extremely limited.

## CONCLUDING REMARKS

In spite of the intimate relation between schools & universities and industry the cooperations between them aren't satisfactory in chemical education. Some Korean experiences in school & university-industry cooperations clearly identifies main obstacles of the cooperative links; mutual disbelief, misunderstanding and a lack of finance to promote the cooperations. Industry's dependence of foreign technology was another special hindrance to university-industry cooperation in a developing country like Korea.

In a developing country realistic university-industry cooperation seems to be established not before but after maturation of both academic research and industrial development. Institutions like The Korean Traders Scholarship Foundation to stimulate academic-industry cooperation might catalyze the maturation of academic researches and industrial developments.

Cooperation of schools with industry related with chemical education is particularly unsatisfactory in Korea and the situations of many developing countries wouldn't be much better. If we realize that the quality of tertiary chemical education is largely determined by the quality of secondary science education a great deal of effort should be made to strengthen the links of chemical education of secondary schools with industry.

## REFERENCES

1. M. Oki, Opening Address to the 8th International Conference on Chemical Education, Tokyo, Japan, 23rd-28th, August 1985.
2. M. J. Frazer, La Chimica EL'industria, 60, 431-434 (1978).
3. C. J. Lee, Korean Traders Scholarship Foudation News, No. 12, 27-30 (1977).
4. R. L. Laslett, R. S. Reeve and R. A. Schulz, J. Chem. Educ., 62, 131-133 (1985).
5. M. J. Frazer, A Paper Presented at the UNESCO Sub-Regional Workshop, Sofia, Bulgaria, 11th-15th, December 1979.
6. H. L. Paige, J. Chem. Educ., 59, 1005 (1982).
7. G. Mattson and J. Gupton, J. Chem. Educ., 60, 124-125 (1983).
8. Ministry of Education, '84 Year Book of Education, The Ministry of Education, Republic of Korea (1985).
9. Ulsan University, University Bulletin (1984).
10. H. Nae, V. Mandler, A. Hofstein and D. Samuel, J. Chem. Educ., 59, 582-583 (1982).
11. W. B. Rah, Korean Traders Scholarship Foundation News, No. 20, 131-133 (1979).
12. KTSF, Ten Years of the Korean Traders Scholarship Foundation (1984).

# Chemistry education for everyone

George C. Pimentel

Department of Chemistry, University of California, Berkeley, California, 94720

In attempting to summarize the rich content of this 8th ICCE Symposium, I find that a suitable place to begin is with a quotation from Oki Sensei's opening plenary lecture:

"Chemistry will be a key science for survival of mankind [because of its importance to] water supply, food, energy, resources, environment, and health."

Yes, we find ourselves in a technological age in which everyone's life is affected by the fruits of science, with chemistry in a central role. These effects extend from issues of comfort and quality of life to economic well-being and even survival. This pervasive impact of technological change is being more and more widely sensed throughout society, perhaps with a significant element of anxiety. This has sharpened and made us more aware of the educational challenges we face.

Professor Malcolm, in his abstract for the third plenary lecture put it rather forcibly:

[We must] "train students of today to be ready in the future to solve problems which have not yet been identified using scientific knowledge which have not yet been formulated and technology which has not yet been invented."

The truth of his statement is displayed in the growth of the chemical literature. This growth is typified in the sheer volume of the 1977-1981 five-year index for the U.S. Chemical Abstracts. This five-year collection fills 4.3 meters of library book shelf and it is only the index. It refers to over 2.6 million scientific papers and patents dealing with chemistry. A second, intimidating measure of the growth of chemical knowledge is found in the number of known compounds. There were fewer than 2 million compounds known in 1950 whereas, in 1980, the total was approaching 7 million. Furthermore, the growth curve is faster than exponential, as shown by the fact that the doubling time was 24 years in the decade of the 1950's while it has shortened to only 11 years in the 1970's. It is sobering to realize that eleven years from today, new Ph.D. chemists will be confronted with twice as many known compounds as there are today when they enter their first high school classrooms to learn what a compound is!

The topics chosen for this symposium display the spectrum of the educational challenges we face--they can be categorized under the following three issues.

1. Are we keeping chemistry education modern and up-to-date?
2. Are we giving the scientists of the next generation the best possible preparation for their professional careers?
3. Are we meeting the educational needs of the public at large?

We began our consideration of the first question with a day devoted to the computer, which has, perhaps, just arrived in time to cope with this burgeoning pool of knowledge. This powerful tool, the computer, places before us a variety of exciting opportunities that we are gradually learning to exploit. Sasaki Sensei gave us a vivid panorama of applications, the most obvious of which are the computer's powerful calculational dimension and its ability to store and make rapidly available the vast store of chemical data that increases every day. Thus, we already have a generation of students who can use neither a set of log tables nor a slide rule. We can look forward to a generation that will not only fail to recognize a Chemistry Handbook, soon they won't know where the library is located because it's all available in their dormitory rooms via their personal computers.

Drs. Moore and Cabrol reinforced Sasaki's theme and Cabrol brought us up to date on the status of artificial intelligence. Then, pervading the poster sessions and demonstration rooms, was an eagerness on the part of the symposium participants to take advantage of this new frontier for education.

At the same time, more than one of our speakers has warned us not to expect too much from or to be beguiled by computers. For example,

Oki-Sensei included in his opening lecture the remarks:

> "Use computers where effective; recognize that computers can't do everything. The computer is not almighty."
>
> He went on to point out effective uses in computer graphics, stereochemistry, and molecular dynamics.

Sasaki-Sensei added a reservation about Computer-Aided-Instruction, CAI.

> "Education by CAI makes one of the important bases of computer education, but it cannot take the place of the human teacher's role in teaching."

However, I believe that the most important of these warnings is concerned with computer simulation of experiments. Professor Sasaki says, "Chemistry is a learning based on experiments." Professor Malcolm agrees, "Computer simulation...gives no indication of the difficulty involved in obtaining accurate experimental results, which are the touchstone of 'truth' in our subject." Those remarks capture one of our most important messages about chemistry--indeed, about all science--that it all begins with measurement. All of our organizing concepts (theories) were ultimately derived from observations and measurements, each with its own intrinsic uncertainty. These intrinsic uncertainties arise from inevitable limitations imposed by our measuring devices. Hence the validity of a given theory is limited by these intrinsic uncertainties and the range of experience over which the measurements are made. This philosophy is embodied in our pedagogic intent when we ask a student to carry out a chemical analysis. It is surely not for the purpose of finding out what is in the sample--it is to give the student the experience of learning <u>how we find out</u> what is in a sample. And as he does so, he also learns that the conclusions to be drawn contain some tentativity because measurements are never perfectly precise.

Let me give you an example. Let us consider the study of the rate of a reaction with mechanism such that for large concentrations of [A], the product [D] grows with the square of the time. In a well-executed experiment, a student might obtain the data and the growth behavior pictured in Figure 1A. With the uncertainty bars shown, however, another student might collect the data shown in Figure 1B. The smoothed curve suggested by these data differs slightly from that obtained before (the dashed curve). Now, suppose that the second set of data are analyzed with the aid of a curve-fitting program designed to represent by a three-term polynomial $[D]=a+bt+ct^2$ the least-squares best-fit to the available data. Figure 1C shows what the computer would conclude, using the data obtained by the second student. With only three points, the curve passes perfectly through the data and it gives a completely false impression of the actual behavior.

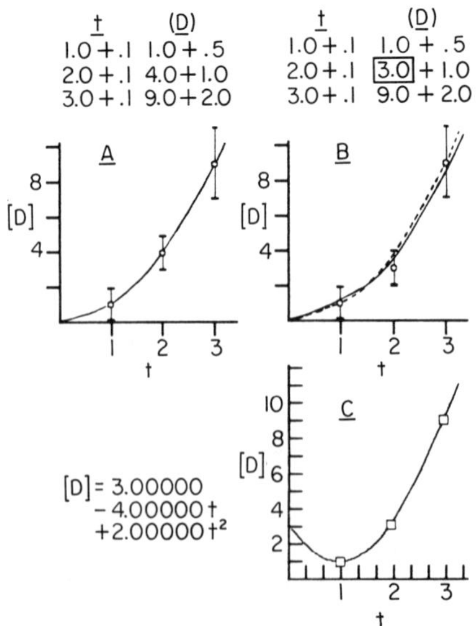


Fig. 1. A kinetic study: computers need judgement

Plainly, a scientist used to dealing with analysis of experimental data that contains experimental uncertainty would not fit three data points with a three-parameter function. Some of our students will, however, and as they do so, they will miss the entire point of the experiment.

Now consider an experiment just conducted at a Summer Institute we conducted at the University of California at Berkeley. The Institute was directed at deepening the knowledge of 50 middle school science teachers. We had access to a sophisticated "thermoprobe" programmed to collect and present thermal data on our PC's. The high school teachers on my instructional staff were quite enthusiastic, at first, about offering this computer-aided approach. As we discussed what we hoped the teacher-students would learn from the experiment and then, what they would take back to their own classrooms, we decided on quite a different experiment. Even though it does not involve computers, it is useful to consider it and see why we preferred it to the more sophisticated approach.

We asked each of the teachers to assemble one of the three thermometers shown in Figure 2. In Figure 2A, the 25ml flask is filled with water, avoiding bubbles and with enough water to extend a few centimeters up into the one mm tubing. In Figure 2B, the liquid is mercury. In Figure 2C, the erlenmeyer contains carbon dioxide and the manometer contains water. Each teacher then calibrated his/her thermometer at two readily reproduced temperatures, the lower of which is called $0^{o}$ and the upper $100^{o}$. We called the three scales $^{o}H_2O$, $^{o}Hg$, and $^{o}CO_2$. The two fixed points we used where ice water and the boiling point of ethanol (which is at $78^{o}C$). Then each teacher used a glass-marking pencil and a ruler to divide the scale into ten equal intervals. Finally, in turn, every teacher used his thermometer to measure the temperature of one and the same water bath, which was at room temperature.

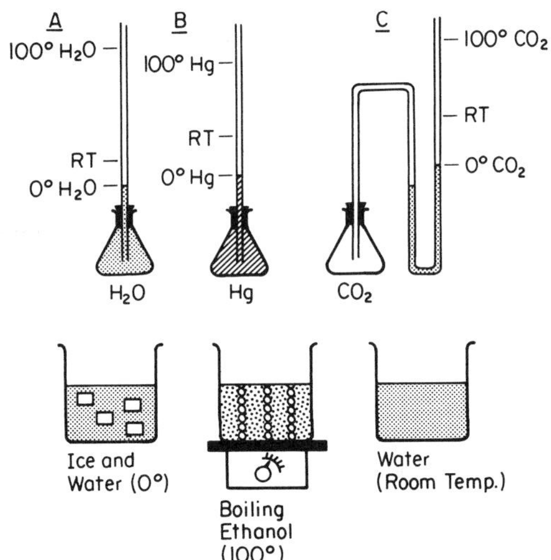


Fig. 2. Simple apparatus can be best for learning.

The teachers were now separated into three groups according to which thermometer they had constructed. They were asked to decide on the "best value" for their particular thermometer and the uncertainty to associate with that "best value." Then all of the teachers came together and one person from each group reported the temperature of the room temperature water bath. The outcome is shown in Table I.

Table I

Room Temperature Measured With Various Thermometers

| | |
|---|---|
| water thermometer | $6.6 \pm 0.8\ ^{o}H_2O$ |
| Mercury thermometer | $20.4 \pm 0.4\ ^{o}Hg$ |
| carbon dioxide thermometer | $19.8 \pm 0.4\ ^{o}CO_2$ |

We spent the next hour discussing why the thermometers gave different results and then, which thermometer to use. One outcome was that other thermometer fluids were tried and it became apparent that gas thermometers all gave the same temperature (within experimental uncertainty). From the known volume per unit length of the tubing and the volume of the erlenmeyer flask, we calculated absolute zero on each temperature scale. Of course, this exercise led to a deeper discussion of the use of an absolute temperature scale defined with a gas thermometer in developing the kinetic theory of gases.

The point of this example is to show that for certain pedagogic goals, simple apparatus can be extremely effective while sophisticated apparatus (including our thermoprobe) would be completely inappropriate.

But let me return to capabilities of computers and useful applications of computers to chemistry. Remember that we are trying to find opportunities to couple these two. For example, Sasaki-Sensei showed us that a computer can draw 217 different structures for a molecule with formula $C_6H_6$ that fit the rules that each carbon forms four bonds and each hydrogen only one. He also used the computer to decide that there are 745 $C_6H_{10}O$ structures and 90,769 $C_{10}H_{18}O$ structures of palytoxin. Place this capability next to a current research question--the structure and synthesis of the nerve poison palytoxin which has the empirical formula $C_{129}H_{223}O_{54}N_3$. Plainly, the result with $C_{10}H_{18}O$ indicates that there is no computer that could handle the problem of drawing the possible structures of palytoxin. Besides, it would be of no use to do so. And, by the way, chemists have already decided what the structure is! They are now dealing with the next problem, which is that there are 63 chiral centers and 7 cis-trans double bonds. We might ask a computer to tell us how many geometric isomers the structure found in nature might have. However, we can handle that with a hand calculator, it is just $(2^{63})\cdot(2^7)=1.2\cdot10^{21}$. Once again, there is no computer in existence that could display for us the $10^{21}$ different geometric isomers.

The intended message is not that the computer cannot help the chemist working on molecules of biological complexity. The point is that we should be seeking ways in which the computer is a real help in teaching chemistry (not teaching computers) or a real help in advancing the frontiers of chemistry. When we ask a student to apply a computer to a problem of chemistry, there are three questions we should think of.

1. Does the computer help us reach the learning goal in chemistry associated with the chemistry problem before us? (Remember, we don't ask the student to determine the composition of an unknown to find out what is in the sample.)

2. Is the use of the computers for the purpose of learning about chemistry or about the computer? (Either is a reasonable teaching goal but we shouldn't pretend we are doing the former when we're actually doing the latter.)

3. Does the postulated use of the computer show how real, contemporary problems of chemistry are being solved by real chemists? (Remember the 217 possible but mostly improbable structures of $C_6H_6$ and the one, unpredicted structure of the natural product, palytoxin.)

Now let us turn to the second question that has been a major theme of this conference. Are we giving the scientists of the next generation the best possible preparation for their professional careers? The answer to this question is complicated greatly in precollege years because students who are likely to be our future chemists are mixed in with a larger group who will find their scientific careers as biologists, physicists, engineers, mathematicians, etc. And all of these are mixed with an even larger group who not only won't go on in science, but may not go on into college. What advice have we heard in this conference that will help us here? What advice have we heard about curriculum?

Professor Malcolm reviewed the history of curricular reform over the last two and a half decades. He criticized the curricula of the '60's as swinging too heavily toward theory and the omission of descriptive material relevant to today's environment and needs. I tend to agree qualitatively with his remarks but I would add a cautionary note. I would remind you of another quotation by Professor Malcolm, that those who fail to learn from history will be forced to live it again. By this, I mean that we must not make the mistake of swinging the pendulum back too far, overcorrecting whatever faults we perceive. To make my point, let me recall the educational situation we faced in high school chemistry teaching in 1958. (Of course, I know specifically about the U.S., but it was probably the same elsewhere.) Chemistry was taught entirely from a descriptive point of view and, in high school and in first year chemistry at the college level, it was characterized as a course in rote memorization. That was quite inappropriate, indeed extremely undesirable, from three points of view.

1. It did not represent a good introduction to science because it omitted the interplay between experiment and unifying principles.

2. It was no longer possible to cope with the burgeoning amount of factual knowledge without the organizational benefit of these guiding principles.

3. It did not present an accurate picture of chemistry as it was practiced at that time.

But we don't have to settle whether the curricular studies of 1960 were appropriate to the problems they addressed--those studies were not tuned to the educational challenges of today. Fortunately, we have had lots of wise guidance from our participants in this conference.

I would begin with the axioms laid down for us by Professor Cole

"Experiences First--Explanations Later"
and, "Theories, concepts, generalizations are important, provided that

- the student is familiar with the knowledge being ordered
- the knowledge being ordered is perceived to be useful
- the theories are not too difficult
- the theories must be honest."

These are, I believe, guidelines we all should attend, although I have a bit of reservation about theories being too honest. What I mean by this is that we must be careful in our zeal to be honest that we don't dwell so much on exceptions that the student loses confidence in, and hence, the benefits of the guiding idealizations. I give three examples. First, the ideal gas concept is not accurate for any gas if we measure carefully enough, yet ideal gas behavior underlies the concept of absolute temperature and, hence, chemical thermodynamics. Second, we know of many non-stoichiometric solids, yet stoichiometry is the cornerstone of chemistry. As a third example, we know of a dozen or so compounds of the inert gases, yet their uniquely inert behavior anchors our understanding of valence.

Reiko Isuyama added to Professor Cole's axioms, "Chemistry is not merely a collection of facts--it is a unified body of knowledge. [We must develop the student's] ability to think logically. [We must remember that] the content of our courses keeps increasing but the duration of the course remains constant."

Professor Malcolm joins in this latter warning:

"[We must avoid] learning overload."
"[We should not be] filling the memory--we should be guiding the mind."

Many of us at this symposium were nervous that our problems of curriculum would become insurmountable by the time we had heard from Professor Kelly about the needs of the biological sciences and Dr. Beek about those for chemical engineers. Instead, their general guidance seems to mesh comfortably into a chemistry curriculum.

Professor Kelly defined the needs for the biology student as:

1) Basic concepts of physical chemistry

2) Bonding, structure, reactions, and synthesis

3) Basic chemistry of the "life" elements: C, H, O, N, P, $K^+$, $Na^+$, $Ca^{+2}$, $Cl^-$, $HCO_3^-$, and $PO_4^{-3}$.

4) Basic chemistry of macromolecules

5) Basic chemical methodology and techniques.

And then, we heard the wisdom of Dr. Beek:

"[We should not end with] a curriculum which consists of too many introductions-into-something, but with one which allows a profound study of a limited number of subjects, combined with a basic experience in experimentation and design."

He goes on to observe that:

"Too much sophistication is bad."
And, "A precooked syllabus takes the life and pleasure out of teaching and learning."

Finally, there is much food for thought in his remarks about what he looks for in a potential scientist employee.

"We expect them to analyze and to judge."

"I look for scientific knowledge and his capability to judge."

Then, thinking ahead to the real leader of the future, Dr. Beek tells us,

"A manager has to be good, really good in at least one discipline."

But a summary of this conference would be incomplete without reference to the importance of the teacher. Speaker after speaker reminded us that the success of the whole educational venture depends upon the quality, enthusiasm, sensitivity and feeling for chemistry displayed by the teacher.

Peter Childs observed

"Only a good teacher can convey his knowledge and love for chemistry." and, he went on--

"You cannot teach a subject well that you do not know and understand thoroughly, but you must also love the subject in order to get it over."

But how can we attract such able teachers to high school and middle school teaching posts when they are grossly underpaid and seriously undervalued by the society they serve? If there is any single goal that this ICCE should try to achieve, it should certainly be to raise the status, prestige, and financial rewards for science teachers, worldwide.

Finally, we should turn to the third theme that was mentioned by more than one of our plenary lecturers, scientific education for the public at large. An increase in the scientific literacy of our entire population is one of today's most pressing needs. It is needed to help society adapt to and benefit from the technological changes that are surely coming. Every citizen has the right to eligibility for meaningful employment in this technological age, the need to live at ease in the presence of these technological changes, and the obligation to participate rationally in the decisions that will decide society's course. Technological issues abound--catastrophic nuclear war, nuclear energy, military actions in space, new drugs, environmental protection, food preservation, herbicides, pesticides, and, now, genetic engineering. Public attitudes, often influenced by media sensationalism, tend to be characterized by anxiety, irrationality, and panic. These attitudes must be replaced by attentative concern, analysis of options, and calm, deliberate choices among realistic options.

This need for better understanding is especially urgent for the field of chemistry. The reason is that even though chemistry responds to vital needs of technological societies, it has assumed a threatening image in the public mind. I can say this with great assurance for the U.S. society, and I believe it is also true here in Japan and in Europe, perhaps throughout the world. The public attention has been so strongly focussed on possible deleterious effects of chemistry technologies--air pollution, being an obvious example--that the public has lost perspective about the benefits of these technologies. I will use urban air pollution as an example. Where does smog come from? Almost all of it is connected with energy consumption, since most of our energy is derived from combustion of fossil fuels. So to cure it, all we need to do is to find the guilty parties who are combusting all of this fuel to make all the smog and make them turn off their engines and boilers. In the U.S. it turns out that two-thirds of the energy use is connected with transportation and the heating, cooling, and lighting of dwellings. Unfortunately, that makes us the guilty parties. We would much rather hear that it is someone else like DuPont Company, or Siemans, or Mitsubishi so that we can make them pay fines and agree to change their ways. But if it is us, it is much more difficult to find a solution--no one could expect me to stop driving my car or not turn on the heat and lights in my home!

This is a good example because it demonstrates the usual outcome--that when there is a risk or cost associated with a use of chemistry, the removal of that risk or cost carries with it its own cost. Inevitably, the use of chemistry involves a benefit to society and if that use is interrupted, its particular benefit will be sacrificed.

What we must do, as educators, is raise the level of understanding of chemistry in the minds of the public at large so that they can more rationally assess the balance between real benefits and possible problems associated with applications of science, including chemistry. This is an educational need shared by all, not just by those who might enter some field of science. To reach this general audience, we must begin with effective science instruction in the middle schools (student ages 12 to 14) and that instruction must include chemistry. There are all sorts of obstacles and problems, but we must remove them. Teacher preparation is, perhaps, the most serious one because there are all too few middle school teachers with strong science backgrounds. We must find ways to increase the depth of science understanding among the teachers we now have in the classrooms. Summer Study Institutes furnish one mechanism. I am pleased to say that such Institutes are again coming into existence in the United States.

A second problem is laboratory equipment and suitable space to give opportunity for student experimentation, particularly in chemistry. We must continue to pursue the ICCE initiative to produce low-cost laboratory kits that are suitable and safe in the hands of children at each educational level. In the U.S., we are searching for ways in which universities and industries can help their local schools through assembly and periodic repair of such kits. We, as science educators, must furnish creative ideas on what these kits should contain.

In this context, let me return once again to the syllabus. What chemistry and what science should be taught to children in their first encounter with science? Should they learn about the structure of the atom? d orbitals? DNA? photosynthesis? genetic engineering? baryons and leptons? f=ma? $E=mc^2$? To me, these are not the right kinds of questions to be asked. For these beginning students, we should not be trying to teach them some particular set of facts or theories of chemistry or physics or biology. <u>We should be trying to affect their attitudes toward science (including chemistry).</u> We must establish a feeling of comfort with how we learn about the world we live in (i.e., with the scientific method). We must be trying to instill in our youngsters the desire to learn more about science, later in their schooling and, indeed, throughout their lives. If we can help them <u>enjoy</u> chemistry and science; if we can keep alive their natural curiosity, we will have done enough.

I will conclude by listing some of the important messages to be taken from this eighth ICCE Conference.

- Chemistry education must reflect today's needs.
- Computers furnish an important new teaching and research tool in chemistry, but their limits must be recognized.
- Good teachers remain the key element in chemistry education. We depend upon teachers who can transfer to and awaken in the student understanding, enthusiasm, enjoyment, and comfort with chemistry.
- We have too much in our curricula

  A modest number of topics should be treated in some depth and with a fundamental approach.

  We must retain flexibility in the syllabus.

  We must not let entrance and exit examinations rigidify our teaching.
- We must provide better science education for the citizen.

# Report on the small group discussions

**M. Chastrette**

Secretary of IUPAC Committee on Teaching of Chemistry
Laboratoire de Chimie Organique Physique
Universite Claude-Bernard Lyon I

Small group discussions were organized to facilitate exchange of information between the participants and to produce brief reports on the present situation in the domain under study, as well as proposals for future action.
The small groups addressed themselves to the four problems corresponding to the themes of the Conference. Four reports were presented independently of the number of groups working on a given theme, on the basis of one report for each theme. The Japanese group, however, discussed the four themes and presented a separate report.

## I. Report of the Japanese Group

Dr. Nakamura, from Nagoya University (Japan) presented a well documented report on the discussion in his group (40 persons). He pointed out how beneficial the meeting between Japanese and overseas teachers was, as well as the difficulties experienced by some Japanese colleagues in adapting to the western discussion style. He made two interesting practical proposals to improve the communication during conferences such as the 8th ICCE:

**1a** To divide the time allocated to a poster session (e.g. 2 hours) into two equal parts. In the first part of the session (e.g. 1 hour) the authors of one half of the posters would have to stay near their posters to answer questions from the participants. During this time the authors of the second half of the posters would be free to circulate and see the other posters. In the second part of the session the roles would be reversed.
**1b** To ask lecturers and presenters to speak slowly and distinctly to ensure better understanding by people whose mother language is not English.

Following these general remarks, Dr. Nakamura addressed the themes under discussion. Quite interestingly the Japanese group produced a set of conclusions very similar to those arrived at by the international groups. Therefore, they will not be dealt with in this part of the report.
The Japanese group discussed an important problem posed by the entrance examinations in Japan and deplored the deformations induced into education as a whole, as well as into chemical education by the severe competition between students for entrance into the best universities.

## II. Report on the use of Computers in Chemical Education

Loretta Jones,from the University of Illinois (USA) gave an excellent concise report on the work of three groups devoted to this theme.
The participants considered computers to be very powerful educational tools that have to be used widely and imaginatively. Four recommendations were made:

**2a** That IUPAC-CTC seek support to set up computer training workshops for chemical educators.
**2b** That IUPAC-CTC support the production of a comprehensive review on the use of computers in Chemical Education. This review should include information on the state of the art in this field and stress the evaluation of past and present projects.
**2c** That participants in the 8th ICCE set up in their countries national or regional information centers on the use of computers in Chemical Education. These centers would distribute information and software, conduct training sessions and maintain links with other centers.
**Note :** Information can already be obtained from three existing centers:
-at Eastern Michigan University, USA,(project SERAPHIM, Dr. J. Moore);

-at the University of Nice, France, (CDCIEC, Dr. D. Cabrol);
-at the University of Ljubljana, Yugoslavia, (UNESCO International Center for Chemical Studies, Dr. A. Kornhauser and Dr. S. Glazar).
**2d** That UNESCO and IUPAC-CTC seek funds to assist in the establishment and operation of these centers.

## III. Report on the Fostering of Chemists of Excellence

A clear and comprehensive report was given by A. Truman Schwartz from Macalester College (USA) on the work of four groups.
He stressed firstly the uniformity of the conclusions drawn by different groups and secondly the necessity, above all, of fostering teachers of excellence. He reviewed the problems encountered and presented some suggestions.

**3a** The status of teachers is unsatisfactory (low pay, low prestige, low professionalism). It was felt useful (but perhaps possibly dangerous) to have a survey on comparative statistics regarding salaries, classes, or hours, class sizes and general teaching conditions in different countries.
**3b** The preparation of teachers should be ameliorated in three areas: practical skills, communication skills and (but less important) educational theories.
**3c** The in-service education should be pursued through workshops, continuing education programs, publications and meetings, cooperation, sabbatical leaves, international exchanges and industrial employment.
**3d** The curricula should be revised, with the following guidelines:
-an inquiry approach at the primary level;
-an approach based on real life at the secondary level;
-an approach taking into account the basic and advanced levels as well as the balance between content and process, for the tertiary teaching.

## IV. Report on the Teaching of Chemistry for Life Science Students

An excellent report was presented by Ms L. M. de Hernandez from CENAMEC (Venezuela).
This group tried to answer four questions:

**4a** How to make chemistry more attractive to life science students? Teachers should
-use analogies, concrete materials and meaningful examples;
-present relevant experiments and demonstrations;
-show the significance of chemistry for the future careers of these students.
**4b** Should the chemistry content taught to life science students be different from that taught to chemistry students?
A consensus was reached on not having a single chemistry course, but there existed a considerable diversity of opinion concerning chemistry courses needed for different life science programs.
**4c** Should life sciences be included in the curriculum of chemistry students?
The answer was generally felt to be 'yes'.
**4d** What to do? (Recommendations.)
-Cooperation and discussion among the different scientists should be encouraged;
-Life sciences teachers should teach their students problem-solving skills.

## V. Report on the Links between Chemical Education and Industry

A rich report was presented by Brock Robertson from the University of Calgary (Canada). This group, like the previous one, sought answers to two questions and made some practical suggestions.

**5a** How can chemical education serve industry? We should provide graduates who:
-read, write, speak and are computer literate;
-are broadly educated and committed to continuous education;
-are innovative;
-are able to design and to work in teams.

More generally, chemical education could help to promote a positive image of chemistry.
<u>5b</u> How can industry serve chemical education? Industry could provide funds, equipment, training facilities for students and teachers, faculty, teaching materials; and more generally, help to develop the scientific literacy of the public.
<u>5c</u> What to do? (Recommendations.)
Cooperation (informal or structured) should be developed by a continuous liason, joint efforts, classroom visits, with a special effort towards teacher training institutions.

## VI. Recommendations of the 6th ICCE

In addition to their debates on the 8th ICCE themes, the discussion groups were asked to examine the 6th ICCE recommendations in the light of the advances made (or not) during the last four years.

### Comments made by M. CHASTRETTE

I was asked to summarize the results presented by the small discussion groups and to make some comments on their suggestions.

It seems to me that there is no need to summarize the reports that you have heard just now. Indeed, they are so well structured and concise that I would just have to repeat what was said. The discussion groups have produced a lot of ideas of great interest, and some practical suggestions as well.
The groups working on the use of computers proposed four recommendations, three of which are directed towards IUPAC-CTC. These are very practical suggestions and IUPAC-CTC is willing to help in this field.
The group discussing the interaction between chemistry and the life sciences recommended that cooperation be developed between chemistry and biology teachers. This is not strictly in the field of the IUPAC, but should be discussed between IUPAC and IUPAB. This discussion could be a simple bilateral action or, perhaps more interestingly, take place within the International Council of Scientific Unions (ICSU).

Concerning the possible action of IUPAC-CTC, I would like to stress that IUPAC-CTC can act only as a catalyst. In fact, the work has to be done by individuals who will conduct a study, write a review, etc. The role of the Committee will be in finding the necessary competent chemical educators and in giving a label of quality to their production. Finally, the Committee will seek, as it has already done many times, money from sources such as UNESCO, which has been very helpful in the past, or from other organizations.

A second task of your groups consisted of examining the recommendations of the 6th ICCE. Four years after the latter conference, the recommendations were found to be quite generally useful and satisfying. This should not be surprising as this set of recommendations defined implicitly a sort of ideal state which could not change much in four years.
The question of the progress induced (or not induced) by these recommendations was often raised. Has progress been made and why ?
The answer is clearly yes and no. For some of the fields related to chemical education little progress was made, but in some important areas remarkable advances can be reported.
The first field is the production of low cost locally-produced equipment. A number of workshops were held in various parts of the world (India, Bangladesh, Brazil or Denmark). Several types of reliable and easily constructed equipment have been produced in significant numbers with the help of UNESCO.
The second area is the organization of national and international cooperation in the use of computers in Chemical Education. The three existing centers have done very important work in this regard.
In these two cases, the work was initiated by some missionaries who managed to convince more and more of their colleagues. They are excellent illustrations of the roles played by national or international institutions in the development of Chemical Education.

Finally, I propose that the participants in this very successful 8th ICCE adopt a single resolution to reaffirm the recommendations of the 6th ICCE.

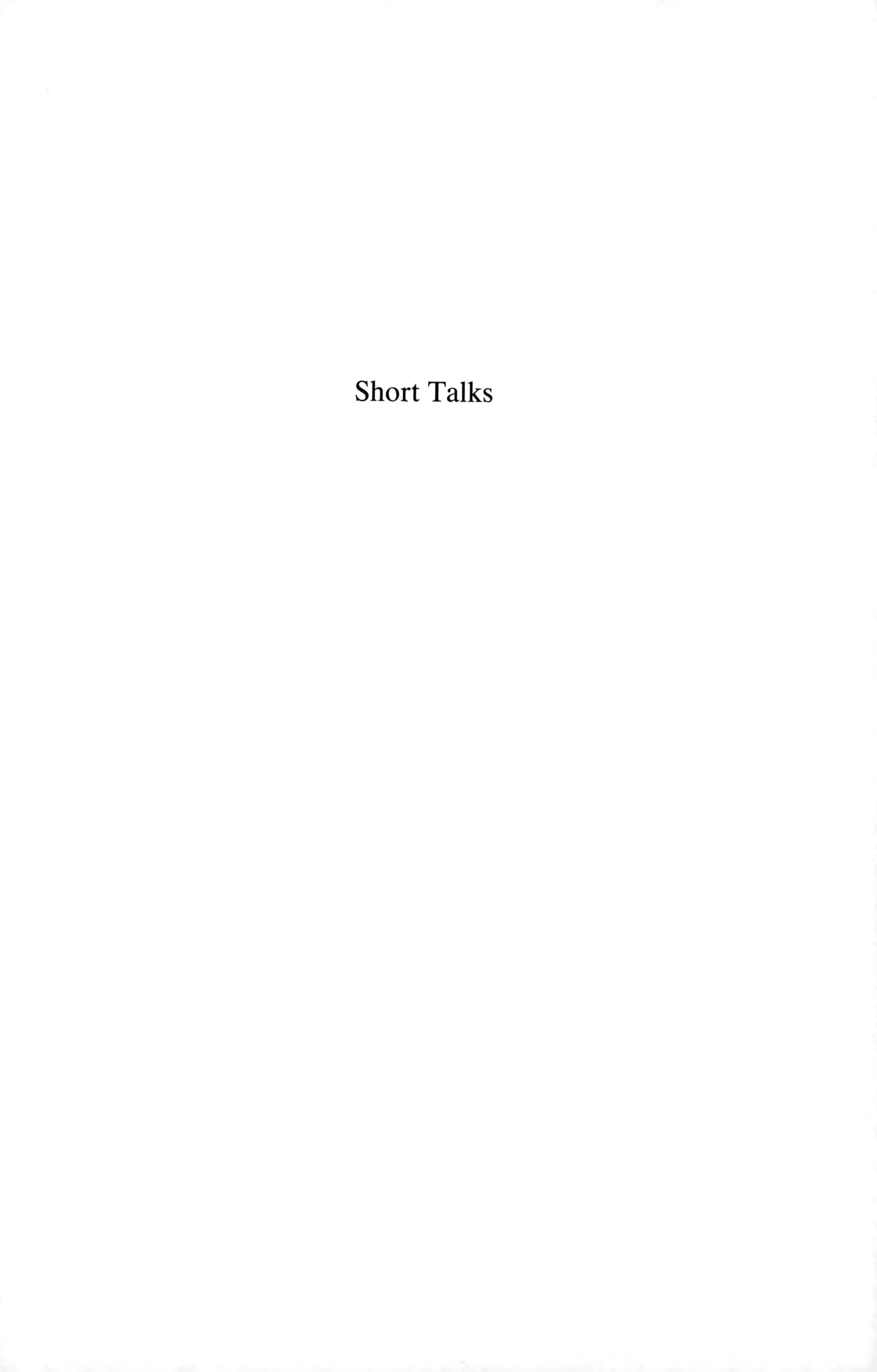

# Short Talks

# Excellence and the accurate use of language, symbols and representations in chemistry

J.D. Bradley, M. Brand and G.C. Gerrans
Department of Chemistry, University of the Witwatersrand Johannesburg, South Africa.

Abstract - Language, symbols and representations are vital tools in conveying meaning. Yet in chemistry, many of them are ill-defined, misleading or inaccurately used. This impedes comprehension of abstract concepts and leads to rote learning and an indifference to accuracy. These outcomes are inconsistent with the aim of fostering future chemists of excellence. It is suggested that IUPAC-CTC should play a role in defining and encouraging the use of appropriate language, symbols and representations.

## INTRODUCTION

Language, symbols and representations are used for purposes of communication and to convey meaning (ref. 1,2). Maintaining precision and accuracy in the use of these optimises the efficiency of knowledge transfer in whatever setting it is conducted. One hallmark of an excellent chemist is an effective concern for this in both research and teaching activities.

As a researcher, the chemist who publishes work in reputable journals, frequently subjects his excellence in this regard to evaluation. In preparing papers for publication the chemist invariably has the assistance of handbooks on nomenclature and symbolism published by bodies such as IUPAC. Peer reviewers and the editors have the responsibility of judging the precision and accuracy of the language he uses. This process serves to maintain excellence in technical communication amongst those active at the advancing frontiers of chemistry.

But away from the frontiers, where newcomers are struggling to enter the field, is this same excellence maintained? As a teacher, does the chemist use language, symbols and representations that have unique, clear and universally-accepted meanings? Does the chemist make his arguments clear and logical and does he carefully distinguish between experimental observation and interpretational model and between cause and effect? We suggest the answer is often 'no' to these questions, and we further suggest that there may be far-reaching consequences of this situation.

In this paper we focus on the basic language, symbols and representations of chemistry and problems in their use in the teaching of chemistry. We exemplify the problems and identify some of their consequences. Finally we suggest ways of ameliorating these problems.

## THREE PROBLEMS OF MEANING IN CHEMISTRY

The focus of our concern is neither systematic nomenclature nor the use of non-technical words (ref. 3): both of these have received attention by others. Our concern is with the technical words, conventional symbols and common representations used in chemistry. Here we have identified three main types of problems. These are presented below with a small selection of examples based very largely on quotations from currently-available college chemistry textbooks.

**(1) Some words, symbols and representations have unique definite meanings, but they are used carelessly.** Often the correct one is used interchangeably with an incorrect one with which it is connected in some way.

Example (a) 'The Lewis diagram for nitrogen is

:N⋮⋮N:'

Here the Lewis diagram represents the nitrogen molecule - not nitrogen, the colourless, odourless gas (for which no Lewis diagram can be given) and not the nitrogen atom (for which a different Lewis diagram can be given). This confusion of macroscopic with microscopic definition is a very widespread kind of mistake (ref. 4,5). It doubtless contributes to the general problem of distinguishing observation and interpretation.

Example (b) 'The total mass of an atom is called its atomic weight'.
The incorrect implication that mass and weight are equivalent in meaning, is coupled with a disregard for the relative character of atomic weights. The vexed question of terminology in this connection has recently been addressed again by IUPAC (ref. 6).

Example (c)

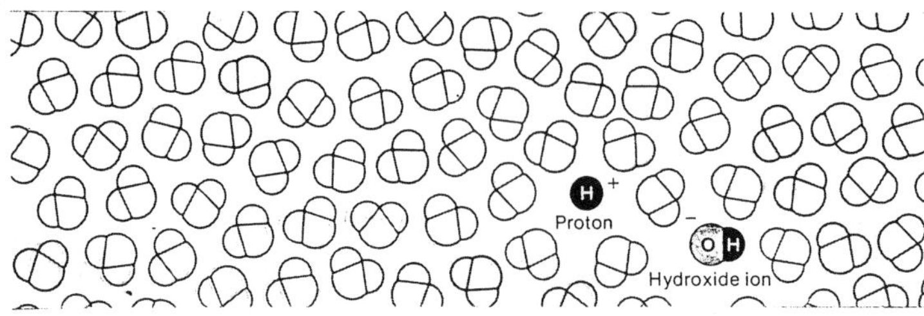


Fig. 1. Pure water dissociates to a small extent into protons ($H^+$) and hydroxide ions ($OH^-$).

In the above microscopic representation of liquid water, the molecules have no contact with one another, (or, indeed, with the proton and hydroxide ion). This is not only grossly misleading but conflicts with other diagrams in the same text, wherein the high degree of contact between the molecules in liquid water is correctly shown.

(2) **Some words, symbols and representations are in common use whose meaning is generally accepted, but which nevertheless appear to conflict with a literal interpretation.**
Example (a) 'The equilibrium moves to the left.' The equilibrium does not, of course, 'move' at all. The statement refers to a change in which the proportions of substances represented on the left hand side of the chemical equation increase at the expense of those represented on the right hand side.
Example (b) 'Ball-and-stick model of water molecule - the H atoms are represented by small balls, while the large ball represents the oxygen atom. Covalent bonds are represented by thick lines.'
Such a drawing of a physical model is an invitation to misinterpretation: spherical atoms are held at a distance by the bonds, rather than being drawn together with a net decrease in volume and, of course, in energy. As others have argued (ref. 7,8), much greater attention should be given by teachers to discussing the meaning of models.
Example (c) 'The point of neutralisation is called the equivalence point.'
At the equivalence point in an acid-base titration the solution is often not neutral. A literal interpretation of 'neutralisation' conflicts with common usage and reality.

(3) **Some words, symbols and representations do not have unique meanings. Different individuals use these in different ways and no authoritative statement of meaning can be found.**
Example (a) 'Element means a simple substance which cannot be changed into simpler substances by any known chemical means.'
'Element - a general name given to each of the 106 different atoms.'
'Element - Matter in its simplest form - a chemical substance that consists of one type of atom only.'
Example (b) 'The smallest neutral particle of a substance capable of independent existence as a gas or as a definite entity in solution is called a molecule. Molecules may be monatomic or polyatomic.'
'Molecule - the smallest possible subunit of a compound - an aggregate of two or more different atoms, or a union of two or more atoms of the same element.
These two examples illustrate how earlier macroscopic concepts associated with the words 'element' and 'molecule', have been overtaken by later microscopic concepts, in some cases producing a confused mixture.
Whilst historic confusion is surely to be expected, the continued existence of this confusion is remarkable (ref. 9,10).
Example (c) Inconsistent symbolism, such as similar formulae to represent grossly different situations (e.g. NaCl and HCl), and grossly different formulae to represent the same situation (e.g. $P_2O_5$ and $P_4O_{10}$). Systematic formulae of substances constitute a rather well-codified part of the language, symbols and representations of chemistry, which has benefitted teachers as

> well as researchers. Nevertheless, although there is this internal consistency, the inconsistency between reality and representation is confusing to the beginner.

In conclusion, it is apparent from these examples that some problems may be attributable to carelessness on the part of the teacher. However, many are embedded in the conventions of the chemical knowledge domain itself. Others (ref. 11) have called for more rational terminology, but we suggest that the scope of the problem is very much greater than has heretofore been identified.

## PROBLEMS OF MEANING IN DOMAINS OTHER THAN CHEMISTRY

The kinds of problems identified above are not a unique feature of chemistry. In everyday experience we are used to encountering:

(1) the misuse of words (e.g. relative and relevant)
(2) the use of words which conflict with a literal interpretation (e.g. a ship sails under its own steam)
(3) different possible meanings of language, symbols and representations (e.g. a mole).

Little difficulty is usually caused, because the meaning intended can be identified from the context (ref. 1). This situation holds, of course, so long as the context is sufficiently familiar and, by implication, the concepts are concrete. Where the knowledge domain is less familiar and/or more abstract - or perhaps where we have to translate from a foreign language - we know from experience that we are often unable to bridge the gaps and we either struggle to extract meaning or abandon the effort altogether.

## SOME CONSEQUENCES OF THE PROBLEMS IN CHEMISTRY

Amongst professional chemists the basic language, symbols and representations of chemistry are mostly agreed, understood and familiar and, just as in the world of everyday intercourse, the context is a sure guide and the concepts are effectively concrete anyway.

The problem in chemistry - as in other knowledge domains - arises when the terrain is unfamiliar and the concepts are abstract. Chemistry probably has a more than its fair share of abstract concepts (ref. 12), and the three types of problems we have identified can therefore have a very serious impact on those still struggling to learn meaningfully. They lack concrete experiences; they lack appropriate conceptual frameworks. They need a well-qualified teacher, a high quality textbook and other learning resources. The teacher needs to use language, symbols and representations correctly, to choose ones least likely to be misinterpreted, and painstakingly to guide the usage and interpretations of the students. The reality, of course, is normally far removed from this best possible scenario. Around the world, there are many more underqualified science teachers than qualified ones. The less-qualified tend to place heavy reliance on textbooks (ref. 13) - if they are available - but, as the earlier example quotations indicate, these often create problems whilst solving others.

One consequence of this state of affairs is that many beginners find chemistry incomprehensible. Some turn away from chemistry and join the ranks of future citizens who are ill-informed and hold antagonistic views about science generally. This outcome reminds one of the situation in the days of the alchemists where obscure and idiosyncratic language, symbols and representations served to preserve their authority and exclude others from poaching on their knowledge. (Perhaps there is something of the alchemist in today's chemists!)

Other students respond by learning by rote and losing interest in searching for meaning. Such responses serve to heighten the confusion between theory and observation and strengthen the disregard for precision and accuracy in thought, word and deed. Regrettably these latter attitudes are very persistent and can be found even amongst graduate students. Clearly these are not chemists of excellence.

## AMELIORATING THE PROBLEMS

We can all contribute to ameliorating these problems and help to foster future chemists of excellence. For a start, we should try to say what we mean and query instances of inappropriate usage by others. We should also try to discontinue the use of some familiar language, symbols and representations that are misleading if literally interpreted. Such individual efforts could be greatly strengthened by the leadership of a body such as IUPAC, and specifically the Committee on Teaching of Chemistry. This would also permit the third of our problems to be addressed, viz. the lack of authoritative statements of meaning in some cases. Precedents for this kind of proposal are abundant: IUPAC has traditionally been recognised as the most important body for defining or recommending terms, symbols and procedures as required by professional chemists. For the average chemist and student of chemistry, the recommendations and the handbooks of IUPAC are its most familiar products. IUPAC should now address the needs of those who are still struggling to learn chemistry and help them become future chemists of excellence.

## REFERENCES


1. J.B. Carroll, Harvard Educ. Rev. 34, 178-202 (1964).
2. C.R. Sutton, School Sci. Rev. 62, 47-56 (1980).
3. J.R.T. Cassels and A.H. Johnstone, Words That Matter in Science, Royal Society of Chemistry, London (1985).
4. J.D. Bradley and M. Brand, J. Chem. Educ. 62, 318 (1985).
5. N.J. Selley, Educ. Chem. 15, 144- 145 (1978)
6. N.E. Holden, Int. Newsletter on Chem. Educ. 16, 13-15 (1981).
7. K. Ehlert and R. Engler, Chemieunterr. 13(4), 27-56 (1982).
8. R.J. Osborne and J.K. Gilbert, School Sci. Rev. 62, 57-67 (1980).
9. P. Huyskens, V.C.V. Tijd. 9, 15-18 (1983).
10. E. Ströker, Angew. Chem. Int. Edtn. 7, 718-724 (1968).
11. R.J. Tykodi, J. Chem. Educ. 62, 241-242 (1985)
12. J.D. Herron, J. Chem. Educ. 52, 146-150 (1975).
13. R.E. Yager, J. Res. Sci. Tchg. 20, 577-588 (1983).

# Children's conceptions in the acquisition of some chemical concepts at middle school level

A. Bargellini, L. Lardicci, M. Mannelli, G. Raspi

Department of Chemistry and Industrial Chemistry, Pisa University, 56100 Pisa, Italy.

Abstract - With the aim of evaluating the real improvement obtained through experimental activity carried out by pupils themselves suitably guided by teachers, in accordance with Ausubel's theory and using suitable questionnaires, an attempt was made to establish what conceptions and misconceptions were held by Italian middle-school pupils (11 - 13 year olds) with regard to the following chemical concepts: Physical states of matter, change of state, solution, element, compound and chemical reaction.

## INTRODUCTION

The acquisition of some fundamental chemical concepts by pupils in the sixth to eighth grades (ages 11 - 14) on an experimental basis, allows them to face present day problems of social and economic relevance. In accordance with Ausubel's theory (ref.1) that the way to improve the level of understanding in learning is "to find out what the learner already knows and teach him accordingly", the following six units of instruction were prepared and experimented in some Italian middle schools: 1) The physical states of matter; 2) Changes of state; 3) Solutions; 4) Elements and compounds; 5) Chemical reactions; 6) Acids and Bases.

The pupils worked in couples, experimentally. The carrying out of the experimental activity for each unit of instruction was preceded by a questionnaire composed of a certain number of open questions by which an attempt was made to test the pupils' conceptions and misconceptions related to concepts including those of a linguistic nature. Each entrance questionnaire was distributed in five middle schools in Lucca and province to a number of students varying from 172 to 86. The answers were examined taking into account the number of exact answers, partly correct answers, that is correct from the conceptual point of view but incomplete, and incorrect answers. Another group of answers were classified: "I don't know". The results obtained from each single questionnaire with regard to the first five topics are reported here.

## QUESTIONNAIRES

Answers obtained from 118 pupils (11-12 years old) to questions on "Physical States of Matter".

1. Chalk is
   a) a solid 99%
   b) a liquid
   c) a gas
2. Air is
   a) a solid
   b) a liquid
   c) a gas 100%
3. Chalk powder is
   a) a solid 94%
   b) a liquid
   c) a gas
4. Sawdust is
   a) a solid 99%
   b) a liquid
   c) a gas
5. Drinking water is
   a) a solid
   b) a liquid 90%
   c) a gas
6. Hail is
   a) a solid 88%
   b) a liquid
   c) a gas

| | | | | | | |
|---|---|---|---|---|---|---|
| 7. Snow is | a) a solid | 90% | 8. Carbon dioxide is | a) a solid | |
| | b) a liquid | | | b) a liquid | |
| | c) a gas | | | c) a gas | 84% |
| 9. Iron is | a) a solid | 90% | 10. Mercury is | a) a solid | |
| | b) a liquid | | | b) a liquid | 91% |
| | c) a gas | | | c) a gas | |
| 11. Water vapour is | a) a solid | | 12. Methane is | a) a solid | 7.2% |
| | b) a liquid | 5% | | b) a liquid | 4% |
| | c) a gas | 92% | | c) a gas | 87% |

13. In what physical states of matter can water be found? -Exact answers: 56%
14. Classification of substances. Inexact answers: 46% including for example "mercury is a solid", "frost is not a solid" and difficulty in the classification of water vapour, copper and sulphur.

<u>Answers obtained from 140 pupils (11-12 years old) to questions on "Changes of state".</u>

| | | | |
|---|---|---|---|
| 1. correct idea of change of state | 75% | 2. heard of changes of state | 88% |
| incorrect | 15 | no | 2 |
| partly correct | 4 | don't know | 10 |
| don't know | 6 | | |
| 3. changes of state known correctly | 50% | 4. correct examples given | 74% |
| incorrect | 25 | incorrect | 13 |
| partly correct | 17 | partly correct | 6 |
| don't know | 8 | don't know | 7 |

5. words that remind of change of state:

| | | | | | | | |
|---|---|---|---|---|---|---|---|
| force | 3% | salt | 7% | condensation | 79% | liquefaction | 83% |
| fusion | 81 | tree | 62 | ether | 7 | | |
| aeriform | 28 | sublimation | 62 | helium | 22 | | |
| boiling | 76 | evaporation | 94 | solid | 52 | | |

6. expressions that remind of change of state:

| | | | |
|---|---|---|---|
| a) a piece of wood burns | 21% | g) ether evaporates | 66% |
| b) water boils | 86 | h) camphor sublimes | 60 |
| c) hail falls | 39 | i) butane, a gas, can liquefy | 61 |
| d) lead melts | 86 | l) solid carbon dioxide changes to aeriform state | 73 |
| e) methane is a gas | 21 | m) lemon juice corrodes marble | 16 |
| f) water vapour mists windows | 78 | | |

<u>Answers obtained from 172 pupils (12-13 years old ) to questions on 'Solutions'.</u>

| | | | | |
|---|---|---|---|---|
| 1. If I speak about solutions what do you think of? | exact | 34% | | |
| | incomplete | 20 | | |
| | incorrect | 34 | | |
| | don't know | 12 | | |
| 2. Have you heard of solutions? Who from? | yes | 72% | family | 2% |
| | no | 16 | school | 23 |
| | no answer | 12 | T.V. | 3 |
| 3. Have you heard of solvent and solute? | yes | 49% | | |
| | no | 32 | | |
| | no answer | 19 | | |
| 4. Familiar solutions | exact | 52% | | |
| | incomplete | 8 | | |
| | incorrect | 22 | | |
| | don't know | 18 | | |
| 5. Solvents you know | exact | 33 | | |
| | incomplete | 10 | | |
| | incorrect | 19 | | |
| | don't know | 38 | | |

6. Which words remind you of a solution?

| | | | | | | | |
|---|---|---|---|---|---|---|---|
| vinegar | 26% | ice | 19% | force | 9% | acetone | 41% |
| lemon juice | 28 | movement | 10 | tea | 37 | salt | 38 |
| ink | 22 | sea | 47 | sugar | 44 | solvent | 34 |

| | | | | | | | |
|---|---|---|---|---|---|---|---|
| bleach | 44 | verdigris | 27 | denatured alcohol | 59 | remove stains | 19 |
| heat | 26 | air | 39 | solute | 26 | | |
| energy | 19 | agitate | 15 | | | | |

7. Which expressions make you think of a solution?

| | | | |
|---|---|---|---|
| a) sugar dissolves in water | 86% | g) salt precipitates in salt-works | 20% |
| b) hail falls | 7 | h) water evaporates | 46 |
| c) acetone dissolves nail-polish | 81 | i) naphthalene sublimes | 22 |
| d) water dissolves salt | 77 | l) ice melts | 39 |
| e) salt is insoluble in alcohol | 22 | m) paint dissolves | 37 |
| f) coffee is soluble in water | 77 | | |

<u>Answers obtained from 115 pupils (12-13 years old) to questions on "Elements and Compounds".</u>

| | | | |
|---|---|---|---|
| 1. correct definition of element and compound | 41% | 2. correct examples of elements given | 60% |
| incorrect | 23 | incorrect | 9 |
| partly correct | 22 | partly correct | 7 |
| don't know | 8 | don't know | 23 |
| 3. correct examples of compounds given | 58% | 4. pupils know correctly what compounds and elements they have seen | 52% |
| incorrect | 6 | incorrect | 7 |
| partly correct | 17 | partly correct | 24 |
| don't know | 18 | don't know | 16 |

5. consider as an element:

| | | | | | | | |
|---|---|---|---|---|---|---|---|
| smoke | 11% | iron | 78% | nitrogen | 76% | marble | 23% |
| cold | 9 | fusion | 3.4 | oxygen | 81 | precipitated | 6 |
| hydrogen | 82 | analysis | 4 | ice | 13 | mulch | 11 |
| water | 10 | aluminium | 65 | boiling | 10 | antibiotic | 10 |
| hail | 5 | medicine | 11 | salt | 30 | | |
| air | 12 | energy | 11 | milk | 11 | | |

6. consider as a compound:

| | | | | | | | |
|---|---|---|---|---|---|---|---|
| marble | 61% | water | 84% | chalk | 62% | paper | 61% |
| rain | 60 | fusion | 9 | glass | 61 | pyrite | 55 |
| salt | 63 | milk | 53 | sand | 52 | | |
| boiling | 8 | hail | 62 | snow | 63 | | |
| aluminium | 24 | lead | 36 | wood | 56 | | |

<u>Answers obtained from 105 pupils (13-14 years old) to questions on "Chemical Reactions".</u>

| | | | |
|---|---|---|---|
| 1. correct definition of chemical reaction | 20% | 2. have heard of chemical reaction | 19% |
| incorrect | 43 | no | 42 |
| don't know | 12 | | |
| 3. chemical reaction seen - correct | 21% | 4. what chemical reaction have you done? - correct | 21% |
| incorrect | 45 | incorrect | 50 |
| don't know | 4 | don't know | 5 |

5. Words which remind of a chemical reaction:

| | | | | | | | |
|---|---|---|---|---|---|---|---|
| explosion | 53% | gas | 40% | poison | 32 | electricity | 24 |
| combustion | 48 | colouring | 36 | energy | 29 | precipitated | 2 |
| solution | 47 | aspirin | 35 | colour | 28 | | |
| pollution | 41 | medicine | 35 | change | 28 | | |
| fusion | 40 | antibiotic | 32 | mulch | 28 | | |

6. Expressions which remind of a chemical reaction:

| | | | |
|---|---|---|---|
| a) a piece of wood burns | 51% | f) a nail rusts | 46% |
| b) a burnt candle | 44 | g) sugar dissolves in coffee | 53 |
| c) water boiling in a saucepan | 57 | h) lemon squeezed into bicarbonate froths | 69 |
| d) must becomes wine | 54 | i) sulphur burns in a tub | 58% |
| e) ice-cream melts | 29% | | |

## CONCLUSIONS

Answers to questions on Physical states of Matter indicate some linguistic difficulty in understanding the meaning of physical state, incorrect ideas on the physical state of $CO_2$ and $CH_4$, and confirm that at this age-level the cognitive approach is of a sensory type.

A good percentage of pupils demonstrate having reasonably clear ideas on changes of state with particular reference to the water cycle, while a certain percentage (21% and 16%) confuse chemical reaction with change of state.

The "Solutions" questionnaire indicated that a considerably high percentage of pupils confuse a change of state such as evaporation of water and the melting of ice with a solution. Solvent and solute are often confused.

While 41% of the pupils can give correct examples of both elements and compounds, we notice a certain degree of misconception regarding water , air, marble and ice considered by 10-12% (23% for marble) as elements because they are found in nature. There are indications of linguistic difficulties.

Only 51% of pupils recognize a chemical reaction in the burning of a piece of wood and a very high percentage (57%,53%) confuse changes of state or phenomena of solution with chemical reaction. Only 21% of pupils remember having seen or done chemical reactions.

## REFERENCES

1. D.P. Ausubel, Educational Psychology - A Cognitive View, Holt, Rinehart and Winston, New York, 1968.

# How to integrate, in chemistry teaching, the learning of fundamental concepts and the acquisition of a scientific attitude

Pierre Pirson

Department of Chemistry, Faculty of Science,
Notre-Damme de la Paix University
NAMUR 5000, BELGIUM.

An ever increasing body of knowledge in Chemistry is often presented to the public and students through media such as television, press, etc. Such information is often delivered loosely and causes the students enormous difficulties in the exercise of organizing and analyzing these data.

Therefore teaching chemistry cannot be a simple transfer of scientific FACTS available from the literature but instead should consist of providing a methodology designed for structuring the information and using it rationally.

In this way, the students will stop being simple listeners recording pieces of information but will become active participants in a process aiming at a real scientific attitude.

This was the problem, some years ago, when scientific teaching and especially chemistry teaching in secondary education was reorganized in Belgium.

At the beginning of chemistry teaching we could have imagined an initiation to the experimental method according to the classical scheme : observation, classification, question, hypothesis, rules and theories. This method would then be used to introduce - generally in an inductive way - the basic concept of chemistry. American and Canadian teachers had already used this approach for their students.

We decided to initiate our students to a scientific attitude by teaching them element by element. We think that each element, each specific know-how, like for example how to observe or how to mesure correctly, needs a real apprenticeship.

You know that for example the teaching of how to measure and give the results with their inherent inaccuracy is rather complicated. That is the reason for proposing a new methodology which no longer consists of in splitting chemistry lessons into notional points (mixture, element, atom, molecule, chemical reactions, etc., ...) but to have points and topics relating to a scientific attitude.

We would like our students who begin their chemistry teaching at the age of fifteen to learn successively :

- observe like a scientist, that is to say learn the rules of how a scientist observes phenomena
- measure and express the results of these measures
- elaborate hypotheses and give experimental proof
- elaborate a model starting from a coherent block of observations and hypotheses
- class using good criteria
- find the significant factors of a phenomenon
- discover and formulate a scientific rule
- simplify the study of the complex reality, that is to say, to be able to imagine an experimental procedure which tends to reproduce an ideal situation.

We think that our students will have adopted a scientific attitude after having learned these eight points.

To give an example : if you compare a scientific attitude to a lemon, you can say that our methodology consists in giving the students the lemon, not at one time and as a whole, but step by step and cut into convenient slices; it is thus an analytical methodology.

The positive side of this methodology is the fact that the students receive a systematic apprenticeship for all the points necessary for a scientific attitude.

The negative side is that they do not have a synthetic view of a scientific attitude. That is why we have introduced after each point of know-how a second phase where the student has the occasion to test and show if he has understood everything so far.

For example, after having learned how to observe, to measure, to formulate and to verify hypotheses, a review point in the form of a laboratory lesson is proposed to our students where they will be able to see if they have understood the points of a scientific attitude already taught. Furthermore, at the end of these apprenticeships, we propose and discuss how a real scientist would approach and organize his research work.

But how to connect this learning of a scientific attitude with the learning of basic chemical concepts?

As this systematic learning of know-how is done on the basis of knowledge, it is therefore clear that these two types of learning have to be done in a coordinated way. I will give one example here to explain what has been said up to now.

EXAMPLE

The aimed know-how is : learn how to elaborate and verify hypotheses experimentally. The aimed knowledge is : what is a chemical reaction.
Under a hood the teacher and his students put together metallic copper and concentrated nitric acid.
The students are then watching what is happening and ask questions concerning these observations. Generally, one of the questions often asked is : "Where has the copper gone?"
Then try to give hypotheses explaining this :
1) The copper has disappeared from the liquid medium
2) The copper is still present in the medium.

If the copper has disappeared from the liquid medium it is:
1a) Either in form of brown smoke
1b) Or in another form.

If the copper is still present in the medium, it is :
2a) Either as a metal dissolved in the acid
2b) Or in another form.

With the assistance of their teacher the students imagine experiments to confront their hypotheses concerning the "disappearance" of the copper.

In general, they throw back the hypothesis "Copper has disappeared in form of brown smoke" by showing that the same smoke can be obtained by pouring concentrated nitric acid on zinc or iron. They also throw back the hypothesis "Copper is dissolved in nitric acid" by showing that after evaporation there is no metallic copper left. Indeed, if they pour concentrated nitric acid on part of the green residue, they can no longer observe brown smoke.

On the other hand, if they redissolve the other part of the residue in water, they obtain a blue solution which ressembles to a solution of sulphate or copper nitrate.
Thus they draw the conclusion that copper is still present, not in metallic form but in a form analogous to that of copper in copper sulphate.

It is now easy to make them understand that every time a substance called a "reagent" at the beginning is found in another form and with other characteristics afterwards and which is then called the "product", chemists call this a chemical reaction.
I hope that my example has explained you this new methodology and proved that the acquisition of the different parts of a scientific attitude and the acquisition of basic chemical concepts can very well be done at the same time.

# The International Center for Chemical Education in French-speaking countries

Daniele CROS and Maurice CHASTRETTE

Centre International Francophone pour l'Education en Chimie
UNIVERSITE DES SCIENCES ET TECHNIQUES DU LANGUEDOC
Place Eugene Bataillon - 34060 MONTPELLIER Cedex (France)

The Centre International Francophone pour l'Education en Chimie (C.I.F.C.) was launched by UNESCO during the year 1984, following a proposal made during the 7th ICCE at Montpellier. The Centre obtained previous help from the Ministere des Relations Exterieures and from national organisations (such as the National Committee for Chemistry and the French Chemical Society) and from regional and local authorities at Montpellier, France.

The Centre is located at the Universite des Sciences et Techniques du Languedoc, at Montpellier.

The Centre was created to help towards the development of Chemical Education in French-speaking countries. It is one of the associate Centres in the International Network for Chemical Education (INCE). INCE has Centre at Ljubljana (Yougoslavia) and several associate Centres all over the world.

## Objectives

The main objectives of C.I.F.E.C. are :

1) Information

a) Diffusion towards the francophone countries of the international research in Chemical Education.
b) Diffusion towards the International Community of research and innovations made in francophone countries.
c) Information of the citizen on the role of chemistry and chemical industry in our societies.

2) Formation

a) In service teacher's training through specialized sessions and workshops.
b) Training of session leaders
c) Training of technicians
d) Production of reference books and documents in French
e) Training by and for the research in Chemical Education.

3) Production

a) Production of low cost locally produced equipment
b) Production of manuals for teaching experimental chemistry
c) Production of manuals for teaching chemistry
d) Research in Chemical Education

4) Cooperation

a) Creation of links between teachers and researchers from different countries
b) Creation of a network of correspondants
c) Creation of international teams of session/workshop leaders

## Recent Actions

We present some significant actions where the role of CIFEC was important.

1) Workshop on the teaching of experimental Chemistry

(TUNIS - September 1984). This workshop was organized by C.I.F.E.C. and the Chemical Society of Tunisia, with the help of UNESCO. It was held at the

Faculty of Sciences of Tunis. Teachers from several francophone countries (many from Tunisia) worked for 15 days and produced a manual for teaching of experimental chemistry (volume VI in the UNESCO series). The experiments described in the Manual were tested during the academic year 1984-1985. A seminar was held at Montpellier to evaluate the impact of these experiments on students in different contexts.

2) Evaluation of curricula and manuals for teaching of sciences at the secondary level. Nabel - Tunisia, September 1984.
This workshop was organized by the Ministere de l'Education National of Tunisia, with the help of ALECSO. C.I.F.E.C. was involved in the organization of the Seminar and was in charge of the teaching throughout the seminar. 110 teachers (from primary and secondary schools) who had experience in designing curricula and writing manuals attented one of the three parts of this Seminar. They produced evaluation tools which are used now in their tasks.

3) Exchange Session on Science Education in francophone countries, Bordeaux, France, June 1984.
This Session was organized in collaboration with the Agence pour la Cooperation Culturelle et Technique (ACCT), 20 participants from 18 countries worked during four weeks to :
- evaluate the role of science education at the secondary level in the nation educational policies in their countries.
- define the needs
- define actions to be undertaken

Following this session a workshop on evaluation and formation to evaluation will be held at Bordeauz (January 1986).

Future Actions (workshops with help from UNESCO)

1) Elaboaration and use of semimicro-materials for teaching experimental Chemistry (Tunis, September 1985).

2) Workshop on low cost locally produced equipment (Morocco, October 1985).

3) Production of a manual for teaching experimental chemistry at the Secondary School level (Togo, November 1985).

4) Workshop on experimental Chemistry Teaching at the University level (to produce a manual, volume VII in UNESCO Series).

Our young Centre has been very active during its first year of existence. We try to develop productive work through cooperation between francophone countries but also between francophone countries and other countries. We are conscious of the enormous amount of work to be done and will gladly accept any help from chemists all over the world.

# The Australian Academy of Science school chemistry course—evolution or revolution?

Robert B. Bucat (Supervising Editor) and Andrew R.H. Cole (Chairman)

School of Chemistry, University of Western Australia, Nedlands 6009, Western Australia.

Abstract - This paper summarises the production of materials for a new-generation secondary school chemistry curriculum, which was carried out by a large team of people on behalf of the Australian Academy of Science during the years 1978 to 1984. These materials are a two-volume textbook entitled *Elements of Chemistry: Earth, Air, Fire and Water* and associated teacher's guides.

## Context for Change

The new course is a reaction to the wave of curriculum materials produced in the 1960's. Perceived weaknesses of these courses are summarised as follows:

- They were discipline-based rather than learner-based and were narrowly representative of "chemists' chemistry".
- These courses were based on theories, principles and organising themes which were both abstract and difficult. Too often the theories presented were based on experimental information which was unknown to the students. The students' first encounters were with abstract theories which seemed to have a "turning off" effect.
- These theories were perceived by students as the "facts" of chemistry, rather than as provisional explanations of observations.
- External examinations for tertiary selection emphasised theoretical concepts and principles to the exclusion of reaction chemistry, laboratory experiences and applications. So, memorisation of facts in older curricula was replaced by memorisation of theories.
- The courses were suitable for tertiary-oriented students only, at the expense of the vast bulk of students.
- A stated commitment to the experimental nature of chemistry was not carried through in the classroom.
- The focus was on the physical states and structures of substances, rather than on reactions of substances to form other substances.
- Students gained little familiarity with common chemicals or chemical reactions. No attempt was made to demonstrate the chemical nature of the substances which people handle every day.
- The influence of chemistry on society and the individual was ignored, so that courses lacked human, social and industrial components.
- Interrelationships with other sciences were scarce.

## Development

The process began in 1978 with an investigation of the state of high school chemical education in all States of Australia by N.S. Bayliss and D.W. Watts. This phase led to the production of a suggested course of study and a critique of the curricula in use in each State at that time. This preliminary document was widely circulated among schools, tertiary institutions and research chemists in both industrial and government laboratories. After revision in the light of comments from all of these sources, the conclusions were published as a recommended course of study (1).

The job of converting the recommended syllabus into a text book was given to an Editorial Panel consisting of D.W. Watts (Director), R.B. Bucat (Supervising Editor), a group of high school teachers and an expert in consumer chemistry. Writing teams commissioned by the Editorial Panel included academic chemists (to ensure chemical accuracy and honesty), teachers (to ensure realistic appraisal of system and student constraints), science education researchers (to minimise inconsistencies with research in student learning) and industrial and other research chemists (to provide current applications).

(1) *Chemistry for Australian Secondary School Students - A Recommended Course of Study*, Australian Academy of Science (1979). Available from the Academy.

## Guiding Philosophies and Design Decisions

The major purpose of this paper is to summarise the philosophies on which the new course is based.

The key philosophies to which the development team continually referred are summarised below. In each case, related decisions regarding the design of the course are listed.

1. *A school chemistry course should be satisfying, meaningful and useful for all students - for those who are not proceeding to tertiary study as well as for those who are.*

   - The emphasis is on the influence of chemistry on our daily lives - not solely on test-tube reactions in the classroom.
   - As much as possible, the chemistry is discussed in relation to local contexts.
   - Theories regarded as too difficult, too abstract or unnecessary for the explanation of observations in the course are not included.

2. *The major emphasis should be on the change of substances into other substances.*

   - From the first pages, the focus is on chemical reactions rather than on theories of a physical nature, such as atomic structure, molecular shapes, hybridisation and ionisation energies.

3. *Chemistry is an experimental science, so school chemistry courses should involve laboratory and field experiences by the students.* Experiments can be the vehicle for arousal of curiosity, familiarity with common substances and reactions, appreciation of aesthetic aspects of chemistry, active participation by the students rather than passive reception, and development from the concrete situation to the abstract idea.

   - Experiment, description and theory are integrated throughout the course. Experiments and/or demonstrations are introduced in the text book at the appropriate moment. There is no separate laboratory manual.
   - Where appropriate and convenient, experiments are used to pose problems and to develop patterns, concepts and theories - rather than simply to illustrate theories previously described.

4. *Theories, concepts and generalisations are important for the organisation of knowledge, but only provided that:*

   (a) *the students are familiar with the knowledge being ordered.*
   (b) *the knowledge to be ordered is useful and significant to the students in their present or potential environments.*
   (c) *the level of sophistication is appropriate to the intellectual abilities of the students.*

   *The text should be at the same time descriptive and yet not lacking in rigour or intellectual content.*

   - Observations are made before generalisations are drawn. For example, the periodic table is not discussed until students have become familiar with the chemistry of representative elements.
   - Unnecessary concepts are not included in the course. Examples of excluded concepts are standard states, standard enthalpy changes, standard reduction potentials, sophisticated models of atomic and molecular structure, electron-pair repulsion theory to predict molecular shapes, enthalpy and entropy changes as factors that govern whether or not reactions will occur.
   - More difficult concepts such as equilibrium and the nature of intermolecular forces are presented late in the course.
   - Periodic classification is based on metal/non-metal character and formulae of compounds, rather than ionisation potentials, electron affinities and atomic radii which are not observable by students.

5. *Chemistry is not an isolated body of knowledge and processes confined to test-tube reactions in the classroom. The interaction of people and chemistry is important. In this context, chemistry is not value-free.*

   - The structure of the course emphasises our dependence on the earth, the atmosphere, energy and water, as well as on synthetic substances.
   - Opportunities are provided in the course for discussion of topics such as air pollution, water pollution, limitations of energy resources and options available, benefits and costs of the minerals industries, costs and prevention of corrosion.

6. *Students should be aware that chemistry is the basis of everything, living and non-living.*

   - Numerous examples are presented of chemistry in the biological and earth sciences.

7. *A chemistry course should provide opportunities for students to become aware of the role of chemistry in industry and technology - especially when such industry is local to the school or home.*

   - Frequent reference is made to industrial applications of chemical principles. The principles and the industrial application are presented as one portion of the course - not as two separate pieces of information to be remembered in isolation. For example, the electrolytic production of sodium hydroxide and chlorine, and the electro-refining of copper are integral parts of the discussion of electrolysis.

8. *We are a consumer society and people should have a basic knowledge of the chemistry of foods, building materials, polymers and other common products.*

   - The course contains frequent discussions of the chemistry of consumer products.

Contents and Sequence

A list of the contents, chapter by chapter, is shown below. A high degree of coherence is obvious in each of the major sections.

# CONTENTS

## VOLUME 1



## VOLUME 2

The sequence of topics differs from most texts in that it does not commence with atomic structure, bonding, the periodic system and molar quantities. In line with the overall philosophy that we should move from the concrete and observable to the abstract and theoretical, we commence with reactions of familiar materials - the metals. This leads naturally to oxidation and reduction, electrochemical cells, electrolysis, corrosion and reduction of ores to metals.

The chemistry of carbon compounds is introduced fairly early since it deals with familiar substances, is mostly descriptive and at this level does not involve theoretical concepts. Acids and bases are used as reagents right from the first chapter, but the formal treatment of acid-base and other equilibria is delayed until near the end of Volume 2. In many instances a "spiral curriculum" is used, whereby many topics are revisited at different levels as the course proceeds and student knowledge grows.

## The Product

The product is attractive and highly readable. The immediate uptake of the course throughout Australia was surprisingly high and considerable interest has been shown from overseas countries. The response of users has been most encouraging and indicates that many students are actually enjoying reading about and doing chemistry (a rare phenomenon in recent decades!).

## Availability

The materials are available from:

The Publications Officer,
Australian Academy of Science,
G.P.O. Box 783,
CANBERRA CITY, A.C.T. 2601,
Australia.

Prices (in Australian dollars) are as follows:

Text, Volume 1: A$21.75 (approx. US$14.50)
Text, Volume 2: A$21.75
Teacher's Guide Volume 1: A$8.25 (approx. US$5.50)
Teacher's Guide Volume 2: A$8.25

# Motivation through everyday chemistry

Jens Josephsen

Institute of Life Sciences and Chemistry, Roskilde University,
P. O. Box 260, DK - 4000 Roskilde, DENMARK

The internal "beauty" of chemistry is seldom the predominant motivating factor among young students, who struggle with chemical concepts and calculations. This is especially true for two types of students:

1. Upper secondary or college students who have not decided on their further education.
2. The vast majority of students doing chemistry as a compulsory tool for their main choice of education.

If chemistry is regarded as boring, difficult or maybe even irrelevant, this situation raises a couple of problems:

- A lack of recruitment of students wanting to become chemists.
- An insufficient chemical education of the future citizen.
- A reduced ability to use chemistry as a tool in other forms of education.

Although great didactic improvements have been made regarding the logical and comprehensible presentation of chemical concepts and their proper use, further motivating factors are necessary. One such factor is the feeling of relevance. In other words, the teaching must meet the students in their world of thinking and at their level of knowledge.

MESSAGE → TRANSFORMATION → RECEIVER

Some important features of the receiving group, the students, can be represented by their intellectual development and by their degree of openness for learning: We experience a greater portion of students in the school and in the college, who are concrete operational thinkers in the Piaget sense (ref. 1) and who show narcissistic trends rather than receptive ones. If we teach those students as if they had developed formal operational skills - and as if they were interested in anything we present to them, we would probably do a poor job.

And about the message and its transformation - chemistry and the teaching: In our teaching we can focus on chemical principles or on the other hand, we can focus on chemical phenomena, which in turn can be more or less academic or taken from everyday life. The treatment of the particular type of chemistry could be deductive or on the contrary based primarily on observation and induction. What we want the students to learn and be able to do at the end of the course is to make observations on phenomena and to rationalize them in chemical terms. If we therefore present the students with focus on deduction from chemical principles, they would probably not come out of the process with any training in handling everyday phenomena or problems for which a chemical description is relevant.

PROJECT - ORGANIZATION

A certain degree of project-organization of introductory chemistry seems to contribute to the motivation of both the above type 1 and type 2 students. At Roskilde University project-organization is a principle. The freshman student enters into our college-like basic studies within the natural sciences and choose only after the completion of this, the subjects he or she wants to graduate in (ref. 2). In this way the student doesn't have to make the final choice at entrance, as is the case at other unversities in Denmark. Half of the time is devoted to introductory courses, i.e. the disciplines within the natural sciences. And the other half of the time, the students work is organized in group-projects, which are problem-oriented, i.e. they deal with: How do natural science handle questions from the "real world". The students are heavily involved in the formulation of the questions, so it is not surprising that questions from everyday life constitute the major number of the starting points for their studies. The questions may have certain narcissistic characteristics, they are rather concrete, and are of

course based on observation.
A rough sketch of the process through the problem-oriented projects can be represented by the following phases:

| | |
|---|---|
| 1 Formulation of a problem: | Suggest, list and define<br>Reduce and choose<br>Formulate |
| 2 Investigations: | Design and find equipment<br>Carry out and obtain data<br>Reproduce and calculate |
| 3 Evaluation: | Interpret and compare<br>Conclude<br>Document |

An example
Among a lot of examples from the basic studies first year program (students aged 19-20), the following example about chemical contraception illustrates how the process works, and since the project ran some years ago, it is also possible to report on what became of the students:
A group of 6 female and 3 male students - all inclined towards biology - decided to work with contraception and its possible risks. The study of birth control pills and claims on their side-effects were, however, out of the question from an experimental point of view. So they turned their interest to contraceptive foam, which is primarily used among young people. Some minor clinical side-effects had been reported among which irritation of the vagina was chosen for detailed study.
At the end of this formulation phase, the question became: "Does a dose of contraceptive foam alter the acidity of the vagina so much that the microflora could change, favouring pathogenic microorganisms?".
Up untill this point, the students had studied the female sex-hormonal cycle, the microbiology, ecology and histology of the vagina and acid-base chemistry, including pH, pK's, titration curves, buffer capacity, etc.
The next phase was to plan and perform the model experiments on three different commercial products. It turned out, that the three products showed quite different values of starting pH and quite different buffer capacities. Finally these numbers were compared to those of the modelled system, the vagina. But since no data on the buffer capacity of the vagina was found in the literature, two different estimates were made, based upon existing data for similar systems. These estimates were in fair agreement with each other. With these results in hand the conclusion was, that neither the acid - base properties nor the contents of benzoic acid as a preservation agent can cause dramatic changes in the microflora of the vagina, to give an increased risk of infectious symptoms.
The students' work was presented in a typed report, including an 18-page section in the shape of a scientific paper. The report also included a number of their own appendixes on relevant basic chemistry and biology for the benefit of their fellow students who for evaluation purposes should also read and understand the "work".

## CONCLUDING REMARKS

This first year project can be considered as a success. Not only did the students formulate a problem - they did also answer the question on the basis of chemical investigations and considerations. In addition, 2 of the 9 students did change their attitude towards chemistry so much that they later studied chemistry successfully. Moreover 4 of the other students took further chemistry courses for their subsequent choice of study, namely biology. Although this has nothing to do with statistics, this example suggests that we can make more students interested in chemistry by taking advantage of their motivation for everyday phenomena.
In conclusion it would be tempting to suggest, that for introductory chemistry we should not hesitate to teach the young students
CHEMISTRY TO SOLVE PROBLEMS which appear in their minds,
in contrast to the often used linkage isomer of teaching the students
TO SOLVE CHEMISTRY PROBLEMS appearing in the textbook.

## REFERENCES

1. J. Dudley Herron, J. Chem. Educ. 52, 146 (1975).
2. J. Josephsen, J. Chem. Educ. 62, 426 (1985).

# The chemistry of everyday life: A short course for junior high school students

Richard Kilker Jr.

Chemistry Department, Drew University, Madison, New Jersey, 07940 USA

Abstract - A course was designed to introduce 8th and 9th grade students to simple organic chemistry and biochemistry with an underlying theme of acid/base reactions. At the beginning of each class, the students were provided with a worksheet covering the subject material of that period. The classroom approach was a lecture/lab format. Every class session had either a demonstration by the teacher, an experiment performed by the students or both. The course was 10 weeks long and met for 2 hours on a saturday morning. Within the course students were introduced to topics and then shown where these fit into their daily lives, in keeping with the title of the course. As well, they were introduced to techniques commonly used in a scientific laboratory.

## INTRODUCTION

A program entitled the Drew University High Ability Program was designed in cooperation with local public school districts, to intellectually challenge a group of 8th, 9th and 10th graders. The disciplines covered in this program were not restricted to the natural sciences. But before detailing the chemistry course, let's look at the High Ability Program in general.

## HISTORY OF THE PROGRAM

The present program evolved from the Drew-Dodge High Ability Program (1980-83). This program drew financial support from the Geraldine R. Dodge Foundation, Drew University and six public school districts in the vicinity of the Drew University campus. The program admitted only selected 8th graders, chosen from the participating public school districts, and exposed them to college level work in mathematics or writing.

The present Drew High Ability Program (1983-84) is open to selected 8th, 9th and 10th graders. The program has been expanded to include school districts other than the original six. The grants supporting the Drew-Dodge program have expired and the present program is supported, in part, by the tuition charge for each student.

## SELECTION OF PARTICIPANTS

A student who participates in the Drew High Ability Program is chosen on the basis of three criteria. First, he/she must demonstrate intellectual ability by achieving a specific score on a national examination. Second, he/she must show achievement in his/her schoolwork as demonstrated by his/her grade point average. Third, he/she must obtain a written recommendation from either his/her school principal or a guidance counselor. A selection committee notifies the students chosen for the program by sending them a formal letter inviting them to participate. This committee makes a concerted effort to insure a broad social and cultural integration of gifted children.

## COURSE OFFERINGS

8th graders chosen for the program choose either a mathematics or a writing course. Each of these courses occupies the two twelve-week sessions comprising the year. 9th and 10th graders choose one course per twelve week session. Courses offered in 1983-84 included archaeology, theater, greek and chemistry. The material was presented at a college level but the courses were for no credit since the rationale for this program lies in challenging these students in a situation where no grades are given.

The classes ran from 9:30 to 11:30 am on a Saturday morning during the normal school year. A 10 minute juice and donut break, occurring at 10:30 am, allows all the students in the program time to meet with each other informally. Each student paid $250/ten week session with a limited number of full scholarships available based on financial need. Class size ranged from 10 to 18.

## CHEMISTRY COURSE

The chemistry course offered was entitled "Chemistry and Everyday Life". As can be seen from the course syllabus below, the content was a mixture of organic chemistry and biochemistry. It was felt that the students might have been exposed to these areas of chemistry in their classroom experience but not at the depth presented in this course. However, I assumed the students knew little of these areas and this proved to be the case. The emphasis of the course was as much on the way in which the subject material was presented, as on the subject matter itself. For example, a solution was determined to be acidic or basic depending upon the color of an added indicator which the students isolated from red cabbage leaves (ref. 1). This pointed out the fact that an acid/base indicator need not come from a mysterious bottle handed out by a teacher, but that a natural source for such an indicator might be found in the crisper drawer of their refrigerator.

## CLASSROOM SET-UP

The course format was lecture/lab. The classroom was a freshman chemistry laboratory with the students assigned to a lab bench space. The students were advised to take notes from the lecture material presented in each class. They appeared to do this and often referred back to the notes of a previous class in order to answer a question posed in class. The textbook used was an issue of Chemmatters entitled "Checking Out Acids and Bases" (ref. 1). The American Chemical Society kindly provided enough copies for this class at no charge.

At the beginning of a each class a worksheet was handed out. The basic format of this worksheet was (1) a few questions which the students should attempt to answer using the material presented in the lecture (e.g.,(week 1) What is the Bronsted-Lowry definition of an acid?) (2) a table for the data collected by each student either from a demonstration performed by the teacher or an experiment carried out by them (e.g.,(week 5) What is the migration distance for D-glucose on the paper chromatogram?). These worksheets helped the students to focus attention on topics discussed that period thus expediting their collection of any data.

As can be seen from the syllabus, every class session had either a demonstration by me, an experiment performed by each student or both. The demonstrations were to familiarize the students with a test they would perform, e.g., Benedict's Test (week 4), or to allow them to see the results of an experiment too unmanageable for an entire class to perform, e.g., blood-buffer simulation (week 3). In the latter case, the students were able to get a close-up view of how a pH meter operated and at the same time come to see how/why the blood acts as a buffer. These demonstrations and experiments were well received by the students and elicited questions and comments which indicated that they were challenged to think beyond each experiment. I feel that some of this enthusiasm was due to their being able to actually do something related to what was presented in the lecture earlier in the period.

TABLE 1. Course Syllabus

| Week | Lecture | Experiment | Demonstration | Student Experiment |
|---|---|---|---|---|
| 1 | Acids, Defn. & Detection | Use of pH paper, Litmus paper & the CRC (ref. 2) | Yes | Yes |
| 2 | Bases, Defn. & Detection, Antacids | Use of Indicators, Comparison of Various Antacids | Yes | Yes |
| 3 | Buffers, Defn. | Blood-Buffer Simulation (ref. 3) | Yes | No |
| 4 | Carbohydrates, What makes a Compd. Sweet? | Benedict's Test, Decomposition of a Sugar (ref. 4) | Yes | Yes |
| 5 | Identification of an Unknown Carbohydrate, Paper Chromatog. | Benedict's Test, Paper Chromatog. | No | Yes |
| 6 | Proteins, Defn., Detection & Denaturation | Fractionation of Milk Components (ref. 5) | Yes | No |
| 7 | Lipids, Defn., Soap, Defn., Micelles | Biuret & Benedict's Test on Milk Components | Yes | No |
| 8 | Process of Refluxing | Prepn. of Soap From Solid Shortening (ref. 6) | Yes | Yes |
| 9 | Mono-, Di-, and Triglycerides | Qual. Tests on Soap & Fat Solns.; Construct Molecular Models | Yes | Yes |
| 10 | Nucleic Acids | Diphenylamine Test for DNA, Construct Molecular Models | Yes | Yes |

Safety was a concern of mine when first starting this class. I cautioned the class concerning basic safety principles. If deemed necessary, I even required them to wear safety goggles. Since the class was so small and I was close at hand at all times, I could observe any potential trouble spots.

Within the course the students were introduced to a topic and then shown where this fits into their daily life, in keeping with the title of the course. Examples are:

(1) How is the taste of a liquid related to its acidity? (week 1)
(2) Is a solution of Drano acidic or basic? What does this mean in terms of safety precautions to follow in using this product? (week 2) How is this related to its action on grease which is clogging a drain? (week 8)
(3) What is an antacid? Is one commercial brand better at neutralizing acid than another? (week 2)
(4) Why is it necessary for the blood to act as a buffer? What physiological states are due to an imbalance of this buffer? How is this imbalance treated clinically? (week 3)
(5) Are all sugar solutions sweet? (week 4)
(6) Can you really turn solid shortening into soap? (week 8)

In addition, they were introduced to techniques that are commonly used in a scientific laboratory. Examples include:

(1) Why might one use a CRC Handbook of Chemistry and Physics (ref. 3)? (week 1 and 9)
(2) How is a pH meter used? (week 3)
(3) How is paper chromatography useful in identifying an unknown sugar from a list of knowns? (week 5)

(4) How is a refrigerated centrifuge used? (week 6)
(5) How does building a model of a molecule help you speculate on its water solubility? (week 9)

Finally, I introduced the usage of a computer in the following way. The students were given samples of 0.1M, 0.01M and 0.001M acetic acid, 0.25M propionic acid and 0.30M formic acid. They determined the pH of each using pH paper and recorded these values on a worksheet. They then were given the dissociation constant values for the three acids. I briefly indicated that there is a mathematical formula which allows one to ***calculate*** the pH of a solution knowing the molarity of the acid in the solution and the dissociation constant of that acid. I had available a computer program which did exactly that. So by entering the appropriate values the computer rapidly ***calculated*** the pH of each solution of acid. The students could then compare this value to the value they obtained using the pH paper and make comments on the accuracy of each method. I wanted them to experience how a computer program could save them time in performing a repetitive calculation. I did not, however, introduce them to the actual calculation the computer was performing. At least one of the students had a microcomputer at home and so was familiar with computers, but there were others who had not seen a computer terminal in the recent past.

## STUDENT INTERACTION

The chemistry class was composed of seven students, of whom three were female. There were three from ninth grade and four from tenth grade. All the students attended the class regularly except one, who eventually dropped the course.

The students' educational backgrounds, on the surface, seemed to be fairly even. The difference that was obvious was the willingness to take a risk by answering a question or better yet, by asking a question or making a verbal observation. By the end of the course, however, all the students were willing to jump-in when given the opportunity.

The students were enthusiastic about most of the topics covered but their interest waned ten minutes before the class was to end. This was not unexpected since these students were giving up *Saturday mornings* to attend this class. There were instances where a student spent considerable time thinking about something discussed in a previous class. One instance relates to the class session where we used the computer in calculating the pH of a solution of an acid. As mentioned earlier, I had not discussed how the computer mathematically calculated the pH of the solution. I had deemed it too difficult to spend any time on it. To my surprise, at the next class session a student presented me with a program written for his microcomputer that did exactly what the classroom computer did *and* he had introduced a 'pH line' such that a pointer positioned itself on this line so as to indicate the pH. I questioned how the student had obtained the formula to carry out this calculation and was told that he had found it in a textbook at home! Needless to say, this student absorbed far more than I could provide within the course, resulting in the teacher being pleasantly challenged by the student!

## CONCLUSIONS

The concept of the High Ability Program was to intellectually stimulate a select group of ninth and tenth grade students. The course size was small so as to foster one-on-one interactions between the students and the teacher. The chemistry course offered attempted to inform the students about the chemistry of everyday life with an underlying theme of acid/base reactions. The subject matter was a mixture of organic chemistry and biochemistry presented at a college level. The lecture/lab format insured that each student had both a theoretical and a practical exposure to the subject material. The worksheets provided in each class allowed the students to focus their efforts on specific topics.

The students indicated that they would take another such course if allowed to do so. A formal student evaluation of this course was not attempted. An effort will be made to locate more textual material to hand out to the students, for the lack of such material proved to be a difficulty for some students.

REFERENCES

(1) Dombrink, K., and Tanis, D. (eds.), ChemMatters, 1 (#2), 7 (1983).
(2) Weast, Robert C., "Handbook of Chemistry and Physics", 49th ed., The Chemical Rubber Co., Cleveland, Ohio, 1968.
(3) Ophardt, Charles, E., J. Chem. Ed., 60, 493 (1983).
(4) Personal correspondence from Maria Christofore, Union High School, Union, N.J.
(5) Armstrong, R. et al, "Laboratory Chemistry: A Life Science Approach", Macmillan Inc., N.Y., pp. 285-290, (1980).
(6) Ibid., pp. 279-282.

# A multidisciplinary approach to the teaching of applied chemistry

M.L. Bouguerra

Faculte des Sciences, Departement de chimie
Campus universitaire, Le Belvedere, 1060 Tunis (Tunisia)

Abstract - A product-centered approach to teach applied chemistry to students majoring in chemistry or education is described. Our aims are : 1) to show that chemistry or education is central science, 2) to teach chemistry "globally" with its numerous ties to the various realms of human activities, 3) to point that chemistry can help fight underdevelopment in Third World countries thanks to a sensible and "utilitarian" approach which yields a better educated citizen in our particular national context. Tunisia is a small country with few mineral and oil resources, many agricultural crops, long shores and the Sahara desert.

-----

Starting from a product or an element placed in the centre of concentric circles, one can develop a method teaching which will be more attractive to students by revealing the various relationship of the compound to the real life and by studying as well its physical and chemical properties, detection and analysis action on environment, economic importance...(fig. I).

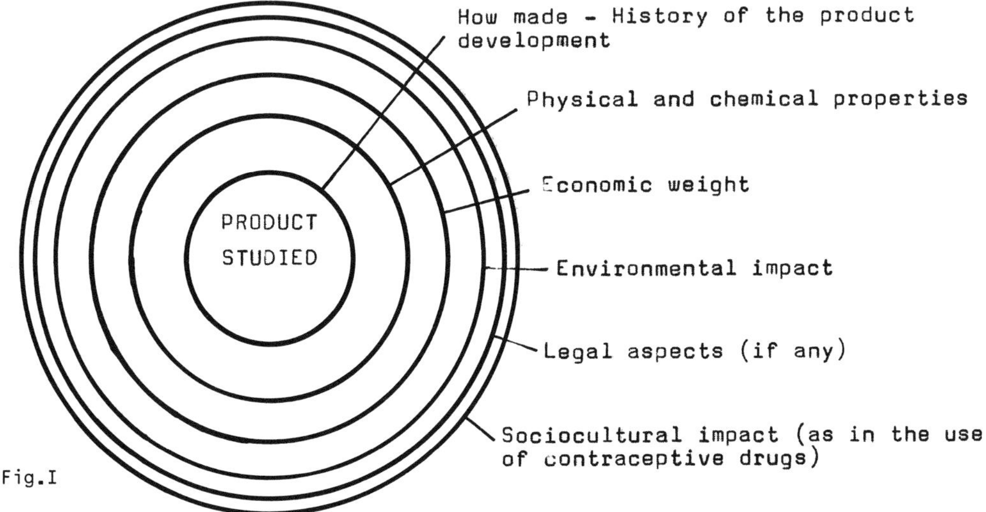


Fig.I

Examples abound such as mercury, sulfur, polyethylene, DDT, tetraethyllead, national products such as olive oil, phosphate rocks, iron ores or flower extracts a traditional undertaking in Tunisia.

We give here, for instance, the broad outline of our treatment for DDT (dichlorodiphenyltrichlorethane). It is far from a definitive study of the subject even if the student is provided with an extensive bibliography as a starter.

1. DDT is a broad spectrum, persistent organochlorine insecticide of the diphenylethane family.
2. First, our students are asked to prepare the insecticide as lab assignment recrystallize it, study its solubility, record its melting point, calculate the yield and the price of one kg, think about its isomers...
3. Pesticides are biocides but DDT is not very poisonous to man : the Merck Index (9th edition) gives an estimated figure of 500mg/kg as a lethal dose.

But soon our students learn that the Merak Report in the USA placed DDT on its group B list - judging it positive for tumor induction because of the findings on mice fed DDT.

Recalling the long controversy about cigarettes and lung cancer, our students realize how extremely difficult it is to prove - scientifically not emotionally - a cause of cancer in man.

4 . We later move to the history of the product which was discovered around the start of World War II and was praised as "miracle compound" and kept as a "war secret" for some time.

DDT was far cheaper and more effective against almost all insects than the then known mineral insecticides. Its uses led to dramatic early successes and saved millions from death from malaria and yellow fever. It was predicted by enthusiastic supporters that all pest insects are going to be completely wiped out in the near future and in 1948 Dr. Paul Mueller of J.R. Geigy AG (Switzerland) was awarded the Nobel Prize for the discovery of the insecticidal properties of DDT which was actually synthetised in 1874 by Zeidler. But within 30 years, the promise of insect free abundance had fallen from grace. Furthermore, significant concentrations of DDT in blood were detected in Hungary. This country then banned the use of this compound and was soon followed by Sweden in 1970. However, in Tunisia DDT was banned only in 1980. Then our students had to tackle with the practical, economic choice facing some societies. DDT is a cheap compound, easy to handle and to use. These are some reasons explaining its success in developing countries where hunger and diseases have to be fought. Nevertheless, one has to question its uses because of some disturbing associated flaws:

Chemical stability : DDT's half life is several years and it resists breakdown by water, air and sunlight. Its main metabolites are DDE and DDD. DDT is now an ubiquitous pollutant and our students perform gas chromatographic analysis (EC detection) of tunisian foodstuffs to appreciate their DDT residue contents and discuss the FAO-WHO daily intake tolerance set up for various food items. (Fig. II for degradation).

Figure II

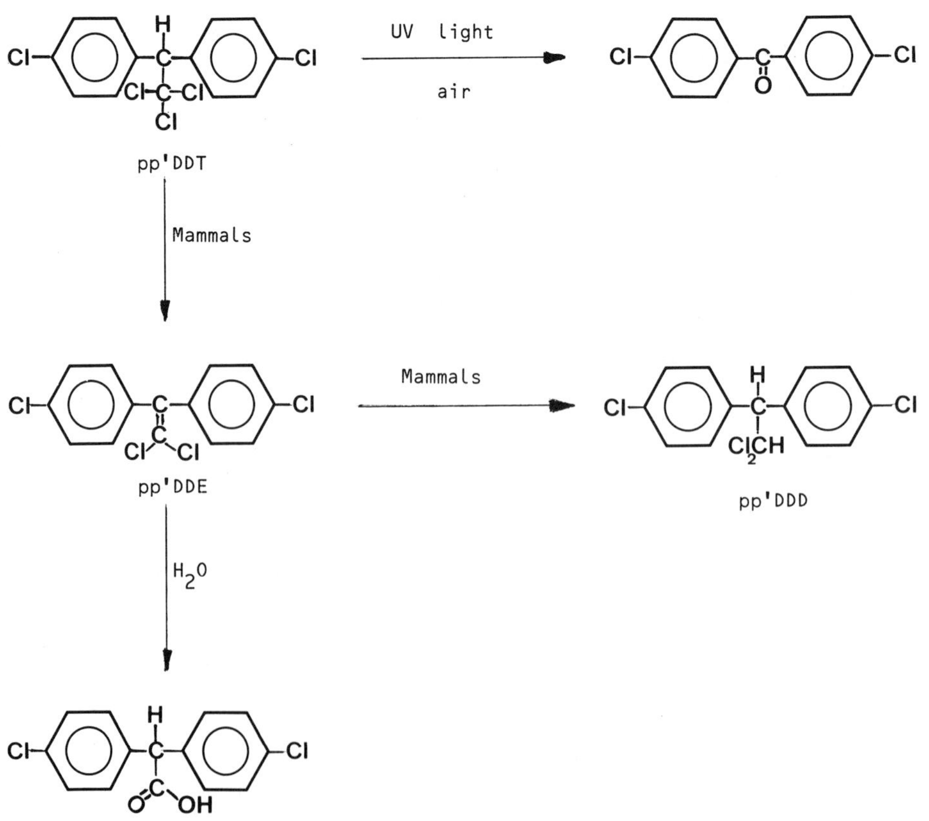

Biological magnification : DDT is not soluble in water, however it is fairly soluble in fatty materials and thus tends to accumulate in lipidic tissues of living organisms. It concentrates along the food chain. A decline in population of flesh eating birds was attributed to the fact that DDT may make the egg shell thinner because the DDT may inhibit some crucial pathways in calcium metabolism.

Ecological threats : DDT has some negative aspects on the environment, on the flora and fauna and creates an ecological imbalance in many developing countries where it was extensively used against vectors of parasitic diseases. DDT was responsible for big losses of marine life in some of these countries.

Pesticide resistance : One major drawback in DDT use is the occurrence of pest resistance. More than 50 different species of malaria carrying mosquito have become resistant to this insecticide and many others world wide and over 400 species of various pests are known now to be resistant to one or more pesticides.

Towards better ways to fight pests : Our students are also made aware to the fact that chemistry is developing new and safer chemicals to fight pests to control parasites and to protect crops such as integrated pest management, pheromones, propesticides, allelochemicals, antifeeding compounds.

Exercises, reports and field visits : Numerical drillings and assignments about DDT and pesticides are made as well as visits to farmers and agrochemical sellers.

Aims achieved : 1) Appreciation of the central role of chemistry in our life,
2) A comprehensive approach to a product through a multidisciplinary approach,
3) Learning of literature searching and selection,
4) Formation of the students'own opinion about a somewhat controversial issue...

BIBLIOGRAPHY

Here is a sample of the literature unearthed by our students.


- P.R. Ehrlish, A.H. Ehrlich and J.P. Holdren, Ecoscience, W.H. Freeman and Cy, San Francisco, 1977;
- T.A. Weil, B.Q. Weil and B.D. Balaustein, J. of Chem. Ed., 51, 198-199 (1974);
- P.S. Corbet, Rev. Zool. Bot. Afr., 57, 73-95, (1958);
-Cleaning our environment. A chemical perspective, Amer. Chem. Soc., Washington DC, (1978);
- J.W. Mellor and R.H. Adams JR, Chem. and Engin. News, 37, (23 April 1984);
- J.B. Tucker, Environment 25, 17-20 (1983);
- R.J. Smith, Science 204, 1391-1394 (1979);
- J. Jeyaratnam, R.S. De Alvis Senéviratne and J.F. Copplestone, Bull. WHO 60
- H.B.N. Hynes, Ann. of Trop. Med. and Parasit. 54, 331-332 (1960);
- G. Matthews, New Scientist 368-372 (12 may 1983);
- C.W. Heckman, Environ. Sci. and Technol. 16, 48A-57A (1982);
- J.A. Nathanson, Science 226, 184-187 (1984);
- G. Chapin and R. Wasserstom, Nature (London) 293, 181-185 (1981);
- D.A. Ratcliffe, J. of Applied Ecology 7, 67-115 (1970);
- R.L. Kalra, R.P. Chawla; M.L. Sharma, R.S. Battu and S.C. Gupta, Environ. Pollut. Ser. B, 6, 195-206 (1983);


Hint : We already published two papers illustrating our multidisciplinary approach : one about the orange fruit and the second one dealing with tetra ethyl lead.


- M.L. Bouguerra, L'Actualite Chimique, 33-34 (January 1982);
- M.L. Bouguerra, Bull. Union des Physiciense (Paris), 668, 165-174 (1984).

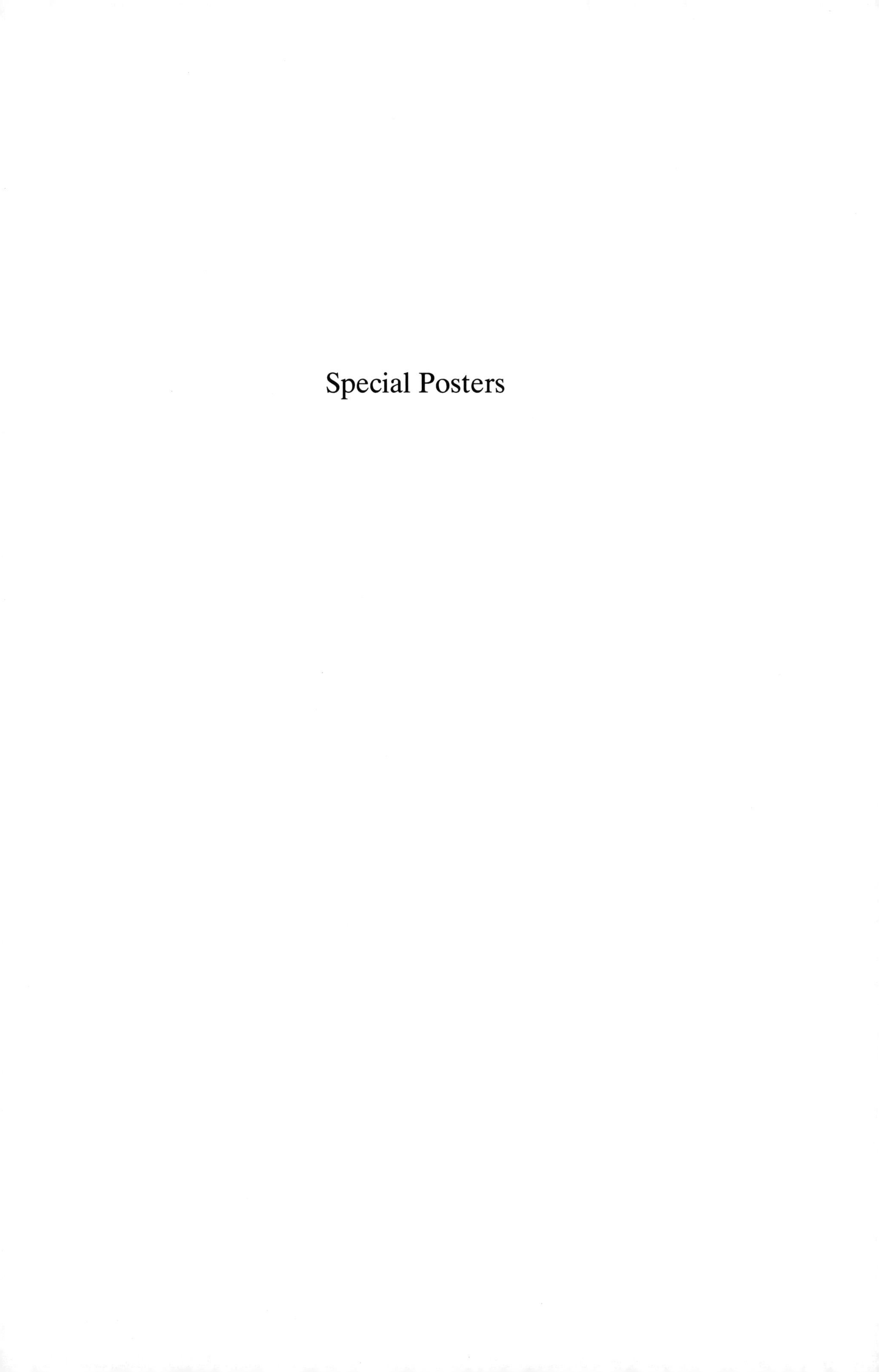

# Special Posters

# Chemical education in an international dimension

M. Chastrette

Laboratoire de Chimie Organique Physique, Universite Claude-Bernard Lyon I, F-69622 Villeurbanne, France
Secretary, IUPAC CTC

D J Waddington

Department of Chemistry, University of York, York, YO1 5DD, UK
Chairman, IUPAC CTC

Abstract - This paper outlines some of the activities of the IUPAC Committee on Teaching of Chemistry over the last 5 years.

## INTRODUCTION

The International Union of Pure and Applied Chemistry promotes cooperation and the furtherance of chemistry worldwide and in 1964, IUPAC formed the Committee on Teaching of Chemistry 'to act as an informational and coordinating body for chemical education throughout the world'. The last 20 years has been marked with many endeavours and opportunities. Some of the current programmes are described below:

## INTERNATIONAL CONFERENCES

CTC sponsors and arranges international conferences on chemical education. These are held every two years, and the most recent have been in Madrid, Spain (1975: Education Technology in the Teaching of Chemistry)(1); in Ljubljana, Yugoslavia (1977: Chemical Education in the Coming Decades) (2); Dublin, Ireland (1979: Interaction Between Secondary and Tertiary Level Teaching) (3); Maryland (1981: Teaching Chemistry in a Diverse World) (4); Montpellier (1983: Chemistry, Education and Society) (5), and now Tokyo (1985: Widening the Scope of Chemistry) (6).

One particular feature is that through these conferences, chemistry teachers from many developed and developing countries can meet - for example teachers from over 70 countries were present at both the Maryland and Montpellier meetings, and over a third of these were from schools. In Tokyo, of the 650 participants, 240 were from 54 countries outside Japan, and 290 gave poster papers and displays.

The next meeting, the 9th International Conference on Chemical Education, will be in Sao Paulo, Brazil in July 1987 (7). The title of the meeting is The Value of Chemistry in our New World.

## INTERNATIONAL NEWSLETTER ON CHEMICAL EDUCATION

CTC produces the *International Newsletter on Chemical Education* (8). Over 3000 copies are printed per issue (twice a year) and distributed free via the national representatives in CTC, UNESCO, and directly from the editor. A recent issue has been devoted to 'Chemistry and Society', in which there were accounts of work going on in nine countries. Another two issues have been concerned with the development of microcomputers in chemical education, and another marks the 150th birthday of Mendeleev with articles on the periodic law.

The *Newsletter* promotes the exchange of ideas and the dissemination of information of chemical education all over the world. The *Newsletter* in the past four years has contained articles about chemical education from, for example Australia, Austria, Bahrain, Canada, Cyprus, Czechoslovakia, Denmark, France, F.R.G., G.D.R., Ghana, Japan, India, Italy, Israel, Kenya, Netherlands, New Zealand, Nigeria, Norway, Papua New Guinea, Portugal, Singapore, Sri Lanka, Switzerland, Togo, U.K., Upper Volta, U.S.A., Yugoslavia, Zimbabwe.

## BOOKS, SURVEYS, PAMPHLETS

This year saw the culmination of an ambitious project in collaboration with Unesco, the publication of a book entitled 'Teaching School Chemistry' (9). This book has 8 chapters, on 'The changing face of chemistry (J.A. Campbell (U.S.A.)), 'Curriculum innovation in school chemistry' (R.B.Ingle (U.K.) and A.M. Ranaweera (Sri Lanka)), 'Some methods in teaching chemistry' (A. Kornhauser (Yugoslavia)), 'Practical work and technology in chemical education' (M.H. Gardner (U.S.A.), J.W. Moore (U.S.A.) and D.J. Waddington (U.K.)), 'Assessment of students' (J.C. Mathews (U.K.)), 'Educators and Training of Teachers' (A.V. Bogatski (U.S.S.R.), D. Cros (France) and J.N. Lazonby (U.K.)), 'Current research in chemical education' (P.J. Fensham (Australia)), 'The future' (M.H. Gardner (U.S.A.)). In turn, these authors were assisted by over 70 distinguished educators worldwide - indeed from over 30 countries. The book has been published by Unesco Press in English, Arabic, French and Spanish.

An anthology, published by CTC, 'Chemical Education in the Seventies' (10), contained contributions from individuals in 40 countries, outlining developments in chemical education in schools, colleges, and universities. It has proved so popular that we have had it reprinted. The book highlights major innovations and new programs in these countries. The text provides a forum for the exchange of information about recent changes in curricula, assessment techniques, and other important aspects of chemistry teaching.

CTC is also involved in other publications. Unesco uses the CTC itself and also its individual members, in their personal capacity, as consultants, writers, and editors. Some recent examples include the series *New Trends in Chemistry Teaching*. The latest volume is an anthology of papers concerned with chemical education which have appeared in international and national journals, (11) grouped into nine sections: Chemical information; What should we teach and how?; Chemistry in space; Chemistry in the future; Controversies in teaching chemistry; Simple experiments; Low-cost equipment; Games and simulations; Chemistry and industry.

CTC has also produced, in the last few years, several surveys. The full reports can be obtained from their authors, abbreviated versions being obtainable in the *International Newsletter*. These include 'Introduction of SI Units in School and University Teaching: Implementation and Difficulties' (12,13), 'Survey of Chemical Education in Developing Countries' (14), and 'Survey on Continuing Education of Chemists in Industry' (15).

The Atomic Weights Commission has produced a table of Atomic Weights specially for schools. It has been recently revised and is available free from IUPAC (16).

## LOCALLY-PRODUCED EQUIPMENT FOR CHEMISTRY TEACHING

CTC has also been involved with the UNESCO-inspired initiative to encourage laboratory work in college and university courses. This has taken the form of workshops, the first being held in Seoul, Korea in 1975 for the Southeast Asian region and later ones being held in other regions (17-20). College and university teachers are brought together to test and devise experiments which can be implemented under the sometimes very difficult conditions under which they have to work (e.g. lack of basic facilities, few resources, large classes, inexperienced teachers). From this work we have concentrated on two aspects, the supply of simple robust equipment for the teaching of chemistry at this level, and its maintenence.

CTC, with the support of Unesco initiated a low-cost locally-produced (LCLP) programme at the University of Delhi under the leadership of Professors K.V. Sane and P.K. Srivistava, of the Departments of Chemistry and Physics. The group has several characteristics which are extremely important.

1. The group has grown out of enthusiasts at the University; it has not been planted there.

2. The group contains academic scientists of wide interest, students and technicians, working together.

3. There is a strong element of self-reliance (whenever possible the work is done within groups using only locally available materials).

4. The latest available technology is used which in turn keeps down costs.

5. There is an insistence that the equipment is as reliable and accurate as that which is commercially available.

6. The equipment is supplied within a training programme. Better still, teachers make the equipment during training, thereby being in the best possible position to maintain it. The

extra pride and motivation the fabrication gives is understandable and vital to the success of the project.

Several regional workshops, concerned with this project have been organised over the last few years, in Madras (India) (21), Dakha (Bangladesh), Talawakalla (Sri Lanka), Sao Paulo (Brazil), Copenhagen (Denmark) (22) and members of the project took part in the recent meetings at Bathurst (Australia) (23), Amman (Jordan), Puerto Rico and Bombay (India).

A variety of institutions have helped with the organisation and funding including the International Council of Scientific Unions (ICSU), Costed (Committee on Science Technology and Education in Development) and the Commonwealth Foundation.

The Commonwealth Foundation awarded Professor Sane with its first Senior Visiting Practitioner Award, in 1984 which enabled him to visit and organise activities in five Commonwealth countries and from which he wrote a booklet entitled 'Science Education through Self-Reliance' (24). The Foundation is also helping with the infrastructure of the unit in Delhi.

The equipment so far developed has been mainly electrochemical (pH meters, conductance bridges). Many versions of these types of equipment are available, to suit the needs of universities, colleges and schools, and to suit the needs of teachers in different countries. A colorimeter and spectrophotometer among other apparatus are now being designed and trialled.

All activities depend upon the community of chemistry teachers at every level, who are willing to give their time, their ideas, and their energy, freely, either working through their national representative or directly through the chairman and secretary of CTC (Professor M. Chastrette, Laboratoire de Chimie Organique Physique, Universite Claude Bernard Lyon I, 69622 Villeurbanne, France). We will be delighted to hear from you if you want to take part and to help CTC to flourish.

## REFERENCES


1. C.N.R. Rao, (Editor), Educational Technology in the Teaching of Chemistry, Proceedings of the International Symposium on Chemical Education, September 1975, Madrid, Spain, 1975.
2. A. Kornhauser, (Editor), Chemical Education in the Coming Decades, Problems and Challenges, Proceedings of the International Symposium on Chemical Education, August 1977, Ljubljana, Yugoslavia, DDI Univerzum, 1979.
3. P. Childs and J.L. Gowan, (Editors), The Teaching of Chemistry - Interaction between Secondary and Tertiary Levels, Proceedings of the International Conference on Chemical Education, August 1979, Dublin, Ireland, 1980.
4. W.T. Lippincott and H. Heikkenen, (Editors), Teaching Chemistry in a Diverse World, Proceedings of the Sixth International Conference in Chemical Education, University of Maryland, 1981.
5. H. Heikkenen and A. Ramboud, (Editors), Chemistry, Education and Society, Proceedings of the Seventh International Conference on Chemical Education, Montpellier, France, 1983.
6. Y. Takeuchi, (Editor), Widening the scope of Chemistry, Proceedings of the Eighth International Conference on Chemical Education, Tokyo, Japan, 1985.
7. For further details, write to Professor R. Isuyama, Instituto de Quimica, Universidade de Sao Paulo, Sao Paulo 20780, Brazil.
8. International Newsletter on Chemical Education. At present the editor is D.J. Waddington, Department of Chemistry, University of York, Heslington, York YO1 5DD, U.K.
9. D.J. Waddington, (Editor), Teaching School Chemistry, Unesco Press, 1984. Also published in Arabic, French and Spanish translations.
10. A. Kornhauser, C.N.R. Rao and D.J. Waddington, Chemical Education in the Seventies, IUPAC CTC, 1980. 2nd ed., Pergammon Press, 1982.
11. M. Laffitte, J.J. Thompson and D.J. Waddington, (Editors), New Trends in Chemistry Teaching, Volume V, UNESCO, 1982 (English and French editions).
12. B.T. Newbold, International Newsletter on Chemical Education, 10, 3 (1978)
13. B.T. Newbold, International Newsletter on Chemical Education, 13, 8 (1980)
14. C.N.R. Rao, International Newsletter on Chemical Education, 8, 2 (1978)
15. W.B. Cook, International Newsletter on Chemical Education, 9, 3 (1978)
16. N.N. Greenwood and H.J. Peiser, Table of Atomic Weights to Four Significant Figures. Prepared for IUPAC CTC by IUPAC Commission on Atomic Weights and Isotopic Abundances. 1983.
17. D.J. Waddington, (Editor), A Sourcebook of Chemical Experiments, Volume 1, UNESCO/IUPAC, 1976. Also available in Spanish translation, Manual de Experimentos Quimicos, Tomo 1.
18. S.I. Bayyuk, M.H. Freemantle and E.C. Watton, (Editors), A Sourcebook of Chemical Experiments, Volume 2, UNESCO 1977. Also available in Spanish translation, Manual de

Experimentos Quimicos, Tomo 2.
19. The Role of Laboratory and its influence on University Chemistry Courses. A Regional Seminar, Universidad Nacional Autonoma, Mexico City, Mexico, 1977. Manual de Experimentos Quimicos. Tomo 3, UNESCO/UNAM, 1981.
20. D. Cros and M. Maurin, (Editors), Manuel d'Experiences de Chimie, Volume 5, UNESCO/ Societe Chimique de France, 1980.
21. K.V. Sane, Locally Produced and Low Cost Equipment An Experiment for Chemistry Teaching. (A manual for teachers). Volume 1 UNESCO/IUPAC, 1981.
22. E.W. Thulstrup and D.J. Waddington, (Editors), Proceedings of the Workshop on Locally Produced laboratory Equipment for Chemical Education, Copenhagen, 1983.
23. C.L. Fogliani, (Editor), Proceedings of the Workshop on Low Cost Locally Produced Equipment, Bathurst, Australia, 1984.
24. K.V. Sane, Science Education through Self-reliance. The Commonwealth Foundation, 1984.

# Activities of the division of chemical education — The Chemical Society of Japan

John T. SHIMOZAWA (*)
Department of Chemistry, Faculty of Science
Saitama University, Urawa 338

## HISTORY OF THE DIVISION OF CHEMICAL EDUCATION

The Division of Chemical Education (hereafter this is abbreviated as DCE) was founded in 1978, reformed from the Committee on Chemical Education which was founded in 1951 in the Chemical Society of Japan (abbreviated as CSJ hereafter). The system of the DCE is similar to that of the American Chemical Society, except that the Journal of Chemical Education (Kagaku Kyoiku, in Japanese) is published directly by the CSJ.

The membership of the DCE is around 2400, of which almost half are affiliated with a university, and the others are secondary school teachers.

The subscription fee is ¥ 4200 and ¥ 4500 for the members of CSJ and non-members of CSJ respectively. Members of DCE receive six issues of the "Kagaku Kyoiku" free, although the Journal is not the official Journal of the DCE.

## COMMITTEES

The DCE has several committees concerned with policies and activities on Chemical Education.

I. The Executive Committee consists of Division Chairman, Vice-Chairman, Heads of Several Sub-Committees and Representatives of Regional Chapters, as will be described below.

II. Sub-Committees.

1. Special Committees on; 1) Industry, 2) University level, 3) General Chemistry Course in Tertiary level, 4) Teachers' Training and 5) Secondary level.
2. Local Committees on; 1) Recommendations for Award candidates, 2) International Relations, 3) Increasing the membership, 4) Planning of the DCE's Congress.
3. Research Committees on; 1) Entrance Examinations, 2) Educational Technology.

In addition to these Committees, there are regional committees in each regional chapter of CSJ, and those committees have representatives.

## MEETINGS

Each year, the DCE hold an Annual Congress at a different venue. Each time, the main topics are selected by the Local Committee, the topics chosen during the last five years were as follows:

1984: How to contribute to the 8-ICCE as Japanese Chemists, at Tokyo.
1983: Requests to Secondary Education in Chemistry from Industry and University, at Sapporo
1982: International Activities of IUPAC-CTC, at Tokyo
1981: How to innovate Chemical Education in the Secondary Level, at Okayama
1980: The Revised Course of Study, at Sendai.

At each Congress, Invited Papers and Panel Discussions were given, and sometimes, contributed papers were also presented.

Furthermore, there is a session on Chemical Education and on the History of Chemistry at the Annual Meetings of CSJ, which are held in April and in autumn every year. The topics are selected by the local commmittee for the Annual Meeting, and Special Lectures are invited in each spring by awardees of Chemical Education.

AWARDS FOR CHEMICAL EDUCATION:

The Award for Chemical Education was founded in 1976, and since 1983, the Awards were divided into two categories; i.e. An award of Chemical Education which goes mainly to University teachers, and 5 Medals for excellence to Teachers at schools. The lists of the recipients of these Awards are as follows:

AWARDS

1976 Bunichi Tamamushi, Nozomu Yamaoka
1977 Teiichi Asahina, Taku Uemura
1978 Hisateru Okuno, Yasuji Takebayashi
1979 Takeji Kashimoto, Matsuji Takebayashi
1980 Tomonari Nishikawa, Seiichiro Hikime
1981 Michinori Oki, Tamon Matsuura
1982 Osamu Shimamura, Keiji Nakanishi
1983 Shunji Kato
1984 Tomohiko Sakaki

MEDALS

1983 H.Ishido, K.Koshino, S.Togari, H.Fukushima, T.Matsuoka
1984 I.Ohtsuki, M.Kijima, S.Noda, H.Matsuo, K.Yago

ACTIVITIES OF THE COMMITTEES:

Every Committee mentioned above has meetings regularly, and the committee heads report their results in the "Kagaku Kyoiku" once a year.

The followings are the special items discussed this year.

1) Industry: The role of university chemistry in developing good chemists useful for industry through the analysis of the University Curriculum.
2) University: FORUM discussion on the curriculum of the Graduate Course was held; it investigated the ideal curriculum for the Masters' Course.
3) Freshmans' Course: The experimental programs used at universities were investigated. The current difficulties were summarized.
4) Teacher's Training: The Questionaires are designed to identify the difficulties to foster teachers especially for primary schools.
5) Secondary level: As the enrollment at high schools has increased up to 97 %, the Course of Study will have to be revised. The members began to make plans to revise the Course of Study as a proposal.
6) International relations: The future plan and also the findings of 8-ICCE were discussed.
7) Future Plan: The budget of DCE is so limited, and the role of chemical education is so important, that CSJ is planning to improve the status of DCE. A tentative plan has been agreed on by the Committee.
8) Entrance Examination: Problems required at the Common Entrance Examination for National Universities were evaluated, and a report was submitted to the National Center for University Entrance Examination.
9) Educational Technology: The software developed at several institutions was evaluated. The comments on each software will be circulated through DCE periodically.

REGIONAL ACTIVITIES

Each Chapter of the CSJ has a Committee on Teaching Chemistry, and some of the bugdets are supported by the Division. It varies chapter by chapter, however, the activities of each are similar; i.e. to have meetings for lectures, and contributed papers, for audio-visual aids; special topics such as how to use the microcomputers in Chemical Education.

As one Chapter covers several Prefectures, the venues of these meetings are rotating every year. The number of participants of these meetings is between 100 and 200. Universities of the region and Educational Authorities of each District are the sponsoring organizations which enable school teachers to attend these meetings.

CHEMICAL EDUCATION FOR THE PUBLIC

The CSJ has been very keen to provide good understanding of chemistry by the general public. According to the plan, there were several examples; "Invitation to Chemistry": Using Department Stores, chemistry exhibitions were held in Osaka, Kanazawa and Nagoya, etc. In Tokyo the National Science Museum was used for the exhibition. The budgets were supported by the Chemical Industries in Japan, and more than a million people attended these presentations. As a result, the popularity of chemistry has increased during the last three years. This is because people better understood the importance of chemistry in our everyday lives and that we cannot survive without chemistry. Moreover it could be shown that there are many problems to be resolved in the field of Chemistry.

The TV commercial messages also emphasized the improved image of chemistry. The Association of Chemical Industry in Japan tried to ask the procedures to improve the TV picture in such a way that the audience sees the scientific aspect of chemistry through the commercial messages.

THE FUTURE SCOPE OF CHEMICAL EDUCATION IN JAPAN

As described above, the activities of the DCE and CSJ for the improvement of chemical education are progressive, however, there are still many problems to be solved. For example, at the primary level, teachers have no concrete idea how to answer the problems raised by pupils which are based upon pupils' daily experiences. For the high school pupils, the problems set at the entrance examination to the university are too difficult, and the students try to remember the right answers to get good scores without having a real understanding of chemistry.

For the university students, there are too many items to be memorized again, therefore they are getting tired of studying their special subjects even though their positions are promised after the graduation. Finally, how to cultivate excellent chemists in future is the main object for members of CSJ, because without having excellent chemists, our nation will not be able to survive.

There are many pessimistic opinions for the future in terms of the fostering of excellent chemists. There are also proposals to improve the situation; however, the social trends indicate that the decreasing demands for materials are effecting the youngsters, and it may be said that these proposals are only just the dream of the established chemists. If this is so, we must have more intensive discussion on this problem, not only by educators but all research chemists must also appreciate the present status of the difficulties of chemical education throughout the world.

---

* Secretary General of 8-ICCE, and Vice-Chairman of the DCE of the CSJ.

# Invitation to chemistry—a public activity of the Chemical Society of Japan

Naoki Toshima*

Department of Industrial Chemistry, Faculty of Engineering, The University of Tokyo, Hongo, Bunkyo-ku, Tokyo, 113 Japan

Abstract - A series of meetings, "Invitation to Chemistry", has been held in Japan as one of a public relations activity of the Chemical Society of Japan since 1981. It was usually composed of a couple of lectures and scientific movies. Recently, however, demonstrations of chemical experiments are often chosen for the program. Themes of the lectures and demonstrations as well as the number of the people in the audience are shown in Table 1. Other public activities of the subcommittee are also mentioned.

## INTRODUCTION

It has been more than a decade since chemistry became an unfavorite science especially among young people. In order to remove this unfortunate situation, we, chemists and chemical engineers, have to make an effort to promote chemistry. The Chemical Society of Japan started "Invitation to Chemistry," a series of lecture meetings, as one of the public relations activities in 1981. (CSJ also held "Chemistry Fair '81" at the National Science Museum in the same year.)

## ARRANGEMENT FOR "INVITATION TO CHEMISTRY"

In the beginning, "Invitation to Chemistry" was held twice a year on the occasions of the National Meetings of CSJ. In this period, the planning subcommittee was prepared for each meeting. In 1984, however, CSJ decided to activate a public activity of the society and asked all the branches to prepare local planning subcommittees for "Invitation to Chemistry" and supported each meeting financially. The subcommittee plans the meeting and executes according to his plan. Each meeting is propagandized though posters at high schools, the organ magazines of CSJ, direct mail to teachers and past participants, commercial magazines, and newspapers.

## RESULTS OF THE MEETINGS

As the results of change in organization, "Invitation to Chemistry" is held more often as shown in Table 1. In 1985, more than ten meetings have been held all over Japan. The contents of the meetings have also changed as shown in Table 1. At the beginning, lectures and short movies were the main program. Now, demonstrations of chemical experiments are often chosen for the program. The number of people in the audience must be more limited in the demonstration than in the lecture, but the demonstration has much more impact on the audience than the lecture.

In 1985, three lecture meetings were held on the scale of about 500 participants for each meeting, while 10 small meetings with demonstrations or plactices were held with about 80 participants for each meeting. The audience of the large lecture meetings includes teachers, students, public servants, house wives, employees and so on. On the other hand, the audience of the small demonstration meetings are often limited to high school students.

According to the response to opinionaires, the audience is interested in practices and experiments which demonstrate chemical phenomena in the surroundings and recent topics which are informed by newspapers or commercial magazines.

The content of most lectures talked about at the meeting is summarized as a long abstract, which is published in Kagaku Kyoiku (Chemical Education), a magazine issued by CSJ. A collective abstract is planned to be published as a book.

---

*Chairman of Planning Subcommittee of "Invitation to Chemistry" (Kanto), Public Committee, The Chemical Society of Japan, Kanda-Surugadai, Chiyoda-ku, Tokyo 101

## OTHER PUBLIC ACTIVITIES OF THE SUBCOMMITTEE

The member of the each local subcommittee often works as a key member in other public activities of each branch of CSJ. For example, CSJ held Chemistry Fairs in large cities. In this case, the subcommittee members should be the main member for the Fair and the meeting "Invitation to Chemistry" held in the same city during the Fair.

Other public activities of the subcommittee in Kanto branch (Tokyo district) include the following: 1) Visiting lecture series; lectures by professors at primary, junior and senior high schools, usually involving demonstrations. 2) "Invitation to Chemistry and Education"; lecture and practices for school teachers. 3) Research meeting of chemistry clubs; presentation of research results by member students of chemistry clubs at high school 4) "Pleasant Chemical Laboratory"; practices on chemical experiments at National Science Museum for teachers or junior high school students.

Table 1. List of "Invitation to Chemistry"

| No. | Date (Place) | Theme (Speaker, Affiliation) (Demonstrations were indicated with * marks) | Participant | | |
|---|---|---|---|---|---|
| | | | Adult | Student | Total No. |
| 1. | 3.27.'81 (Tokyo) | 1)"Science and Taste" (Y. Omata, Ajinomoto Co.) 2)"Space and Chemistry" (S. Murayama, Nat. Sci. Museum) | 54% | 46% | ~330 |
| 2. | 10.11.'81 (Okayama) | 1)"Life Science and Chemical Industry in Tomorrow" (T. Yasui, Kuraray Co.) 2)"Space Development and Chemistry" (S. Saito, Tokyo Inst. Tech.) | 19% | 81% | 507 |
| 3. | 4.3.'82 (Tokyo) | 1)"Cancer and Chemistry" (C. Nagata, Nat. Cancer Center) 2)"Electronics and Materials" (Y. Iida, Matsushita Elec. Co.) | 54% | 46% | 513 |
| 4. | 10.2.'82 (Niigata) | 1)"The Birth of Ceramic Age" (H. Yanagida, Univ. Tokyo) 2)"Life Science and Chemical Industry in Tomorrow" (T. Yasui, Kuraray Co.) | 0% | 100% | ~230 |
| 5. | 3.19.'83 (Tokyo) | 1)"Birth and Evolution of Atmosphere and Sea" (Y. Kitano, Nagoya Univ.) 2)"Expectation for Life Science" (K. Nakamura, Mitsubishi Inst. Life Sci.) | 59% | 41% | ~400 |
| 6. | 3.20.'83 (Kyoto) | 1)"Chemistry and I" (K. Fukui, Kyoto Inst. Tech.) 2)"Antibiotics and Challenge of Carring Cancer" (H. Umezawa, Inst. Micro Org. Chem.) | 68% | 32% | ~350 |
| 7. | 8.28.'83 (Sapporo) | 1)"Surroundings of Chemistry" (K. Fukui, Kyoto Inst. Tech.) | 5% | 95% | ~1400 |
| 8. | 3.24.'84 (Tokyo) | 1)"Single Crystal of Pure Silicon" (Y. Abe, Shin-etsu Semicond. Co.) 2)"New Prospective in Synthetic Organic Chemistry" (T. Mukaiyama, Univ. Tokyo) | 79% | 1% | 545 |
| 9. | 4.2.'84 (Tokyo) | Part I 1)*"Mysterious Surface Chemistry—Easy and Pleasant Experiments—" (T. Sasaki, Tokai Univ.) 2)*"Nobel Syntheses of Nylon" (N. Ogata, Sophia Univ.) | 2% | 98% | 86 |
| | | Part II 1)*"Materials for Displaying Images from Electric Signals" (I. Shimizu, Tokyo Inst. Tech.) 2)*"Molecules Which Adsorb or Release Light" (K. Mutai, Univ. Tokyo) | 2% | 98% | 87 |
| 10. | 8.7~8.'84 (Osaka) | Part I 1)"Figure and Architecture of Biomolecule" (M. Kakudo, Himeji Inst. Tech.) 2)*"Visualized Organic Chemistry" (M. Kozuka, Osaka City Univ.) | 50% | 50% | 121 |
| | | Part II 1)"Sanshisuimei-Ron—Fundamentals in Environmental Chemistry" (T. Fujinaga, Nara Tech. Coll.) 2)*"Looking Surroundings by Chemical Eye" (S. Kato, Kobe Gakuin Univ.) | 52% | 48% | 120 |
| 11. | 10.20.'84 (Kyoto) | 1)"Chemistry of Natural Dyes—An Ancient Purple" (T. Yoshioka, Osaka Art. Coll.) 2)"Chemistry of Dyeing" (Y. Kuroki, Osaka Pref. Univ.) 3) Practices in dyeing | 100% | 0% | ~40 |
| 12. | 11.23.'84 (Tokyo) | 1)"Chemistry Assists Electronics Breakthrough" (T. Takamura, Toshiba Res.) 2)*"Seaching Life through Chemical Experiments" (M. Karube, Tokyo Inst. Tech.) | 2% | 98% | 67 |

(Continued)

Table 1. (Continued) List of "Invitation to Chemistry"

| No. | Date (Place) | Thema (Speaker, Affiliation) | Participant Adult | Student | Total No. |
|---|---|---|---|---|---|
| 13. | 3.21.'85 (Tokyo) | 1)"Evolution of Chemistry" (Y. Oshima, Tokyo Inst. Tech.) 2)"Recommendation for Chemistry" (K. Fukui, Kyoto Inst. Tech.) | 66% | 34% | 439 |
| 14. | 4.3~4.'85 (Tokyo) | Part I 1)*"Electroresponsive Plastics" (S. Miyata, Tokyo Coll. Agric. Tech.) 2)*"Relationship between Light and Materials" (N. Toshima, Univ. Tokyo) | 2% | 98% | 93 |
| | | Part II 1)*"Light and Color—Pleasant Chemical World" (I. Ikemoto, Tokyo Metro. Univ.) 2)*"Shape and Property of Organic Compounds" (M. Saburi, Univ. Tokyo) | 2% | 98% | 78 |
| 15. | 7.25.'85 (Hiroshima) | 1)"Looking Surrounding by Chemical Eye" (S. Kato, Kobe Gakuin Univ.) 2) Training course 3) Free talking | 6% | 94% | 250 |
| 16. | 7.27~28.'85 (Sendai) | 1)*"The First Stage of Glass Blowing" (Y. Takahashi, Tohoku Univ.) 2)*"Let's Look the Degree of Reaction by Eye—Follow-up of Esterification Reaction by Gas Chromatography" (S. Yamaguchi, Tohoku Univ.) 3)*"Electrolysis of an Aqueous Solution—Making Pendants of Metal Leaves" (H. Itabashi, Miyagi Teach. Coll.) | 8% | 92% | 158 |
| 17. | 8.3.'85 (Nagoya) | 1)"Chemistry and Life" (K. Fukui, Kyoto Inst. Tech.) 2)"Life and Material" (S. Iijima, Nagoya Univ.) | 94% | 6% | 415 |
| 18. | 8.5.'85 (Osaka) | 1)"Chemical Materials in Living" (K. Isagawa, Osaka Pref. Univ.) 2) Discussion meeting, "Q and A" | 46% | 54% | 132 |
| 19. | 8.5.'85 (Sapporo) | 1)"Syntheses and Applications of Diamond" (S. Yatsu, Sumitomo Elec. Ind.) 2)"Development and Application of Ceramics for Electronics" (S. Hayakawa, Matsushita Elec. Co.) 3) Social meeting | 100% | 0% | 136 |
| 20. | 8.19.'85 (Morioka) | 1)*"Plastics and Heat" (K. Mori, Iwate Univ.) 2)*"Variety of Chemistry in the Surroundings" (K. Sato, Iwate Univ.) 3)*"Separation"(with practice)(Y. Takigawa, Iwate Univ.) 4)*"Making a Blue Photograph"(with practice)(T. Murakami, Iwate Univ.) | 21% | 79% | 47 |
| 21. | 8.19~20.'85 (Fukuoka) | 1)*"Iodine-Starch Reaction—What Happens When a Molecule Enters into Another Molecule Containing Large Cavity?" (T. Inatsu, Kyushu Univ.) 2)*"Enzyme in Radish" (Y. Kato, Kyushu Inst. Tech.) 3)*"Preparation of Paints"(with practice)(Y. Moriguchi, Fukuoka Teach. Coll.) 4)*"How Does Coal Burn?"(with practice)(I. Mochida, Kyushu Univ.) 5)*"Synthesis of Nylon"(with practice)(C. Kajiyama, Kyushu Univ.) | 0% | 100% | 79 |
| 22. | 8.20.'85 (Tokyo) | 1)"The Beginning to Chemistry—Looking for Isomerism" (S. Yoshikawa, Univ. Tokyo) 2)*"Plating without Electricity" (T. Osaka, Waseda Univ.) | 3% | 97% | 129 |
| 23. | 10.5.'85 (Kanazawa) | 1)"Chemistry and My Step in Life" (S. Yoneda, Konohana-cho Primary School) 2)"Chemistry in the Dimensions of Molecules" (M. Imoto, Osaka City Univ.) 3) Panel Discussion "Let's Talk on Chemistry" (T. Shiba, Osaka Univ.; H. Seki, Sumitomo Chem. Ind. Co.; H. Fukami, Kyoto Univ.; R. Noyori, Nagoya Univ.) 4) Exhibition, "Local history of chemistry" | 40% | 60% | ~500 |
| 24. | 11.24.'85 (Tokyo) | 1)"How to Prepare Optical Fibers" (M. Watanabe, Sumitomo Elec. Ind.) 2)*"Chemistry of Iron—Iron Compounds in the Surroundings" (H. Sano, Tokyo Metro. Univ.) | 0% | 100% | 69 |
| 25. | 12.7.'85 (Sapporo) | 1)"Hydrogen Energy System" (T. Ohta, Yokohama Nat. Univ.) 2)"Usage of Solar Energy" (K. Wakamatsu, New Energy Dev. Org.) | 100% | 0% | 70 |

# Educational programs currently broadcast by NHK

Nippon Hoso Kyokai (Japan Broadcasting Corporation)
2-2-1 Jinnan, Shibuya-ku, Tokyo 150, Japan

INTRODUCTION

The letters NHK stand for Nippon Hoso Kyokai - in English: the Japan Broadcasting Corporation, founded in 1925.

NHK operates two television channels, two AM radio channels and one FM radio channel in its domestic services, together with an international short-wave broadcasting service for overseas listeners. NHK is the sole public broadcaster in Japan and conducts its entire range of operations on the financial basis of the receiving fees paid by 30 million households. Standing in this way on a nationwide foundation, it can operate free from political influences and the various social and economic pressures of the day.

While NHK has spread its broadcasting networks to cover the entire country, it broadcasts programs that meet the requirements of the Japanese people, while at the same time assuming the mission to contribute to the improvement of the nation's cultural standards.

The weekly transmission hours and numbers of programs are as below.
"General TV and Radio 1 Network": In these services news, cultural and entertainment programs intended for the general audience are leading items. While presenting such programs under well-balanced scheduling, these two networks also take care of local services.
"Educational TV and Radio 2 Networks": These two services deal mainly with educational programs, especially for school broadcasting. The details of educational programs will be discussed in a later session.
"FM Broadcasting Networks": NHK conducts FM broadcasting on both national and local networks, taking maximum advantage of the high sound quality that frequency modulation offers and laying stress on programs on music, and language training.

Percentages of programs of different categories in an average week and their daily broadcasting hours are as follows (Table I).

TABLE I. Weekly Programs Categorized ( % )

General TV (Daily Average 18 hours)
News: 36.7 , Cultural: 27.6, Entertainment: 21.4, Educational: 14.3
Educational TV (Daily 18 hours)
Educational: 78.2, Cultural:19.7, News:2.1
Radio 1 (Daily 19 hours)
News: 42.8, Cultural: 30.7, Entertainment: 25.1, Educational: 1.4
Radio 2 (Daily 18.5 hours)
Educational: 71.5, Cultural:16.1, News: 12.4
FM (Daily 18 hours)
Cultural:43.5, Entertainment:35.8, News: 13.8, Educational: 6.9

EDUCATIONAL PROGRAMS

"Broadcasts for Schools" : Broadcasts for schools have a long history of some 50 years in Japan, stretching from the early days of radio down to the present. NHK sends out programs to kindergartens, primary schools, lower and upper secondary schools through its Educational TV(ETV) and Radio 2 networks, as shown above. The number of programs carried every week and the number of broadcasts, are 115 programs and 31 hours 35 minutes for ETV and 52 programs and 50 hours 20 minutes for Radio 2.

The utilization rate of these programs from kindergartens through to upper secondary high schools, is very high as shown in the Table II.

TABLE II. Ratio of School Program Utilization (%)

| | Television | | Radio | |
|---|---|---|---|---|
| | Utilization Rate | No. of Schools (Estimated) | Utilization Rate | No. of Schools (Estimated) |
| Kindergartens | 71.5 | 10,834 | 10.9 | 1,652 |
| Day Nurseries | 78.5 | 17,860 | 7.8 | 1,775 |
| Primary Schools | 89.8 | 22,489 | 16.6 | 4,157 |
| Lower Secondary | 65.4 | 7,115 | 21.0 | 2,284 |
| Upper Sec. Schools | 66.1 | 3,279 | 12.7 | 630 |

This means that practically everyone has had some experience of broadcasts to schools during his or her education. Visual materials that cannot be had in the classroom are being supported and those cover the entire curriculum: i.e. Japanese language studies, science, mathematics, social studies, English, music, art, physical education, ethics, home economics, and other subjects.

By using these, teachers are able to stimulate students' interest in their studies and a desire to work; consequently they are also able to develop their students' independent studies. There is an organization called All Japan Teachers Federation for studying the use of Radio and Television in Education. This was organized voluntarily by teachers, representing 83 % of all schools. Its members concern themselves actively with the question of what forms broadcast education should take:in other words, the ways in which school broadcasts should be included in lessons in order to heighten the effectiveness of class- room teaching. The activities of these teachers assist the wider introduction of broadcast programs into the classroom, and the study reports they prepare lead directly to the production of better programs.

These curricula follow the contents of the "Course of Study", and the programs receive the advice not only of the school teachers but also of university professors who have a good understanding of the Course of Study.

PROGRAMS FOR CHEMISTRY TEACHING:

The Programs relating to school teaching follow the "Course of Study" which describes the official curriculum for pre-university courses. Therefore, the texts used in the on-air-programs conform with the contents of the Course of Study.

1) Primary school Course: There are 20 minute programs for each grade per week, in day time. However, pure chemistry courses are not required in the lower grades in primary school, there are programs to teach chemical concepts using marionettes to present children who have discovered something interesting in their daily lives. The chemistry related items appear in the higher grades of primary school courses, such as the neutralization of acid and base, and the oxidation process in burning. Water solutions are used as the examples to understand the concept of small particles which are sometimes invisible and the ion model in the solution is displayed in the TV screen using animation. These patterns help the pupils to understand the small particles constituting whole materials.

2) Lower Secondary School(Middle School) Course: There are 30 minute programs for each grade per week, in day time. These include the Chemistry Courses required in the First Field of Science Course in which Physics is also combined. The Course of Study involves the chemical concepts of burning and solubility treated quantitatively and also the symbols of the elements to formulate chemical reactions. The programs broadcasted demonstrate these through models to describe molecular and ionic concepts. Since these programs are used as color dispays, the understanding of these concepts by the pupils is much improved. The environmental course of our lives is also explained through the program of the combined science.

3) Upper Secondary School (High School) Course: As the Course of Study requires both "Science I", which is compulsory and the combined science, and also "Chemistry", the programs are divided into two series. In addition to these, there are programs on Radio 2, which are called as "Teachers' Hour" to explain how to teach pupils newly developed fields such as molecular dynamics, semi-conductors, bio-technology, and computer literacy, etc.

In "Science I", a distinguished professor illustrates the scientific approach to look at nature from his point of view. This is highly appreciated not only by the pupils but also by the teachers, because the combined science course is difficult to deal with for teachers who are not trained in general science courses.

For the Chemistry Course, the programs on TV are always used for colorful displays of experiment. Simulation of molecular and ionic models are used to illustrate the more difficult concepts of chemistry.

All of these school programs are used either on real time or video-taped versions to the class, however, it is not popular to use them as a library; i.e. as individual after-school studies.

EDUCATIONAL BROADCASTS FOR ADULTS:

The desire to study has been growing year by year among adults. It can be said that we live in an age when leisure is increasing, while the tendency of the time is toward a highly information-oriented society and an older population. In these circumstances, people look to broadcasting to provide them with information useful for the enrichment of human values in their lives and to enable them to satisfy their various intellectual interests.

In "Hobby Guide", for example, the number of programs has been increased to two a week, each dealing with a different subjct, such as how to use the video camera, or new scientific technologies such as biotechnology. There are also general information programs on medicine or health improvement which are useful for the family, and programs giving the latest information on various industries and useful for business management, such as "Business Network", and "Farmers' (Fishermans') Day". There are also programs related to social welfare, such as "Welfare Tommorow" and "Speech Development".

The "NHK Culture Seminar" has been established in response to the demands from adults for intellectual satisfaction in various fields, including social studies, history and science. It has received a strongly favorable response from viewers.

---

This article was written by the Secretary General of 8-ICCE. The article refers to the pamphlet, partly, published by the Audience and Public Relations Bureau, NHK.

# International Chemistry Olympiad: U.S. preparation and participation

Michael D. Hampton, Mary Beth Key, James Wright, Patricia Smith

Department of Chemistry, University of Central Florida, Orlando FL U.S.A.

Abstract - In July of 1984 the United States competed for the first time in the International Chemistry Olympiad. This was the XVI International Chemistry Olympiad and it took place in Frankfurt, West Germany. Preparation for the competition began with the selection of the top twenty high school chemistry students. These twenty students then participated in a two week study camp during which they were involved in an intense program of classroom and laboratory sessions. At the end of the study camp the top four students were chosen for the United States team for the International Chemistry Olympiad. The U.S. team fared quite well at the Olympiad, earning one silver medal, two bronze medals, and one certificate. More importantly, however, the students and the coaches had an extremely enjoyable and expanding experience. In 1985 the United States participated in the XVII International Chemistry Olympiad in Bratislava, Czechoslovakia. Once again, the students and coaches had extremely enjoyable and expanding experiences. To make it even better, the U.S. team earned two silver medals and two bronze medals.

## INTRODUCTION

The International Chemistry Olympiad (IChO) is an event which began in 1968 when students from Hungary, Poland, and Czechoslovakia met at the First IChO in Prague, Czechoslovakia. The following year Bulgaria joined the competition. A year later the number of participating nations rose to seven when the German Democratic Republic, Soviet Union, and Romania joined the competition. In 1976 the competition had grown to the point that formal organizational rules had to be established. The Olympiad has been held annually since 1968 and has shown steady growth and refinement. In 1984, when the United States entered the competition, participation had increased to twenty-two nations. In addition several nations, including Brazil, Kuwait, Venezuala, and Canada, sent observers, anticipating competition in future Olympiads. Twenty-two nations also participated in the XVII IChO in 1985.

In 1982 the American Chemical Society (ACS) decided that the United States should participate in the International Chemistry Olympiads. As a result, the U.S. sent an observer, Dr. Marjorie Gardiner, to the Olympiad in 1982 and another observer, Ms. Patricia Smith, to the Olympiad in 1983.

In 1984 the U.S. prepared and sent a team to compete in the XVI IChO. The XVI IChO was such an excellent experience that the U.S. prepared another team in 1985 and competed in the XVII IChO and will continue to participate in future Olympiads. The student selection and preparation and the Olympiad competitions will be discussed below.

## SELECTION OF STUDENTS FOR STUDY CAMP

The process by which students were selected for participation in the study camp was the same in both 1984 and 1985. The selection process was as follows. Each local section of the ACS (roughly one to three local sections per state) chose a number of students to take a national examination. The number of students chosen by each local section was based on the population of that section. The method by which these students were selected was left to the discretion of each section; many used a written exam prepared by the ACS. Based on the results of the national exam, the top student from each of the six regions of the ACS, along with the fourteen next highest scoring students, was selected to attend the study camp.

In 1984 two thousand students were screened by fifty local ACS sections and 192 students were selected to take the national exam. One hundred local ACS sections screened 5000 students and 473 were chosen to take the national exam in 1985. The national exam consisted of an objective portion, made up of multiple choice questions, and an essay section. The two sections were given equal weight in the grading. The exam was developed, administered and graded by the ACS. This exam tested students' knowledge of all the basic areas of chemistry. The essay portion provided an opportunity to evaluate the students' problem-solving abilities and writing skills.

## STUDY CAMP

The study camps in both 1984 and 1985 were two weeks in duration and were held at the U.S. Air Force Academy in early June. The camps were sponsored by the ACS and the U.S. Air Force Academy. Instruction was provided by three mentors (one university professor and two high school teachers in 1984, two university professors and one high school teacher in 1985), the study camp coordinator, and various faculty from the Academy and from universities near the Academy. The three mentors lived with the students in the dormitory.

The goals of the study camps were to provide the students a chance to learn chemistry and to enhance their enthusiasm for the subject, to expose students to learning in an intensive university environment, to give exceptional students the chance to interact with their peers from all over the U.S., to promote cooperation between university and high school faculties, to prepare students for competition in the IChO, and to choose the four students for the U.S. team for the [illegible]

A typical day at the study camp began with a wake-up call at 0600 followed by breakfast at 0615. The morning session, [illegible] then commenced. This session was classroom time used for lectures, [illegible] problem solving and test taking. Lunch was served at 1115. At 1300 the laboratory session began. Dinner began at 1615 and from 1800 to 2300 the evening session occurred. This time was used for homework and lab reports, tutorial sessions, added lectures, study, and sometimes free time. Saturday morning was occupied with a horseback ride and Sunday morning was free time.

During classroom sessi[illegible] following subjects were covered: Analytical Chemistry - [illegible]ons, Beer-Lambert Law; Physical Chemistry - thermodynamics[illegible]on, graphical methods; Inorganic Chemistry - coordination c[illegible] analysis, reactions; Organic Chemistry - synthesis, reaction [illegible] acids, proteins, sequencing, saccharides, fatty acids, nu[illegible]

In the 1985 study camp [illegible] However, more depth was added to the Organic and Biochemistry [illegible]ion had conductivity and non-aqueous acid-base chemistry added, the emphasis [illegible] chemistry was shifted more towards kinetics, and the Inorganic [illegible] descriptive.

The laboratory sessions in both 1984 and 1985 were designed to provide the students with experience of a wide variety of laboratory techniques and apparatus. The experiments done covered qualitative and quantitative analysis and synthesis. For some experiments detailed instructions were provided while virtually no instructions were provided for other experiments. Thus the students were forced to follow instructions as well as to organize their work plans, anticipate experimental requirements, and manage data independently. Written reports were required for every experiment.

The evaluation of the students in 1984 and in 1985 was based 60% on theoretical knowledge and 40% on laboratory abilities. Subjective evaluation was used in borderline cases. The theoretical knowledge was evaluated by means of two three-hour written exams. The first exam covered Analytical, Physical, and Inorganic Chemistry. The second exam was comprehensive over the entire study camp. The laboratory exam (3 hours) involved qualitative inorganic analysis in 1984 and inorganic synthesis in 1985. By this evaluation the top four students were chosen to represent the U.S. in the IChO.

## INTERNATIONAL CHEMISTRY OLYMPIAD

The International Chemistry Olympiad is an international competition for students taking chemistry at the secondary level. Nations send teams of up to four members each and up to two adult coaches. The students compete individually on a five hour written theoretical exam and a five hour practical (laboratory) exam. The coaches work together as the International Jury to decide the final form of the exams, translate them as necessary, grade the exams, decide award distribution, and plan the next year's Olympiad. The International Jury is chaired by the country hosting the Olympiad. The official language of the Olympiad is generally English, though the proceedings are quite often held in the language of the host country.

The country hosting the Olympiad sends preparatory problems to all of the nations which are to compete. These problems are sent 3-4 months before the Olympiad to help the entrants prepare for the competition. The preparatory problems contain both problems and experiments which indicate the emphasis of the upcoming Olympiad exams. It is the responsibility of the host country to prepare the exams and present them to the International Jury for approval.

The exams are graded separately by both the host nation and the International Jury to provide double checks and reduce bias. Further, the coaches are separated from the students by significant geographical distances while the exams are administered. The host country provides all room and board, laboratory space, manpower, and even spending money for the participants during the Olympiad. Each participating nation is responsible for transportation of their team and coaches to and from the Olympiad. The ACS covered the travel expenses of the U.S. delegation.

The Olympiad is typically 9-10 days in duration. Official opening ceremonies are held on the first day. The students and coaches are then separated, and the International Jury is convened to approve and translate the theoretical exam while the students are given free time. The next day the students take the theoretical exam while the coaches attend tours and/or presentations. The Jury then reconvenes to approve and translate the practical exam while the students attend tours and/or presentations. The following day the students take the practical exam while the coaches tour the host country. The remaining days of the competition are spent by both students and coaches touring the host country, visiting schools, plants, and factories, and attending lectures and parties. During this time the Jury receives copies of the students' tests to grade, convenes numerous times for planning the upcoming Olympiad, compares grades with those of the graders from the host country, and decides award distribution. The awards ceremony is held the last day followed by a formal party to close the Olympiad.

The XVI IChO was held in Frankfurt, West Germany in July 1984. Seventy-six students from twenty-two nations competed. The theoretical exam covered first order radioactive decay, coordination chemistry (reactions and structures), stereospecific organic reactions, organic synthesis and reaction mechanisms, absorptiometric analysis, gas laws and determination of chemical formulas, DNA and nucleotide pairing, and proteins (amino acids, sequencing,enzymes). The practical exam had two parts - nitration of phenacetin and quantitation of phosphoric acid in a cola.

The U.S. students did very well on the inorganic, physical, and biochemistry problems but very poorly on the practical exam and problems involving organic chemistry. They did very well in the overall rankings. Seth Brown placed 13th earning a silver medal, Keith Rickert and Aaron DiAntonio both earned bronze medals, placing 36th and 50th respectively, Peter Capofreddi placed 51st and was awarded a certificate.

The XVII IChO was held in Bratislava, Czechoslovakia in July, 1985. Eighty-three students from 22 nations competed. In this Olympiad the theoretical exam covered analytical separation and titrimetric analysis of aluminum in an alloy, electron configuration, paramagnetism, molecular structure, bond order and length, simple and competitive reactions of EDTA with calcium ions, kinetics of a racemization reaction, thermodynamics of electrochemical cell, organic synthesis, kinetics and mechanism of an organic reaction, structural formulas of monosaccharides, and stoichiometric calculations involving interconversion of these monosaccharides. The practical exam involved the determination of the relative molecular mass of a weak acid by titration in a non-aqueous solvent.

The U.S. team fared very well in this Olympiad. The only consistent weaknesses shown were in organic synthesis and monosaccharide structures. The U.S. team members scored very highly on the practical exam this time. Keith Rickert and David Maymudes earned silver medals, placing 12th and 25th respectively. Glenn Whitney and Eric Nelson earned bronze medals, placing 44th and 54th respectively.

While the competition was exciting and challenging, the U.S. students and coaches found that the real value of the Olympiad went well beyond the competition. The best parts of the IChO were the lasting friendships formed with people from other nations, the chance to learn about and experience other cultures and lifestyles, the opportunity to freely exchange ideas with chemists from other nations, and the chance to mingle with one's peers from other nations.

# If water had no hydrogen bonds . . .

Mariana P. B. A. Pereira

Department of Education, University of Lisbon, Lisbon, Portugal

Abstract - In a creative event, students were asked to foresee what would happen if water had no hydrogen bonds. The answers given illustrate relationships between chemistry and biology, physics, geology, other areas, as well as extrapolation of the knowledge of chemistry.

## INTRODUCTION

The chemistry Olympiad in Portugal includes a creative event. In 1982, seventeen teams of four students each, answered the following creative question:

Suppose that because of some accident in Nature, there were no hydrogen bonds between water molecules. Foresee the consequences of such an accident.

The answers given by the participants were more numerous than those anticipated by the jury, and are illustrated below.

## RELATIONSHIPS BETWEEN CHEMISTRY AND BIOLOGY

1- If water had no hydrogen bonds, the water molecule would not have its dissolving character, and hence there would be no life on earth, since there would be neither mineral salts, nor any dissociated compounds necessary to life.
2- Life would not exist; with no hydrogen bonds DNA molecules would not exist.
3- Each living organism would be obliged to live in an isolated system, as, for instance, inside a glass campanule.
4- Probably different forms of life would exist.
5- Life which is dependent on liquid water would not exist.

## RELATIONSHIPS BETWEEN CHEMISTRY AND PHYSICS

1- The conductivity of water would be zero.
2- Water would be less dense.
3- Solid water would be more dense than liquid water; so, it would not float.
4- Water would be gaseous and not liquid, since there would be no interaction between water molecules, and they would not be free and chaotic like in the gaseous state.

## RELATIONSHIPS BETWEEN CHEMISTRY AND GEOLOGY

1- With no liquid water, that keeps the temperature more or less constant, the thermic amplitudes would be very high.
2- There would be no oceans, no lakes, no rivers, because water would be in the gaseous state.

## EXTRAPOLATION OF KNOWLEDGE OF CHEMISTRY

1- The chemical bonds between oxygen and hydrogen atoms would be covalent non -polar.
2- The intermolecular forces would be reduced to the van der Waals forces, considerably weaker.
3- The particle $H_3O^+$ would not be established, since the $H^+$ would not be formed to establish the necessary bond.
4- Less energy would be needed to break the intermolecular bonds; consequently, the boiling and the melting point would be considerably below.
5- The concepts that we have for acids and bases would not hold.
6- Aqueous solids would not exist.
7- Hydrated salts would not exist.
8- Ionic compounds would not be dissolved in water.
9- Dissociation would fail to exist.
10- Covalent compounds that form H-bonds in water would not dissolve in water.
11- The acids would not dissociate, since $H_3O^+$ would not exist.

12- The ionic product of water would be zero.
13- $K_w$ would have no meaning, since the water molecules would be ionised.
14- Electrolites would not be formed.
15- pH would have no meaning.

RELATIONSHIPS BETWEEN CHEMISTRY AND THE REST OF LIFE IN THE WORLD

1- In order to drink water, people would have to buy gas cylinders
2- There would be no snow flakes.
3- Icebergs would not exist.
4- The conservation of food in ice would be impossible.
5- There would be no ice creams.
6- Winter sports would not exist.
7- Ice-hockey players would have no job.
8- The process of dyeing would not exist.
9- There was no need to learn the topic on hydrogen bonds in school.

# Analysis of item characteristics in the JFSAT: chemistry

Fumiyasu YAMADA, Yujiro NOMURA, Tomoichi ISHIZUKA,

Research Division, The National Center For University Entrance Examination, 2-19-23 Komaba, Meguro-ku, Tokyo 153 Japan

The contents and the item characteristics of the chemistry test in the JFSAT were analysed. The aim of this paper is to clarify what kind of scholastic attainments the JFSAT can measure and to investigate the relation between the scholastic attainments which an item can measure and the item characteristics of the item.

## THE JOINT FIRST STAGE ACHIEVEMENT TEST

The Joint First Stage Achievement Test(JFSAT) has been administered every January since 1979 in Japan. Every candidate who aims to enroll himself in any national or local public university has to sit for the JFSAT which is conducted by the National Center for University Entrance Examination. Approximately 350 thousand candidates take the examination every year. The JFSAT is constituted by five subjects: Japanese, social studies, mathematics, natural science, and foreign language; chemistry is one branch of natural science. The test is constructed by multiple choice items and the items are to be answered based on the mark sheet system.

Multiple choice tests have been criticized as that they only measure the examinees' knowledge while there may be a variety of scholastic attainments to be measured. This criticism is a part of the motivation which leads us to the content-oriented analysis of the JFSAT.

The chemistry test of the JFSAT in 1982(see Appendix) has been analysed here. The test was constituted by four large questions each of which was constructed by combination of several small items. On the whole, 36 small items were contained in the chemistry test.

## EVALUATION OF TEST ITEMS

Out of the many aspects of examinees' scholastic attainments, the following five aspects were taken up: Knowledge, comprehension, skill and experience of scientific observation (observation), reasoning, and interest and attitude towards nature (attitude). Three hundred samples were drawn from the population of all the Japanese high-school chemistry teachers. They were requested to evaluate each of 36 items from the point of view how well it measures the five aspects by using the four point scale as follows.

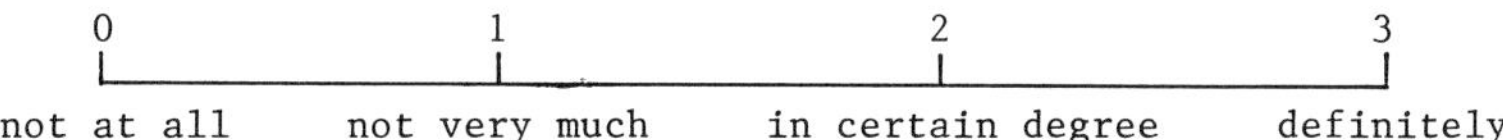


From the results of their evaluation, average scale point of each aspect for each item was calculated. Applying a technique of cluster analysis to these average scale points, five clusters were identified, i.e. 36 items were divided into the five clusters as is shown in TABLE 1.

TABLE 1. The clusters of items based on average scale points.

| Cluster Number | Number of Items | Item Number |
|---|---|---|
| 1 | 2 | (1) (2) |
| 2 | 7 | (3) (4) (5) (6) (7) (8) (15) |
| 3 | 10 | (9) (10) (11) (12) (14) (16) (17) (18) (19) (30) |
| 4 | 5 | (13) (28) (29) (31) (36) |
| 5 | 12 | (20) (21) (22) (23) (24) (25) (26) (27) (32) (33) (34) (35) |

Fig.1 shows the average scale point of each aspect for each cluster by star-chart. According to the figure, cluster-1 is constructed by the items which are biased toward knowledge. The items included in cluster-2 can measure comprehension in addition to knowledge; in addition to these aspects the items in cluster-3 or cluster-4 can measure also reasoning. Finally the items in cluster-5 are evaluated to be able to measure almost all of the aspects. Therefore, if an item can measure observation or attitude, it can measure also knowledge, comprehension and reasoning. This fact shows that contrary to the criticism the items of the JFSAT can measure not only knowledge but also various aspects.

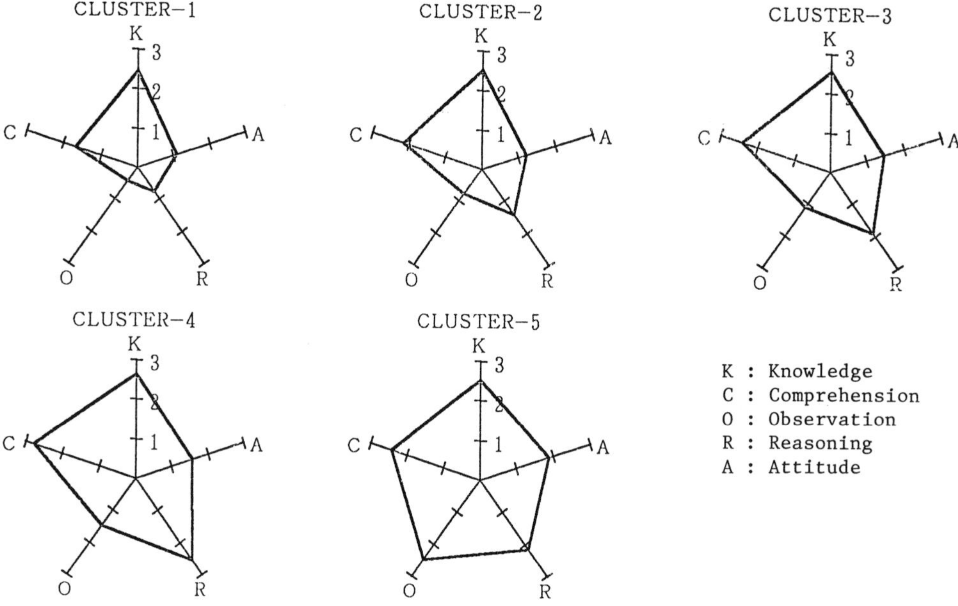


Fig. 1. The average scale point of each aspect for each cluster.

## ITEM TEST REGRESSION

The item characteristics of the item which is included in each cluster was analysed by the method of item test regression. Fig.2 shows the item characteristic curves in five clusters. The figure shows the percent of the examinees passing the item for five groups, namely L, ML, M, MH and H. The letter L, ML, M, MH and H represent the groups of the

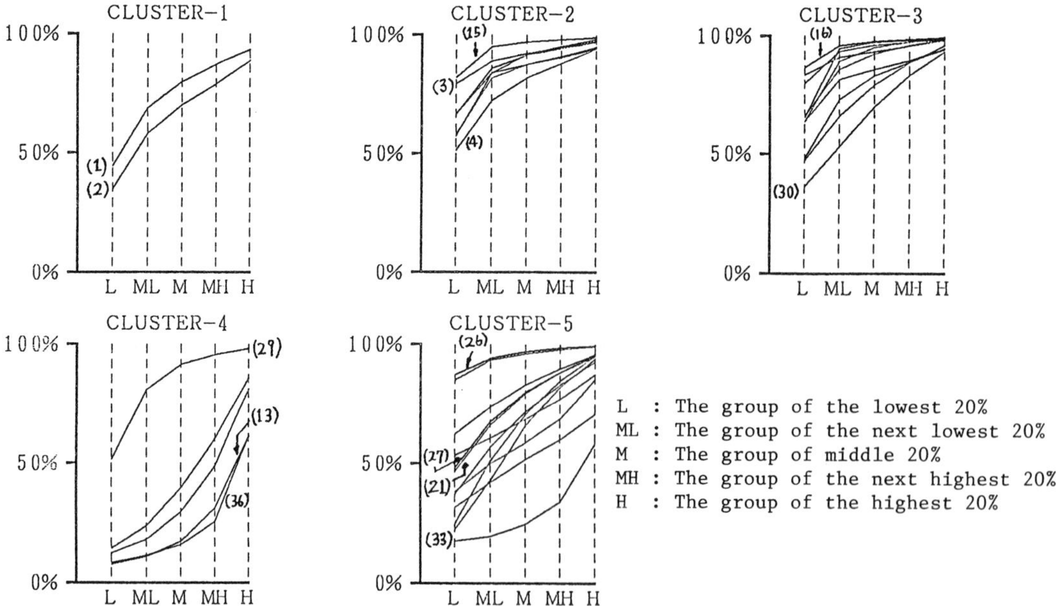


Fig. 2. The item characteristic curves in five clusters. The numbers in parentheses represents the number of typical item in the respective cluster.

examinees which are constituted by dividing the total examinees into the five subsets of same size according to their scores on chemistry test. The group H is the highest and the group L is the lowest. The main findings were as follows:

* The items which measure mainly knowledge and comprehension (cluster-2) tend to be easier items for the examinees of the JFSAT and to have low discriminating power, i.e. the slope of the item characteristic curves are flat (item (3),(15)).
* The items which measure also reasoning (cluster-3 or 4) tend to be rather difficult items and to have high discriminating power especially among the groups of examinees with high scores (item (13),(36)).
* The item characteristics of the items which measure each aspects equally well (cluster-5) are rich in variety; some items have high discriminating power among the groups with high scores (item (33)), some items, on the other hand, have high discriminating power among the groups with low scores (item (21),(27)).

## APPENDIX

The chemistry test of the JFSAT in 1982 (approx. 60min.).

Use the following atomic weights when necessary.

H 1.0  O 16.0  Mg 24.3

I. Answer Q1 and Q2.

Q1. Choose one most appropriate name, word or number from the answers for each of (1)-(10) in the following sentences (a-c).

a The original version of the Periodic Table of Elements was invented by (1) in 1869. Later, various alternatives have been proposed, and now (2) period type in which A and B subgroups of each group of elements are separately placed is used.

b In the Periodic Table of (2) period type, elements are arranged in the order of (3), and the elements of (4) series and these located close to (5) of each series are non-metals. (6) elements are located near to the center in this Periodic Table.

c The number of electrons of outermost electron-shell in the atoms of the chemically most stable (7) group elements is either (8) or (8'). If one election is transferred from the atom with one more electron than those element to the atom of (9) group element, with one electron less than these, the two atoms combine each other by forming the (10) bonding.

Answers (1) 1) Mendel 2) van't Hoff 3) Arrhenius 4) Mendeleev 5) Bohr 6) Lavoisier 7) Dalton 8) Boyle 9) Gay-Lussac 0) Avogadro

(2)(5) 1) upper edge 2) lower edge 3) right edge 4) left edge 5) long 6) short 7) large 8) small 9) vertical 0) horizontal

(3) 1) atomic weight 2) molecular weight 3) oxidation number 4) atomic number 5) mass number 6) valence

(6)(10) 1) active 2) inactive 3) typical 4) transition 5) metallic 6) coordination 7) covalent 8) hydrogen 9) intermolecular 0) ionic

(4)(7)(8)(8')(9) 1) 1 2) 2 3) 3 4) 4 5) 5 6) 6 7) 7 8) 8 9) 9 0) 0

Q2. The table given below is a part of the Periodic Table described in Q1a. The elements belonging to the first to third series are designated as A,B,...,R. Choose the set which contains only the gaseous elements at normal temperature and pressure from 1)-6) given below. (11)

Answers 1) CEHJL 2) DFHJL 3) AGIQR 4) EGILN 5) AILOR 6) BHJKR

| series \ group | 1A | 2A | | | | | | | | | | 3B | 4B | 5B | 6B | 7B | 0 |
|---|---|---|---|---|---|---|---|---|---|---|---|---|---|---|---|---|---|
| 1 | A | | | | | | | | | | | | | | | | B |
| 2 | C | D | | | | | | | | | | E | F | G | H | I | J |
| 3 | K | L | | | | | | | | | | M | N | O | P | Q | R |
| 4 | | | | | | | | | | | | | | | | | |

II. Answer Q1-Q3.

Q1. Choose one appropriate figure from 1)-0) given below for (12) and (13) in the following sentences.

The saturated vapor pressure of water increases, as is shown in the Figure, as the temperature rises. The saturated vapor pressure at point A is (12)mmHg.

The air in which the partial pressure of vapor is 17.0mmHg was kept in a 10L flask at

1 atm. and 25°C. When the temperature lowers to 0°C, a part of vapor condenses and the pressure of the gas in the flask decreases to (13)mmHg.

Answers 1) 470 2) 556 3) 681 4) 685 5) 700 6) 743 7) 760 8) 777 9) 845 0) 930

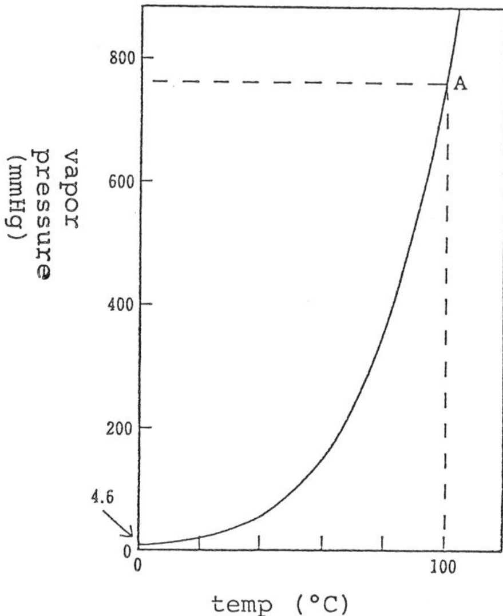


Q2. Choose one proper word for each of (14)-(19) from the answers. Answers may be used more than once.

The volume of n moles of gas is given by the following equation.

(volume) = n x (14) x (14') x (14")

This equation is derived from the equation of (15) of the gases. Accurate measurements, however, indicate that the above relation does not strictly hold for many gases such as oxygen or nitrogen. The reason is that it is impossible to neglect the (17) between (16) of gases, or the volume of (18) for (19) gas.

Answers (14)(14')(14") 1) molecular weight 2) reciprocal of molecular weight 3) pressure 4) reciprocal of pressure 5) the absolute temperature 6) reciprocal of absolute temperature 7) gas constant 8) reciprocal of gas constant 9) Avogadro's number 10) reciprocal of Avogadro's number

(15)(16)(17)(18)(19) 1) ideal 2) real 3) state 4) reaction 5) kinetic 6) attraction 7) pressure 8) electrons 9) molecules 0) gas constant

Q3. In order to determine the volume (at 20°C, 1 atm) of hydrogen evolved when magnesium is dissolved in hydrochloric acid, an apparatus shown in the Figure was employed. 0.12g of magnesium and 20mL of hydrochloric acid (1.00mol/L) are placed in A and B, respectively. Hydrochloric acid in B is then transferred to A, and the gas evolved was introduced into a graduated cylinder filled with water and set upside down above the surface of water, and its volume was determined. Comments (20)-(23) were raised concerning this experiment.

Mark 1) for proper comments while mark 2) for improper ones.

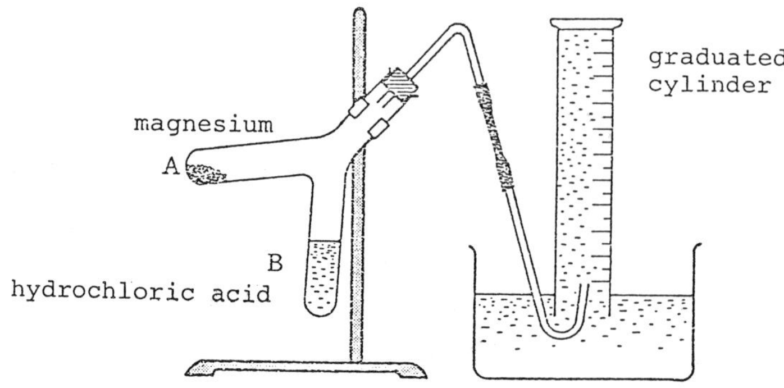


(20) The graduated cylinder is too small to determine the volume of gas evolved.

(21) Hydrochloric acid is not enough to dissolve all of the magnesium.

(22) Since hydrogen easily dissolves into water, the collection of hydrogen over water is an improper method.

(23) Since oxygen in air reacts with hydrogen, the air in the container should be replaced with hydrogen beforehand.

III. Answer Q1-Q3.

Q1. Choose one word from the answers for (24)-(27) in the following sentence.
When a small amount of saturated aqueous solution of iron(III)chloride to hot water heated nearly to the boiling point, iron(III)(25) is formed by (24) reaction, and the solution became transparent and exhibited a beautiful red color. If this solution was (26) in water by means of cellophane bag, (27) which is larger than the diameter of pore of cellophane sheet cannot pass through the membrane.

Answers (24) 1) oxidation 2) reduction 3) hydrolysis 4) neutralization 5) substitution
(25) 1) oxide 2) hydroxide 3) ionized 4) carbonate 5) metal
(26) 1) oxidized 2) reduced 3) subjected to salting out 4) evaporated 5) dialysed
(27) 1) chloride ion 2) colloidal particle 3) water molecule 4) hydroxide ion 5) dissolved gas

Q2. Find the order of the hydrogen ion concentration of the mixed solution obtained by mixing an equal amount of two solutions described in a-c. Choose the correct one from 1)-6).(28)

a 0.2mol/L hydrochloric acid and 0.1mol/L aqueous ammonia
b 0.1mol/L hydrochloric acid and 0.2 mol/L aqueous barium chloride
c 0.02mol/L hydrochloric acid and 0.2 mol/L acetic acid (the degree of electrolytic dissociation of acetic acid is assumed to be 0.01).

Answers 1) a>b>c 2) b>c>a 3) c>a>b 4) c>b>a 5) a>c>b 6) b>a>c

Q3. The following is the thermochemical equation for the oxidation of sulfur dioxide.

$$2SO_2 + O_2 = 2SO_3 + 46\text{kcal}$$

Answer the following questions (29)-(31) in connection with the above equation.

(29) Choose the incorrect one from the following descriptions 1)-4).
1) A decrease in the temperature of mixed gas leads to a forward reaction.
2) An increase of the pressure of mixed gas leads to a reverse reaction.
3) The reverse reaction is endothermic.
4) Addition of oxygen at constant volume and temperature leads to a forward reaction.

(30) Choose out of 1)-3) the atom whose oxidation number decreases by 2 when the reaction above proceeds forward.

Answers 1) S of $SO_2$ 2) O of $O_2$ 3) O of $SO_2$

(31) Suppose the heat of formation of $SO_2$ and $SO_3$ is $Q_1$ kcal/mol and $Q_2$ kcal/mol, respectively. Choose the one out of 1)-0) which is the closest value to that of $Q_1$-$Q_2$.

Answers 1) 11 2) 23 3) 34 4) 46 5) 92 6) -11 7) -23 8) -34 9) -46 0) -92

IV. Answer Q1 and Q2.

Q1. There is a mixture of a small amount of aluminum, copper, table salt, sodium carbonate naphthalene and cane sugar (all in powdered form). To the mixture in a test-tube, an equal amount of water and benzene is added. Then the test-tube is tightly stoppered, vigorously shaken, and left standing a while. After the benzene layer and aqueous layer are separated, the content of the test-tube is examined as described in a-d. Determine compounds A,B,C and D described in a-d. Choose one for each from 1)-0).

a The benzene layer is transferred by means of a dropping pipette to the evaporating dish, and benzene is evaporated. Only solid substance A(32) which has a strong odor and readily sublimes is left.
b The aqueous layer is transferred by means of a dropping pipette to an another test-tube to which dilute nitric acid is added. Gas B(33) is evolved.
c When aqueous silver nitrate solution is further added to the solution b, white precipitate is formed. This is formed from substance C(34) in the mixture.
d To the substance which remains at the bottom of the initial test-tube with added water and benzene, hydrochloric acid is added. One component is dissolved to evolve gas, but there remains a substance D(35) which is not dissolved even after sufficient hydrochloric acid was added.

Answers 1) aluminum 2) copper 3) table salt 4) sodium carbonate 5) naphthalene 6) cane sugar 7) hydrogen 8) oxygen 9) nitrogen monoxide 0) carbon dioxide

Q2. 200mL of 0.10mol/L aqueous copper sulfate was electrolysed using a platinum anode and cathode. Calculate the concentrations of copper sulfate and sulfuric acid in the solution after 1930 C(0.02F) of electricity was passed. Choose the correct one from 1)-8). (36)

| Concentration(mol/L) | 1) | 2) | 3) | 4) | 5) | 6) | 7) | 8) |
|---|---|---|---|---|---|---|---|---|
| copper sulfate | 0 | 0 | 0.01 | 0.01 | 0.05 | 0.05 | 0.08 | 0.10 |
| sulfuric acid | 0.01 | 0.02 | 0.02 | 0.05 | 0.05 | 0.10 | 0.02 | 0 |

# Introduction of in-service training for chemistry teachers of senior high schools in Tokyo, Japan

Kazuo Suzuki and Toshiaki Hasegawa*
Tokyo Metropolitan Institute for Educational Research and In-Service Training;
1-14, Meguro 1-chome, Meguro-ku, Tokyo, Japan.
*Shimura High School;
41-10, Nishidai 1-chome, Itabashi-ku, Tokyo, Japan.

1 The goals of in-service training in 1983 、1984

In-service training in the institute aims to develop chemistry teachers' abilities. In particular 、 it aims to contribute to the guidance of making students understand the principles and laws of chemistry by carrying out observations and experiments 、 thereby developing students' abilities as well as positive attitudes toward the fundamentals of chemistry.

2 The number of participants

The number of in-service training participants in 1983 amounted to 30 and it was equivalent to 75% of the maximum capacity 、 40.

The number of participants with in-service training in 1984 amounted to 28 and it was equivalent to 70% of the maximum capacity、 40.

3 The records of participants' impressions in 1983、 1984.

1983

- I thought that the contents and methods were suitably distributed in this in-service training.
- I hope that newly developed experiments will be introduced.

1984

- Because we saw the factory of the Plastics Recycle Center 、 I came to understand the intricacies of making products.

In-Service Training for Chemistry Teachers of Senior High Schools in Tokyo — 1983、1984 -
1983

| Inning、Date、Time 、 Contents(Methods),Lecturers |
|---|
| 1,2 7/22 Fri. 9:00 ～ 16:00<br>・Recent Subjects at Chemical Education Sakuramachi High School Yûji Yamazaki<br>(Lecture, Conference)<br>・Chemical Properties of Substances — Organic Compounds —<br>(Experiments, Conference) Koishikawa High School Mitsugi Anraku<br>Experiments<br>(Substitutions of Hydrocarbons)<br>cyclohexane + $Br_2$ $\xrightarrow{h\nu}$ , benzene + $Br_2$ $\xrightarrow{h\nu}$ , toluene ($CH_3$) + $Br_2$ $\xrightarrow{h\nu}$<br>(Detections of Functional Groups) R-OH, R-CO-R', R-CHO etc.<br>(Make of Mirrors)<br>(Oxidations of Toluene)<br>toluene ($CH_3$) + $MnO_2$ → benzaldehyde ($CHO$) etc. |
| 3 7/25 Mon. 13:30 ～ 16:30<br>・Visit to a Chemical Factory - Sodium Factor - Asahi Glass Corporation Chiba Factory<br>(Conference) Asahi Glass Corporation Chiba Factory Chief Masao Yokomizo |
| 4 7/26 Tue. 13:00 ～ 16:00<br>・Chemical Reactions -Rates of Reactions- Tôkyô Gakugei University Takeo Kawaguchi<br>(Experiments, Conference) Mita High School Seiji Kurokui<br>Experiments<br>( Rates of Hydrolysis)<br>$CH_3COOC_2H_5 + H_2O \xrightarrow{\text{(Neutralization)}} CH_3COOH + C_2H_5OH$ etc.<br>Ethyl acetate |
| 5 10/6 Tue. 13:15 ～ 16:15<br>・In-Service Training Class Research-Teaching for Making the most of students' abilities<br>(Conference) Mukogaoka High School Kaoru Idei, Hideo Yoshida<br>Relations among Rates of Reactions, Concentrations, Temperature, Catalyzer<br>$2H_2O_2 \xrightarrow{(MnO_2)} 2H_2O + O_2$ |

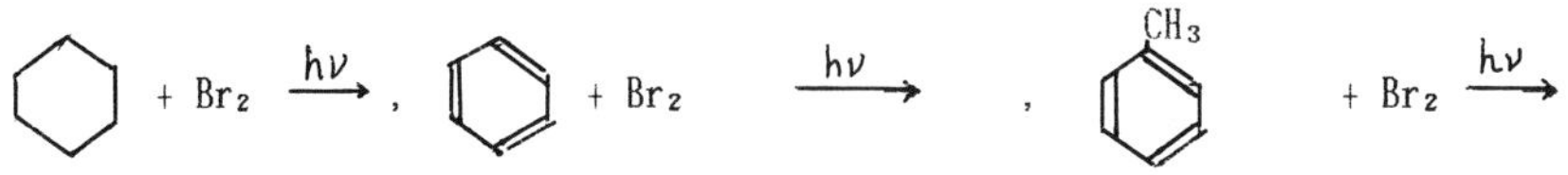

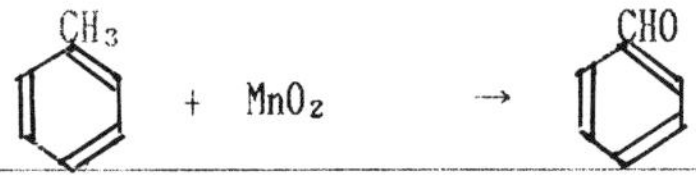

1984

| Inning、Date、Time 、 Contents(Methods),Lecturers |
|---|
| 1,2 7/31 Tue. 9:00 ～ 16:00<br>・Chemical Reactions — Reactions between Acids and Bases,Syntheses of Polymers —<br>(Lecture,Experiments,Conference) The University of Tôkyô Kiyoshi Mutai<br>High School attached to The University of Tsukuba Yoshio Yoshida<br>Nishi High School Yukyu Nakashita<br>Lecture<br>(Reactions between Acids and Bases,Definitions of Acids and Bases,Strength of Acids and Bases, Acids and Bases in Organic Compounds—etc.)<br>Experiments<br>(Syntheses of Viscose Rayon and Nylon—etc.) |
| 3 8/1 Wed. 13:00 ～ 16:00<br>・Relation between Science I and Chemistry The University of Tôkyô Michinori Ôki<br>(Lecture,Conference) |
| 4 8/2 Thu. 9:30 ～ 17:00<br>・Industry and Chemistry — Visit to The Nodono Plastics Recycle Center —<br>(Conference) The Nodono Plastics Recycle Center Chief Saburô Nishimura |
| 5 9/18 Tue. 13:15 ～ 16:15<br>・Teaching for Developing Students' Abilities of Research<br>(Studies of Teaching, Conference) Chitose High School Yuji Yamazaki,Seisuke Hara<br>Cycle of Copper Reactions<br>$Cu \xrightarrow{H_2O_2} CuO \xrightarrow{H_2SO_4} CuSO_4 \xrightarrow{NaOH} Cu(OH)_2 \xrightarrow{NH_3} [Cu(NH_3)_4]^{2+} \xrightarrow{H_2SO_4} CuSO_4$<br>$\xrightarrow{Fe(Steel\ Wool)} Cu$<br>$(Fe + Cu^{2+} \rightarrow Fe^{2+} + Cu)$ |

# Science teaching in Canadian schools and chemical education

Brian T. Newbold

Department de chimie et biochimie, Universite de Moncton, Moncton, N.B., Canada.

Abstract - This paper describes some of the findings of a science education study recently carried out by the Science Council of Canada and mentions certain of its recommendations aimed at bringing about renewal of science education in the schools without delay. A number of implications for chemical education are identified.

## INTRODUCTION

Science education in Canada has received considerable attention and especially over the last few years during which the Science Council of Canada carried out a detailed study of science education in both elementary and secondary schools across Canada. Education in the schools in Canada is the responsibility of the provinces and it is important to note that the four-year science education study was conducted in cooperation with the Council of Ministers of Education(each provincial minister of education being a member of that organization), the Federal Government, and the Science teaching profession.

## SCIENCE EDUCATION REPORT

Following the study, the Science Council produced Report 36 (ref. 1) entitled "Science for Every Student : Educating Canadians for Tomorrow's World", which concluded that immediate renewal in Canadian science education is essential, and made a series of recommendations aimed at bringing about this renewal.

The Report states that Canadians in school today "will face a world made daily more complex by rapid scientific and technological developments" and that "Canadians must be literate not only in the traditional basics of language and mathematics, but also in the new basics of contemporary society : science and technology". The science education study was embarked upon in order to answer questions such as : "How well is Canada's educational system equipped to meet the need for scientific literacy for all ? Do students receive enough science education ? - is it appropriate to individual needs ?, Are some groups - girls, for instance, - neglected ? What science should students be taught and how should it be taught to them ? and what indeed are the aims of science education ?".

Before describing some of the findings of the Report, it should be mentioned that the science education study had three phases, the first being the identification of issues via critical reviews of contemporary science education by well-informed persons. The second part, research phase, consisted of four major projects, namely : "an analysis of science curriculum policies from all provinces and territories; an analysis of 33 commonly used science textbooks; a survey of more than 4000 science teachers; and 8 case studies of science teaching in schools in all parts of Canada". This research program gave rise to a considerable database on Canadian science education in schools which will be useful for future reference and the results have been published in three volumes (refs. 2-4).

The third, and crucial phase, was a series of 11 deliberative conferences for discussion of future orientations for science education in each province or territory. These meetings brought together more than 300 ministry officials, school administrators, school teachers, university professors, trade union representatives, employers, students, and parents; and provided the Science Council with advice and suggestions regarding what should be done to renew

Science education in Canadian schools. I attended the New Brunswick conference which lasted a full day and assembled 30 individuals from all of the interested sectors who debated the science education needs in the English- and French-language schools of the Province, and made some constructive proposals. All three phases of the study provided the Science Council with essential input for the production of its report. Later on, further meetings of the same type were organized to discuss the implementation of the recommendations of the Report.

The study consultations showed that there was general agreement that science education can be of benefit to all students. However, the requirement was for a science education suitable for individual needs and designed to let the students : "participate fully in a technological society as informed citizens ; pursue further studies in science and technology; enter the world of work; and develop intellectually and morally". Report 36 considers Science for the informed citizen to be very important and underlines the science-technology-society connection. It says that "ideally, science education is a preparation and encouragement for students to learn about science throughout their lives". Under the heading 'Science for the world of work', the Report says that science education must not be limited to a presentation of information but should provide students with real problems that can be solved by processing information in a creative fashion. It also states that science education can contribute to the development of rationality and the ability to think in a critical way.

The study examined the 'Intended curriculum'(prescribed by ministries of education), the 'Planned curriculum'(developed by School Boards and individual teachers), and the 'Taught curriculum'(experienced by students in the classroom). It was found that curriculum resources and evaluation methods linked to objectives concerning the relation of science and technology to society and the world of work are often unavailable; and that School Boards and teachers choose the objectives to be emphasized in the classroom. In general, in the early and middle school years teachers stress development of scientific skills and attitudes, and in senior years the learning of science content. Technology, the social context of science or the history of science in Canada are not systematically dealt with in science courses.

Regarding the 'Taught curriculum', it was found that most textbooks emphasize only the learning of science content and acquisition of scientific skills; and that practical work prescribed by texts is highly structured and there are few opportunities for class discussion. The case studies of science teaching revealed a number of problems : there are significant differences between ministry guidelines and classroom practice; science in the early years is often taught without adequate facilities; and only rarely is a timetable period set aside for science at that level. The study also showed that most science teachers in elementary schools are inadequately prepared (75% not having taken a science course since high school); in-service education opportunities for most science teachers are nonexistent or of little value; many girls drop sciences as soon as possible; and students interested in science or particularly high achievers complain of lack of challenge in their courses. On the other hand, "many examples of dedicated, innovative, and successful science teaching, curriculum development and teacher education" were found in every region of Canada.

Based on the study findings, the Science Council proposed eight ways in which renewal of science education in Canada can be initiated : 1) Elementary schools must provide science education for all students; 2) Girls must be encouraged to continue with science throughout their schooling; 3) High achievers and science enthusiasts must receive greater challenge; 4) Science education must provide a more accurate view of the practice, uses and limitations of science; 5) Science education must include study of how science, technology and society interact; 6) Students must be taught how Canadians have contributed to science and how science has affected Canadian society; 7) Technology courses must be included in the secondary school curriculum; 8) Teachers and curriculum planners must evaluate students' progress towards all the goals of science education, not just their learning of science content. The Council pointed out that the responsibility for education lies with the ministers of education and that any process of renewal must be sanctioned and encouraged by them. Besides proposing the eight initiatives, the Science Council made 47 specific suggestions(strategies for implementation) aimed at facilitating the necessary renewal. The limited space available here does not allow the citing of each suggestion, but mention must be made of some of them.

Under the heading of 'Curriculum leadership', the Science Council suggests the following actions to the Ministries of Education : increasing the amount of time devoted to science in elementary school to 15%(i.e. 45 minutes per day); developing technology courses for secondary schools; requiring all students to take science until grade 11. Among proposals aimed at School Boards and schools, the Council recommends the setting-up, where numbers warrant it, of high schools of science and technology. The Science Council recognizes that the number and quality of science teachers are of crucial importance to science education and recommends the following with regard to 'Human resources' : subject-specific teaching certificates; school-focussed inservice education programs; improved planning of pre-service teacher education by universities; and special summer institutes to upgrade elementary school teachers.

With regard to 'Instructional resources', Report 36 includes the following recommendations : the setting up of a Canadian foundation for science and technology education to help develop new curriculum materials; and the creation of a centre for research and development in computer-aided learning. Since numerous organizations and groups outside the school can support the work of science teachers the Council included some suggestions concerning 'External resources': such as the establishment of a federally sponsored national information program stressing the need for young women to participate in science and technology; and the creation of a program of awards for excellence in science education to be offered by the Royal Society of Canada. Under the heading 'Research resources', the Science Council also made proposals because some of its recommendations require further research, and one of these was : the setting up of an interprovincial network of researchers who study methods of teaching about science, technology and society.

The Science Council believes that the recommendations in Report 36 are "concrete, practical and immediately applicable" and that their "implementation is essential if science education is to succeed in preparing Canadian students for tomorrow's world". It estimates that the total cost (over 5 years) of implementing its proposals will be about 155 million dollars (0.154% of the total cost of education) which works out at a total cost per student per year of only ¥6.28. Copies of Report 36 and the Background Study may be obtained from the Canadian Government Publishing Centre (ref. 5).

## IMPLICATIONS FOR CHEMICAL EDUCATION

I have no doubt that if the recommendations in Report 36 are implemented across Canada without too much delay there will be a considerable positive impact on science education in Canadian schools, and this includes the teaching and learning of chemistry. Increasing the amount of time spent on science at the elementary school level to 15% would give pupils a better preparation to proceed later on to further studies in the sciences (e.g. chemistry). The recommendations councerning science teaching and teacher education are particularly important since the teacher is the key component that determines the quality of the system. The measures suggested would improve the qualifications of the science teachers and this would give them increased confidence and more enthusiasm for their subjects (chemistry, physics and biology). This in turn would favour student motivation. Better school chemical education would result from the recommended intiatives and this would be beneficial for the fostering of future chemists of excellence (one of the four themes of the Eighth International Conference on Chemical Education).

## ACKNOWLEDGEMENT

The author thanks the Science Council of Canada for permission to reproduce material from Repot 36.

## REFERENCES

1. Report 36, Sci. Council Can., 85 pp. (1984).
2. G.W. F. Orpwood and J.-P. Souque, background Study 52, Sci Council can., I, 227 pp. (1984).
3. G.W.F. Orpwood and I. Alam, Ibid., II, 122 pp. (1984)
4. J. Olson and T. russell, Ibid., III, 297 pp. (1984).
5. Canadian government Publishing Centre, Supply and Service Canada, Hull, Quebec, KLA 0S, Canada.

# Chemical education and industry in Nigeria

Israel T. Eshiet

Dean, Faculty of Education (Chemical Education) University of Cross River State, Uyo, Nigeria

Abstract - This paper examines the educational system in Nigeria and highlights the place of Chemistry in the school's curriculum and in the overall educational framework of the country. The transition rate from school/university to industry however is not stated but the various Chemistry-related Industries have been listed covering chemicals/chemical products manufactured. Comparison is made between imported and locally manufactured chemicals/chemical products and it is concluded that for the country to be self-reliant, local sources of raw materials should be explored and developed. The current efforts of the Federal and State governments in this direction are mentioned and appreciated.

Among the Natural Sciences of Biology, Physics and Chemistry, the study of Chemistry has increasingly developed and has assumed a unique place in the scientific world partly because of its very close relevance to industry especially the Chemical Industry and largely because of its specific role in Chemical Engineering. Chemistry is thus the only subject among the natural sciences whose name is directly linked to industry. In the Chemical Industry, chemicals are manufactured and from these chemicals, chemical products including drugs, fertilizers, dyes, synthetic fibres and several others are manufactured.

Chemicals and chemical products are commonly grouped into three categories (ref. 1) - basic chemicals such as acids, alkalis, salts, organic chemicals; special chemicals used for the manufacture of chemical products such as pigments, plastics, synthetic fibres; and finished chemical products which include drugs, soaps/detergents, cosmetics, fertilizers, paints, etc. The unique place of Chemistry in Industry becomes therefore more revealing as the manufacture of basic chemicals, as an industry, can be sustained by another industry which manufactures allied chemical products such as synthetic fibres, paints, etc. or finished chemical products such as drugs, etc. The place of Chemistry in industry is even more pronounced in the area of petro-chemicals, refining of crude petroleum, manufacture of synthetic rubber and plastics, glass fibres, paper and pulp.

## CHEMICAL EDUCATION IN NIGERIA

Education in chemistry in Nigeria commences pronouncely at the level of post-secondary schools mainly in the Senior Secondary Schools and Technical Colleges which are two of the three branches of the Post-secondary Education in the Revised National Policy on Education. In this new system, the Junior Secondary School (JSS) offers Integrated Science (ages 13 - 15). In the 1982/83 academic year, 956,918 (ref. 2) students (ages 12 - 13) enrolled in Post-primary Institutions in the country representing 0.99% of a population of 96,127,700 (1985) (ref. 3) inhabitants.

Although the four basic elements for Teaching/Learning process namely - students, curriculum, teachers and facilities are much recognised, there is an increasing awareness of the shortfalls in two areas of Teachers and Facilities. In 1983/84 academic year, there were 47,000 serving teachers in Forms 1 - 3 of the Secondary Schools in Nigeria and by this current academic year, (1984/85) about 125,000 (ref. 4) serving teachers would be needed based on a 70% transition rate from Primary to Secondary Schools. It is estimated that, inspite of the efforts to solve teacher shortage, by 1986/87 and 1987/88, the country would be short by 14,247 and 33,123 (ref. 5) teachers respectively in Secondary Schools. These shortfalls notwithstanding, academic programmes continue to develop from year to year with an obvious bias towards science and technology. The overall educational Structure in Nigeria follows the following framework (ref. 6) - Fig. 1.

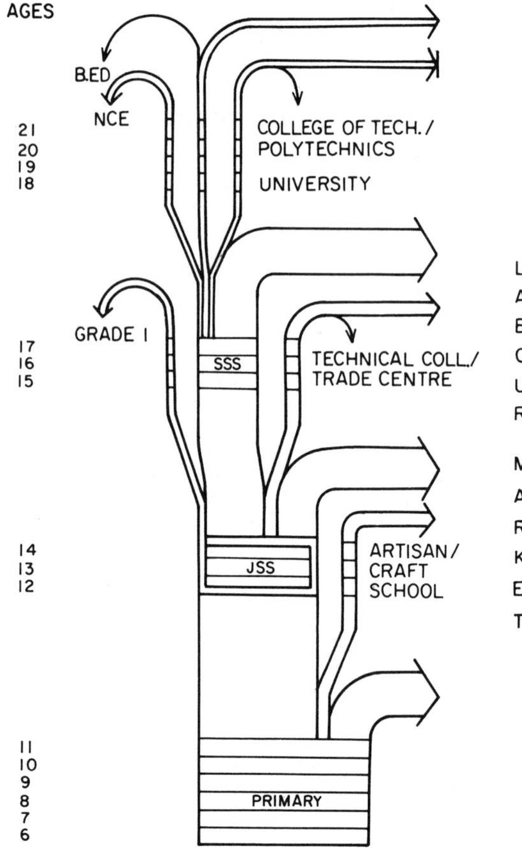


Fig. 1. Educational Structure in Nigeria.

Source: Federal Ministry of Education, Lagos, Nigeria.

JSS = Junior Secondary School
SSS = Senior Secondary School
GRADE I = Grade I Teachers' Cert.
NCE = Nigeria Certificate of Education
B.ED. = Bachelor of Education

Currently, there are 24 Universities in the country all of which offer programmes in all three basic science areas of Chemistry, Physics, and Biology including Mathematics. In addition to these basic Science subjects, programmes are mounted in some of these universities in the areas of Agricultural Science, Engineering, Environmental Science, Medicine, Industrial Chemistry, Pharmacy, Veterinary Medicine, Applied Earth/Mineral Science and Agricultural Technology.

Table 1 shows the extent of Science/Technical Programme in the universities in terms of percentage of such programmes with reference to the number of universities offering each programme (24 = 100%).

| Chemistry | Medicine | Pharmacy | Industrial Chemistry | Vet. Medicine | Applied Earth & Mineral Science | Agric. | Engineering | Environmental Sciences |
|---|---|---|---|---|---|---|---|---|
| 100 | 58.3 | 29.2 | 50.0 | 16.6 | 16.6 | 58.3 | 58.3 | 45.8 |

TABLE 1. Science/Technological Programmes in Universities.

Nearly all the universities are owned by the Federal and State Governments each university having defined objectives as spelt out in the Instrument establishing it. In general, every university in the country has objectives which fall in line with the overall national educational objectives.

According to Aliyu (1985) (Ref. 7),

> Universities should see themselves as real instruments for the actualisation of national objectives and goals. In the third world in general, private universities have not developed in good enough numbers to make any appreciable impact. Most universities are public institutions developed and maintained by Governments as the apex of educational systems and for the purpose of contribution to the positive development of their communities.

## CHEMISTRY AND INDUSTRIES IN NIGERIA

The transition rate in terms of graduates from the University to Industry in the areas of Chemical Manufacture, Chemical Engineering and Industrial Chemistry has not been studied. Records however show that from 1975 - 1978, the following chemistry-related industries have been established in the country (Table 2). The table also shows the number of Industries by year of each item together with the yearly gross output in Naira, (the country's currency).

TABLE 2.

| | 1975 | | 1976 | | 1977 | | 1978 | |
|---|---|---|---|---|---|---|---|---|
| | No. of Indus-ries | Gross Output | No. of Indus-ries | Gross Output | No. of Indus-ries | Gross Output | No. of Indus-ries | Gross Output |
| 1. Basic Industrial Chemicals Fertilizers, and Pesticides | 5 | 15,680 | 8 | 24,160 | 6 | 13,809 | 11 | 46,105 |
| 2. Paints, vanish & lacquers | 6 | 29,263 | | | 6 | 35,802 | 6 | 57,802 |
| 3. Drugs and Medi-cine | 14 | 35,232 | 16 | 41,309 | 17 | 71,673 | 12 | 90,294 |
| 4. Soaps, perfumes comestics, and cleansing agents | 16 | 139,126 | 17 | 225,008 | 19 | 294,675 | 19 | 379,676 |
| 5. Coal & petroleum products & other chemical products | 15* | 151,617 | 3 | 319,721 | 3 | 210,750 | 10 | 312,049 |
| 6. Tyres and Tubes | 7 | 26,780 | 9 | 41,702 | 9 | 46,477 | 3 | 86,737 |
| 7. Plastic products | 33 | 54,902 | 31 | 81,848 | 36 | 100,660 | 38 | 116,611 |
| 8. Cement | 5 | 14,017 | 5 | 54,373 | 8 | 58,102 | 10 | 59,587 |
| 9. Sugar Refinery | 5 | 113,980 | 5 | 64,563 | 4 | 48,776 | 3 | 64,196 |
| 10. Chocolate, Cocoa & Sugar Confec-tionery | 12 | 65,160 | 11 | 69,545 | 13 | 132,579 | 12 | 97,798 |
| 11. Fruit Canning & Preserving | 4 | 5,948 | 3 | 5,937 | 4 | 6,036 | 3 | 781 |
| 12. Animal Feeds | 4 | 1,861 | 6 | 6,692 | 6 | 6,693 | 5 | 149,239 |
| 13. Spirit distillery and Beer | 8 | 161,044 | 10 | 194,615 | 9 | 217,098 | 8 | 345,218 |
| 14. Soft drinks | 9 | 60,382** | 5 | 40,439 | 8 | 68,105 | 7 | 102,545 |

| | 1975 | | 1976 | | 1977 | | 1978 | |
|---|---|---|---|---|---|---|---|---|
| | No. of Indus-ries | Gross Output | No. of Indus-ries | Gross Output | No. of Indus-ries | Gross Output | No. of Indus-ries | Gross Output |
| 15. Tanneries and Leather | 9 | 10,074 | 11 | 18,099 | 13 | 18,830 | 10 | 24,506 |
| 16. Glass Products | 10 | 6,273 | 13 | 45,192 | 14 | 28,058 | 9 | 23,394 |
| 17. Primary Iron and Steel | 6 | 29,007 | 4 | 5,751 | 4 | 8,848 | 9 | 37,430 |
| 18. Non-ferrous Metals | *** | *** | 6 | 52,189 | 7 | 58,238 | 6 | 90,451 |
| | | 920,346 | | 1312,043 | | 1425,209 | | 2094,419 |

* Combined with other chemical products
** Combined with Tobacco
*** Combined with Iron and Steel.

Source: Industrial Survey of Nigeria (1975 - 1978)
Federal Office of Statistics, Lagos - Nigeria.
Data for 1979 - 1984 not yet available.

Comparison of the gross output over the years (1975 - 1978) without taking into account the effect of inflation, is as shown in Fig. 2 which represents 32.7% average growth rate over the years.

Fig. 2. Gross Output (Naira) on Chemicals/Chemical Products

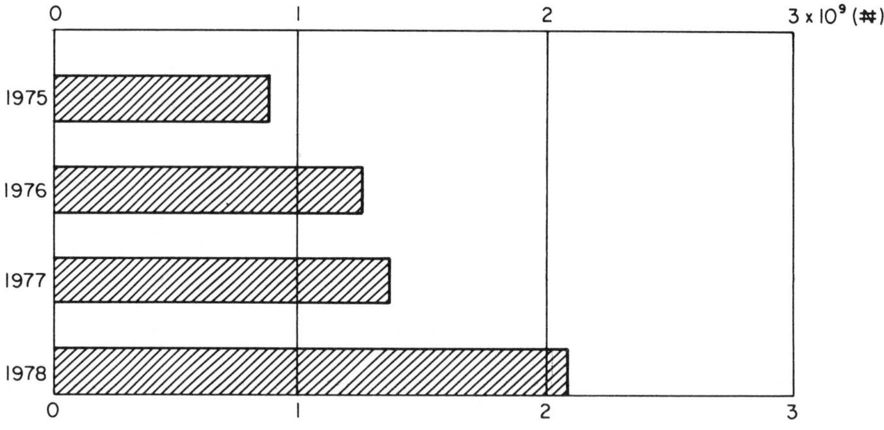


Between 1978 and 1981, the following chemicals/chemical products were imported into Nigeria - Chemical Elements and compounds, Mineral tot and crude chemicals from coal, petroleum and natural gas, Dyeing, Tanning and Colouring materials, Medical and Pharmaceutical products, Essential oils and Perfume Materials - Toilet, polishing and cleaning preparations, Fertilizers Manufactured, Explosives and Pyrotechnic Products, Plastic Materials, Regenerated Cellulose and artificial Resins, Chemical Materials and Products. Table 3 shows the growth of import (in Naira) on these items over these years.

TABLE 3.

(₦ x 1000)

| | 1978 | 1979 | 1980 | 1981 |
|---|---|---|---|---|
| Chemical Products | 640,200 | 647,030 | 881,040 | 1,220,402 |

**Source:** Nigeria Trade Summary, Federal Office of Statistics, Lagos.

THE CURRENT MOOD OF CHEMICAL INDUSTRIES

Partly because of the need for self-reliance and mostly because of the current import restrictions on many items including certain chemicals and chemical products, new industries are springing up while existing ones are expanding. Research is going on into the suitability of locally available raw materials which could replace imported ones for the manufacture of most items. Research is also going on for the manufacture of beer, wine and beverages from our local cereals, production and preservation of food products, and production of dyestuffs. Liquefaction of natural gas, and establishment of petro-chemical industries are on the priority list of government investment ventures.

In conclusion, at present, Nigeria cannot boast of any appreciable level of success in Chemical Industries when compared to developed countries. It is however hoped that with existing import restrictions and current interest in promoting private investments and development of raw material based on local resources, the situation will change for the better.

REFERENCES

1. Bailey, F. E. (1974) - Economic Aspects of the Chemical Industry. Rieget's Handbook of Industrial Chemistry (7th Edition) - Van Nostrand Reinhold Co.
2. Educational Planning (2), (1985), Federal Ministry of Education, Science and Technology, p. 9.
3. National Population Bureau, Lagos, Nigeria.
4. Educational Planning - Special Issue: Federal Ministry of Education, Science & Technology (1984), p. 3.
5. Op. Lit. - p. 19.
6. Op. Lit. - (Cover)
7. NUC Report - Bulletin of the National Universities Commission Lagos, Nigeria, (March 1985), p. 2.

# Some trends in hydrocarbon-producing crops in Rwanda

Pierre C. KARENZI and Joseph MUNGARULLIRA*

Natural Products Laboratory, CRAFOP, National University of Rwanda, B.P. 56, Butare, Rwanda and National University of Rwanda, P.O.B. 117 Burare, Rwanda*

Abstract - Many species have been identified throughout the world as potential hydrocarbon crops. In Rwanda we have recorded and collected latex-bearing local species and tested some of them for the health value of their extracts. This work now under progress is promising especially in providing a basis for chemical education based on local ressources.

## INTRODUCTION

Much work has been carried out so far on hydrocarbon - producing crops by M. Calvin *et al.* (ref. 1 - 2), Buchanan et al. (ref.3) and many others (ref.4-5). However, more research work is needed on most promising plant species and on the chemistry and biochemistry of reduced solvent extracts from available species.

In Rwanda we have started such work on latex-bearing species from the country. A thorough review of the literature on rwandeese botanical species has been made (ref. 6). Some of them have been collected and grown. Solvent extraction has been made and the extracts tested for heat value. The rubber content of latex has been evaluated for two species. More chemical analysis is underway.

## RWANDEESE LATEX SPECIES

Fourteen families have been reported in Rwanda as having latex-bearing species. The Moraceae are the most numerous with 42 species, most of which are various Ficus. Next are Asclepiadaceae with 28 species mainly found in eastern dry area of the country. Then come Euphorbiaceae with 21 species among which are very popular ones like *Euphorbia tirucalli*; *Synadenium grantii*. Other latex plants are found among Apocinaceae (14), Lobeliaceae (12), Asteraceae (9) and some others. Most of these species are native but some (14%) are ornamental and have been recently introduced in the country (ref.6). Among the most recent is *Euphorbia crotonifolia* which grows very well in the country and whose leaves are rich in reduced compounds.

## HEAT VALUES OF SOME HEPTANE-METHYL ALCOHOL EXTRACTS

A thorough screening on rwandeese species for the heat value of their solvent extracts has not yet been done. However some of the most common ones have been tested. Table 1 give heptane and methyl alcohol extracts in percent of dry weight of the whole plant's part (bark, stems, leaves or roots) while Table 2 shows the heat value of these solvent extracts.

## SOME FUTHER WORK

It is worthwhile to look for how the extract-yield and composition are dependant on the age and growing conditions of the plant. Such work is under progress. The physicochemical characterization of rubber from *Synadenium grantii* latex is being carried out as well as thorough chemical analysis of latex and whole plant composition. Some preliminary experiments have been made in cracking heptane extracts of *Euphorbia tirucalli* into hydrocarbons. Last, the biosynthesis pathway of reduced compounds and the solar energy conversion of these species are also under study.

TABLE 1. Heptane and Methyl Alcohol (MeOH) Extracts

| Family | Species | Part | Label | Heptane Extract % d.w. | MeOH Extract % d.w. |
|---|---|---|---|---|---|
| Euphorbiaceae | Euphorbia tirucalli | Bark | E.T.B. | 9.7 | 23.1 |
| | Synadenium grantii | Stems | S.G.S. | 10.2 | 16.5 |
| | | Leaves | S.G.L. | 8.8 | 4.3 |
| | Euphorbia crotonifolia | Stems | E.C.S. | 3.0 | 10.4 |
| | | Leaves | E.C.L. | 10.0 | 16.6 |
| Apocynaceae | Thevetia neriifolia | Stems | T.N.S. | 2.0 | 17.7 |
| | | Leaves | T.N.L. | 6.2 | |
| Asclepiadaceae | Sarcostema viminale | Stems | S.V.S. | 4.9 | 9.5 |
| | | Roots | S.V.R. | 4.2 | 5.8 |

TABLE 2. Heat values and C,H,O ratios of Heptane and Methyl Alcohol Extracts

| Sample | Heptane Extract | | MeOH Extract | |
|---|---|---|---|---|
| | Heat value MJ/kg | Formula | Heat value MJ/kg | Formula |
| E.T.B. | 40.97 | $CH_{1.72}O_{0.07}$ | 15.53 | $CH_{2.04}O_{0.85}$ |
| S.G.S. | 38.80 | $CH_{1.67}O_{0.10}$ | 16.09 | $CH_{1.57}O_{0.74}$ |
| S.G.L. | 39.85 | $CH_{1.72}O_{0.09}$ | 20.19 | $CH_{1.72}O_{0.76}$ |
| E.C.S. | 39.92 | $CH_{1.72}O_{0.09}$ | 15.25 | $CH_{2.11}O_{0.61}$ |
| E.C.L. | 41.57 | $CH_{1.73}O_{0.06}$ | 17.33 | $CH_{1.81}O_{0.65}$ |
| T.N.S. | 38.59 | $CH_{1.70}O_{0.07}$ | 14.58 | $CH_{2.16}O_{0.91}$ |
| T.N.L. | 39.68 | $CH_{1.69}O_{0.08}$ | 15.02 | $CH_{2.23}O_{0.99}$ |
| S.V.S. | 39.15 | $CH_{1.72}O_{0.09}$ | 15.63 | $CH_{1.92}O_{0.78}$ |
| S.V.R. | 39.68 | $CH_{1.42}O_{0.07}$ | 18.42 | $CH_{1.65}O_{0.64}$ |

ACKNOWLEDGMENT

Part of this work has been supported by the U.S.Council for International Exchange of Scholars and U.S.Department of Energy under the guidance of Professor Melvin Calvin. The work under development is supported by the National University of Rwanda, Applied Research and Continuing Education Center (CRAFOP).

REFERENCES

1. M. Calvin, Science, 184, 375-381 (1974); En.Res., 1, 299-327 (1977); Pure and a Appl.Chem.,50, 957-960 (1981).
2. E.K. Nemethy, J. Otvos, M. Calvin, Pure and Appl. Chem. 53, 1101-1107 (1981).
3. R.A. Buchanan , IM. Cull, F.A. Otey and C.R. Russel, Econ. Botany, 32, 131, 146 (1978); J. Am. Oil Chem. Soc. 55, 657, 662 (1978).
4. P.C. Karenzi, in Alternative Energy, vol. 4, p. 1-5, Nejat Veziroglu, ed., Ann Arbor Science Publishers, Ann Arbor (1982).
5. G.A. Stewart, J.S. Hawker, H.A. Nix, W.H.M. Rawlins and L.R. Williams, The potential for production of "hydrocarbon" fuel from crops in Australia, Australian Government Publishing Service (1981).
6. F. Ayobangira, F.X. Habiyaremye and P.C. Karenzi, Bulletin Agricole du Rwanda, 18, July, 167-178 (1985).

# University leadership in scientific and technological education for development in developing countries

Brahim Elouadi

Faculty of Science, University Mohammed V Charia Ibn Batota Rabat Morocco

Abstract - It is suggested that a leading role for Universities and research Institutions in the transfer of technology, should be vigorously encouraged by Developing Countries. Greater attention should be given to the role these vital institutions can play as the most appropriate, if not unique locomotives for their social welfare progress and national technological development.

## INTRODUCTION

Developing Countries in 1985 face a common dilemma—they all import basic technology, the indispensable tool for the development of their fragile economies. Nevertheless, these countries have not all reached the same level of development in the cultural, scientific and technological fields: some have already acquired nuclear power technology, while others are still struggling with illiteracy.

Scientific and technological education in the fields of agriculture, health, natural resource exploitation, etc. is the key element in facing problems of development in these countries.

## PERSPECTIVES

Although the technological background is not uniform throughout all these nations, many of them have succeeded in creating indigenous qualified personnel (engineers, scientists, physicians, economists, ...) in such large numbers that unemployment is beginning to be a serious problem for the great majority of the countries. It may take more than four years for a newly graduated student to find a job in Egypt for example, where the Government is actively seeking a new way of utilizing all university graduates (ref. 1). Yet it becomes urgent everywhere to find immediate solutions for this grave problem which represents the greatest danger for the whole strategy of technology transfer, even in industrialized countries (ref. 2 and 3).

It should be pointed out that the situation of the world economy is less favorable now than it was in recent decades and the design of any strategy should avoid any direct confrontation with the strongest international economical forces; and it is thus the immediate responsibility of all indigenous qualified personnel to offer solutions and to participate actively in the battle against under-development.

## THE LEADING ROLE OF UNIVERSITIES AND RESEARCH INSTITUTIONS

It is now obvious to everyone, that great changes are occurring very rapidly in various scientific, engineering and technological fields and the amount of technical information available is increasing every day. Any strategy for technology transfer must start by getting and selecting the appropriate information needed. Since the continuously increasing amount of information may be a promise as well as a burden, the countries aspiring for technology transfer must put in place institutions which will be in charge of the selection, appropriation and integration of technical information in order to integrate it into the already existing or embryonic industries. The best institutions to be elected for this fundamental step already exist and these are the universities. Universities should be regarded by developing countries as their best if not unique locomotive for a real technology transfer and social progress. By their activities of higher education and scientific research these institutions are called to seek the appropriate solutions for all problems concerning development in the country: technology, economy, sociology, culture, .... This role has already been proved to be a very successful one in all developed countries where the close interaction between university-industry and more generally between university and society, has allowed a substantial technological and social progress. The contribution of university to the technological and social improvements have been more important in countries where this interaction is the highest, like the USA, Germany .... In France where there is evidence of substantial communication between academic and industrial personnel, advanced technology must be transferred through normal commercial contacts (ref. 3). In should however, be noticed that the French higher educational system is one of the most theoretical and rigid systems existing in industrialised countries, and there is a real need to make it more flexible in order to meet the requirements of rapid change in science and technology. All the countries (France included), actually adopting this system, should seriously think about its readaptation in order to strengthen the relationship between universities and all active forces in the society: industry, financial institutions decision makers, etc. The close relationship will certainly give universities and research institutions a major role in being the motor of technological and social progress, instead of remaining, like nowadays, just ivory towers of knowledge, theoretically oriented and without any evident practical uses for the society. The government authorities in charge of industry and higher education should work hand in hand in order to bridge the gap existing between universities and industry, and make higher education institutions more dynamic than they actually are. New strategy and curriculum should be designed in order to meet the following objectives:

*high standard basic academic education

*facilitate the recycling of personnel in industry and all sectors of activity by providing sandwich courses, and organizing short but frequent training (evening courses, weekend courses) for these persons who will not be able to leave their work for a long period of time without any prejudice to normal functioning of their company.

*the materials to be taught should be displayed in modules (credits) which must be updated every two years in order to meet the new needs of the society, as well as science and technology evolution. A great place in this curriculum, should be reserved for computer science and electronics which should be taught to all univeristy students no matter what their major is, because these two sciences have already substantially affected our everyday life and their influence will certainly be much higher in the next century.

*the university should provide a varied menu of studies in a flexible manner for graduation. This broad curriculum should be able to help students developing gradually their personal skills as they advance in age and knowledge. For this purpose many strategies could be designed by the university concerned, depending on the local needs and realities. For example, the higher educational system actually adopted in almost all Francophone countries needs to be redefined because it is not flexible at all: once a student is engaged in a course he has no choice but to go until the end of the program, if he wants to change avenue he must restart from the beginning, which represents a very high new investment in time, intellectual effort, etc. Furthermore the official way followed in many of these countries for the adoption of the curriculum, involves such a complicated administrative path, that it is almost impossible to adapt any part of this curriculum to such change in an efficient and rapid manner. In Morocco for example, the procedure is so long that it may take years before a new curriculum can be officially printed in the government official bulletin. This does not help to make the university very dynamic in building new courses for particular needs. It seems obvious that great reform is necessary and this responsibility is to be transferred to the university itself whose management and organization should be redesigned and adapted to the local realities, in order to meet the national needs.

*the university and research institutions should be elected to more responsibility in higher education and scientific research for development. This necessitates a climate of intellectual freedom because if they are asked to search at least theoretical solutions for crucial problems, they have to feel free to consider all the possibilities. In conjunction with this freedom of expression and management, universities must plainly feel very responsible for the best use of public funds they are receiving and for the impact their conclusions may have on all levels of the society: political, economical, cultural, ... They must also accept control afterwards. Autonomy in management style has proved to be very successful in almost all developed countries, and there is no reason why it should not succeed for universities in developing countries after a slight adaptation which should take into account local realities.

A rapid glance at universities in most developing countries shows that they all suffer from a lack of equipment and technicians, even though their teaching staff has actually a relatively good international standard. The research activity is extremely hard to carry successfully under these very unfavorable conditions. It should be pointed out that these circumstances will certainly affect, in the near future, the quality of education there if appropriate measures are not carried out now. The governments who are really willing to overcome underdevelopment should pay real attention to their universities and research institutions. They can already measure their future chances of success by the dynamism of their intelligentsia in these establishments, because the latter represents future leadership in all fields, and the carriers for their "in gestation" technology. Any weakness which might strike these institutions will undoubtedly compromise their chance of succeeding in real technology transfer. There is a great need for urgent action, in order to make universities and research institutions more directly involved in technology transfer and social development by allocating them more funds and assigning them more precise projects of research. A system of consultancy between university and industry should be initiated and supported by the authorities in charge of industry and higher education. It is a pity that many private and public institutions in developing countries are still continuing to ask nearly exclusively foreign universities and research institutions for elementary chemical analysis, for instance, and ignore their national universities which are certainly as able as the foreign ones to do the job with the same scientific accuracy. There is a great need for self-confidence and trust in their young scientists. In order to gain the esteem of their country which will then allow them more responsibility, these scientists must prove quickly their high capability of designing appropriate solutions for the urgent problems of their community. The authority in charge of higher education should be aware of giving these young people, the opportunity and material means to keep up research, because after one or two years of inactivity these bright scientists will lose enthusiasm for creativity and certainly will be overtaken by very fast technological development; since they are young they will keep their position for many decades and get left more and more behind by scientific progress. Due to their ignorance of the new trends of technology, their role will become consequently more and more an obstacle to the technological and scientific progress; this will certainly contribute to the cultural and scientific asphyxia of their university, which is evidently not suitable. To minimize the scientific regression resulting from equipment shortage and lack of research activity, it is absolutely essential to update the university library. This institution should be regarded by all governments, university personnel and students, as the keystone of higher education and scientific research. At least one library in the country must have every paper printed in the world. This is already the case in all developed countries. Acquisition of knowledge should be considered now, and this will be more true in the 21st century, as a question of life or death for a nation. To be informed about other scientific studies and experiments is just as important as one's own experiments, because it is obvious that bibliography can save time and money, ... It is unfortunate that nearly all university libraries are not adequately furnished in developing countries. It is suggested here that the university library should be the first building to be constructed in a university. Furthermore, this institution should be considered as an independent faculty with its own budget for books and journals acquisition, but also for personnel training in the modern techniques of communication and information supply and storage. We should be aware that the 21st century library will certainly be more audio-visual and computerized than it is now, and we have to start getting ready for this new technique of information diffusion.

It is obvious that developing countries who have to face a great number of priorities such as food, health, natural resources exploitation, deep international crises .., and who are already spending more than 20 to 25% of their budget on education may feel disarmed in following the continuously accelerating technology progress. But one is tempted to advise: you have no choice, fast and study, because the battleground of your future survival as an independent nation is certainly dependant on your success in the race for technology transfer. Investment in higher education scientific research is just as important (if not more so) as investment in providing food. For

TABLE 1. Some Research and Development Programs Supported by Governments in Industrialized Countries (after ref. 3).

| Country | Nature of the Programme | Budget |
|---|---|---|
| USA | *Department of Defence is expected to invest for 5 years program in support of advanced manufacturing technology | 1.2 billion dollars |
| | *Federal funding for the conduct of research and development for the year 1984 (after Ref. 4) | (estimate 45.8 billion dollars) |
| W. Germany | Investment in Research | 2 to 3% of GNP |
| France | One national programme for research and development on robotics and automation supervised by Ministry of Industry and Research (for perid 1982-85) | 3 billion French Francs |
| Italy | 5 years programme on flexible manufacturing systems | 200 million dollars |
| Japan | *One programme supported by Ministry of International Trade and Industry (MITI) on Flexible Manufacturing System | 42 million dollars |
| | *One programme supported by MITI, for research on high performance materials: high performance ceramics, synthetic membranes for new osmotic techniques, composite materials, conductor polyners | 175 million dollars |

TABLE 2. Examples of Private Industries Investments in R & D Programs (Ref. 3).

| Country | Company | Programme | Budget |
|---|---|---|---|
| USA | IBM Cor. | Research and Development | 3.1 billion $ |
| | Grneral Motors | Research and Development | 3.07 billion $ |
| | AT&T | Research and Development | 2.37 billion $ |
| France | Renault | Investments in research | 4.3% turnover |
| Italy | Fiat | Automated System for building engines at Mirafiori plant | 14 million $ |
| Japan | Toyota | New materials and electronics | 380 million $ |

conviction of the necessity of conducting research and development (R & D), one has to refer to amounts of investments made in R & D by the already developed and leading technological countries, as shown in Table 1 (ref. 3). Beside the various government funds allocated to R & D, in developed nations, many private companies are conducting their own research programmes, where they are investing millions of dollars as shown in Table 2 (ref. 3). When these firms, who have the first objective to make profits, invest huge funds in R & D, developing countries should understand this is the way to make more profits. It has recently been demonstrated by all world-wide private companies that they can remain competitive, only if they can continuously improve their technology and products. This improvement is first experienced in their R & D department, and no serious company can now afford to go long without a dynamic department of R & D. This department becomes more and more the central heart of every well-stood company.

The analysis of funds allocated to research and development by governments as well as by private companies in the industrialized world, if they are certainly beyond the financial possibilities of nearly all developing countries (except those with oil income), they demonstrate clearly the seriousness attached by all developed countries to research and development. How could developing countries, who are in a bigger need of technology improvements, neglect this vital activity for technology transfer? It is a big mistake to consider scientific and technological research as a low priority.

This mistake will certainly come back to haunt them in the near future in the form of total dependency from abroad for food, security, and a marginal role in the world. Very urgent action must be carried out in order to recover part of the already wasted time and one to three percent of the GNP should be allocated for the revival of research and development. Since 1961 until now, western industrialized countries have always reserved at least 1.4 to 3% of GNP to R & D, while USSR has reserved between 2.7 to 3.7% og GNP, during the same period (ref. 5)

## CHALLENGE FOR THE THIRD WORLD

A Rapid glance at Tables 1 and 2 shows that the leading technology countries haved already started on a high speed race for the acquisition of the best and most economical technology. They all have clearly understood the main role of R & D in the amelioration of their technology and consequently their ability to face the competitivity of their adversaries. They all also have invested a lot of energy and money in new technologies: advanced manufacturing technology, computing systems, electronics and robotics, etc. (ref. 3).

One question remains, what is or what will be the position of the Third World, during the next decades, in this technological race? Is it already too late for all these nations? The answer depends partly on how strong is their own willingness to overcome their underdevelopment and to participate in the universal human endeavor.

## REFERENCES

1. Alahram, Cairo, Egypt 35980, (June 13, 1985) and 35982. (June 15, 1985)
2. C.F. Von Weizsacker. "Designing the Next Century", World Press Review, 27-29 (Feb, 1985).
3. "New Opportunities in Manufacturing - The Management of Technology". Her Majesty's Stationary Office, Government Bookshops, London (1983).
4. Annual Science and Technology Report to the Congress: National Science Foundation, Washington DC, USA (1982).
5. Science Indicators 1982. Report of the National Science Board 1983, National Science Foundation, Washington, DC, USA.

# The enrolment and training of chemists in China

Tong-Wen Hua

Department of Chemistry, Peking University, Beijing, China

Abstract - In the People's Republic of China, most of students are enrolled according to the grades they get in the unified entrance examination, while some are admitted through recommendation and oral test. The Ministry of Education has a unified training programme which ensures the students enrolled to be brought up to a certain high level. In order to foster more excellent chemists, some measures are taken, recently.

The People's Republic of China has a large population. Her education system is quite complex and diverse. There are general high schools as well as vocational schools in the secondary education level; and universities, colleges, institutes, evening colleges, correspondence universities, etc. in the higher education level. About two million students will graduate from general high school every year, among whom only about four hundred thousand are admitted to universities or colleges. Most of them are enrolled according to the grades they get in the unified entrance examination, while some are admitted through recommendation and oral test. Better results have been achieved through this method.

The Ministry of Education organizes the Entrance Examination Committee which administers the unified admission examination of the whole country. The Committee will set the examination papers according to the published school syllabus. All the test-takers join in the entrance examination on the same date (usually in July). Examination rooms are set in all the cities and counties throughout the country. There are about 40,000 rooms in the whole country and about 1300 in Beijing. All the examination papers are marked according to the unified score standard. The institutions select students among applicants on the basis of their grades. The famous universities and colleges have the privilege to enroll the students preferentially.

As a result of unified entrance examination, which has of course the advantage of saving manpower and expenses, the level of freshmen will nearly be the same and most of the top students can surely be enrolled. In addition, the high school teachers are urged to improve their teaching quality. However there are also some deficiencies, such as, with only one examination some of the good students with special abilities may not be discovered, and in order to get good marks in the entrance examination some of the high school teachers fail to train their students in a balanced way.

In recent years, some measures have been taken to remedy these deficiencies, some acadamic organizations have held science competitions for youth, or scientific activities in summer camps. Through these activities students with special abilities will be discovered and recommended to universities. In the "Chemistry Competition for Youth" sponsored by The Chinese Chemical Society in 1984, 88 students were chosen as winners, among them those who were secondary school graduates were enrolled by some famous universities. Apart from the competition the head masters of some key schools are qualified to recommened their excellent students to universities without taking the entrance examination.

As for those under the age of 16, who are bright enough to accept education of higher learning, they can enter juvenile classes in some universities. Peking University will take 20 students into juvenile class this year.

Adult education is for those under the age of 35 who have not entered universities or colleges. It has become one of the main commonest of higher education in China. Some universities set up evening classes or correspondence sections. Some large factories set up professional classes. CCTV set up TV University. Most of them study in spare time. In 1985, there are 73 evening classes, 15 correspondence classes, and 6 TV classes in Beijing, among which 13 classes are training chemists, such as the chemical evening class by the Department of Chemistry of Peking University, the chemical reagent class by Beijing Chemical Bureau and the chemical engineering evening class by The Beijing Chemical Engineering College.

Universities, colleges and research institutes can enroll graduate students and grant master and doctor degrees to the post-graduates. In 1985, 371 universities and 212 research institutes will enroll about 40,000 graduate students. They are admitted according to the grades they get in the entrance examination as well as some being admitted through recommendation. In 1985 the Chemistry Department of Peking University will enroll 60 graduate students, among whom 25 are recommended. The test papers of professional courses are developed by the concerned universities, while those of political science and foreign language are unifiedly set by the Entrance Examination Committee.

The Ministry of Education has issued unified training programme for the under-graduates and arranged to publish various text books, thus ensuring that the under-graduates be brought up to a certain level. In order to foster more excellent chemists, the following measures are taken:

(1) The elective credits are increased with the decrease in the required credits. The required credits in the Chemistry Department of Peking University are 77% of total credits in 1982, while they are decreased to 67% in 1984.
(2) The decrease in lecture hours encourages students to study on their own. In 1978, the lecture hours for general chemistry in Peking University were 160, while in 1981 they were only 124 and in 1984, 108.
(3) Honor of students are given the privilege doing research work with their professors.
(4) Students are organized to go into society, for instance, to participate in technical work in factories.
(5) Special seminars and extracurricula activities are offered, more outside readings are assigned.
(6) The "advisor system" is introduced, so as to guide students according to their abilities.

China needs to foster many excellent chemists for her own construction.

# Towards excellence in chemical education

Lawrence H L Chia

Department of Chemistry, National University of Singapore, Lower Kent Ridge Road, Singapore 0511

Abstract - It is timely that the 8th ICCE adopted the theme "Chemical Education for Fostering Future Chemists of Excellence". For Singapore, the challenge is not only to produce chemists of excellence, but also to ensure excellence in chemical education for all concerned. Manpower needs are such that some able and committed people who could otherwise be pursuing successful and satisfying careers in Chemistry are required in other areas. That these people have received grounding in chemistry is an asset. This paper refers to the aspirations, background and culture of the people, and the challenges and constraints of an external examination system. The contributions to chemical education by educational and professional bodies are also discussed.

Chemists could not agree more with Nobel Laureate Sir Peter Medawar when he described Science as a great and glorious enterprise. One needs only to look at the inroads and impact, influence and innovations, that Chemistry has had on society, to know how true that is.[1-3] There is, thus, the constant challenge to pass on all that is important in this growing body of knowledge to succeeding generations. Traditionally, modern Chemistry has been communicated in or through the universitites and much has already been written on this.

Emerging from Chemistry itself, Chemical Education, concerned with the teaching and learning of Chemistry,[4-9] has developed as a distinct discipline. Put simply, it is all too easy to assume that because of his expertise, the chemist can effectively impart his knowledge to others. This view is compounded by the concept which some hold that "anyone can teach". Chemical Education seeks to bring to the teaching and learning of Chemistry the very best resources available from education and communication, while at the same time thoroughly looking at Chemistry itself to see how this fascinating subject can best be taught and learnt.

Developing countries like Singapore need both the chemists who can apply Chemistry to good advantage, and also chemists who can teach it effectively. This latter aspect would of course be considerably enhanced with the growing influence of Chemical Education and the input, influence and impact of the IUPAC Conferences in Chemical Education. It is truly timely that the 8th ICCE adopted the theme "Chemical Education for Fostering Future Chemists of Excellence". The knowledge explosion associated with Chemistry, an exciting as well as exacting Science, comes to us at a time when there are disturbing trends that a growing number of our youth may opt for softer options or the line of least resistance even in their studies. The clarion call for excellence will thus ensure that we not only sustain but also increase the momentum gained by Chemistry thus far to carry our civilization into the 21st Century.

In Singapore, having Chemistry courses and graduates of excellence is almost axiomatic, given the Republic's philosophy and policy for excellence and meritocracy. It is however important to note that the country's manpower needs are such that some able and committed people, who could otherwise be pursuing successful and satisfying careers in Chemistry, are required in other areas of society as well. We believe that the Chemistry acquired, be it at school or university, would help such a person in his later pursuits or career. We are convinced that Chemistry by its very nature, ranging from the theoretical or abstract aspects of Physical Chemistry to the more qualitative aspects of Inorganic and Organic Chemistry, prepares the student to have both the broad as well as the integrated view of situations and problems. It trains the student to be both observant as well as analytical. The wide-ranging scope of Chemistry interfaces, if not embraces, on the one hand the disciplines of Biochemistry and Physics and Engineering on the other. It prepares our students to be multi-disciplinary in approach and outlook. Furthermore, Chemistry would be useful to students in Architecture, Engineering, Dentistry, Medicine, etc. Even for those in Accountancy or Arts, a feel for and understanding of Chemistry would be useful in our increasingly sophisticated and complex society. In addition the fascination that Chemistry may have on students, has provided them with a discipline which serves as a motivator for both excellence and self-confidence.

In Singapore where there is no lack of things scientific, the average student has an early start in Science.[10] He is in a community where the people appreciate the place and contribution of Science and Technology in their individual and communal well-being and progress. The applications and benefits of Science can be seen all around. By heritage, tradition and environment, the people realise the value and importance of learning and scholarship. At an early age, children are exposed to scientific phenomena, applications and ideas, especially through the mass media. At home and at school, television and video cassette recorders provide interesting and relevant presentations. The telephone, various home appliances, and toys all contribute to the children's appreciation of Science. Even before going to primary school at the age of six, the average child would have spent two to three years in some form of pre-school education. These formative years are useful in giving the child a good foundation in basic literacy and numeracy skills besides providing a school setting with all the necessary social interactions and adjustments.

In the primary school itself, Science is presented as General Science. Although the term "Chemistry" is not used until much later, its presence is evident when the children are introduced to the more basic or obvious properties of materials like air, water, rubber, wood, paper, plastics, glass and metals. The properties introduced, eg density, magnetic properties and thermal conductivity, are easily related to everyday experience. This whole approach is also consistent with the notion that scientific concepts should be built around practical work and the students' own experience. Pupils are encouraged to inquire, to explore and to discover. Text books are locally written by local authors and compare favourably with the best abroad.

The opportunities for learning more Science increase as pupils continue into secondary school. At the lower secondary level (aged approximately 12-14) all children are taught General Science with Chemistry and Physics taking about 30% each of the syllabus with Biology, the remaining 40%.

After these common courses, students may specialise in the upper secondary school (14-16 years). Many aspire to qualify for the Pure Science stream, taking Biology, Chemistry and Physics as separate subjects. Others may choose to take Physical Science and/or Biology. All others offer Combined Science or Human and Social Biology. Thus at the end of their fourth year in secondary school, students sit their 'O'-level examinations. Most of the Science students who pass their 'O'-level examinations hope to prepare for their Cambridge 'A'-level examinations over the next 2-3 years. Some however prefer to go directly to one of the two local Polytechnics.

A special feature in the 'A'-level syllabus is the section dealing with the attractive and varied options available. 'A'-level candidates are expected to study at least one option. The two suggested thus far have been Phase Equilibria and Further Transistion Metal Chemistry. The other options (Biochemistry, Chemical Engineering, Soil Chemistry, Food Chemistry, Polymers and Spectroscopy) are not only alternatives but also challenging opportunities for those students to develop additional areas and expertise.

At the tertiary level, students could continue their Chemistry studies at the University. Chemistry—related studies may also be pursued by those who enroll at the Institute of Education, the Polytechnics or the Technical Institutes.

While it is possible for local authorities and experts to have our own National School Leaving Examinations, it is felt that to enhance our creditability and standards on the international scene, we should continue to have these Examinations as external examinations associated with the Examinations Syndicate of Cambridge.

While the basic 3-year degree pass structure with an additional year for the Honours degree has evolved from our early experiences of the UK pattern, our staff are always on the alert to monitor, revise and innovate to ensure an academically respectable and relevant Chemistry degree. An example is the availability of Chemistry as a second subject with topics that are associated more with Applied/Industrial Chemistry and Chemical Technology - a popular combination with 3rd year Chemistry.

A Direct Honours Degree was also recently introduced in NUS. At the end of the first year, outstanding students are admitted to the course which extends over a period of two years. During this time only one subject will be studied. At the end of these three years an Honours Degree is conferred.

Agencies like the Curriculum Development Institute of Singapore, Science Council, Singapore National Academy of Science, Singapore National Institute of Chemistry (SNIC), Singapore Science Centre and Science Teachers Association of Singapore (STAS), also contribute to Chemical Education by organising and conducting out-of-class Chemistry activities, eg television programmes, exhibitions, quiz and speech contests, Youth Science Fortnight.

Given its influence, television could indeed be regarded as a major contributor to out-of-class Chemistry activities. The Singapore Broadcasting Corporation (SBC), consistent with the Government's objective to develop Singapore into a modern industrial economy based on Science and Technology, has given adequate time to educational and scientific programmes where the Chemistry content is not insignificant.

The Science Centre aims at sharing and passing on to the general public some of the exciting concepts, findings and practice of Science and Technology. Its impact and value lie in its participational element and the use of models, especially working-models.

The Singapore National Institute of Chemistry contributes by organising speech contests. It aims at giving students the experience of presenting chemical concepts and information in public. The contests have received wide coverage by the mass media.

Another recent innovation was the introduction of the Youth Science Fortnight, jointly organised by Science Centre and STAS and sponsored by Shell. This is a national event with mass participation by students of all ages. It has the aim of attracting the imaginative and creative youth in our society. The Fortnight includes Science Fair, Science Circus, Science Photo Competition, the Science Olympiad, Science Fiction, Film Festival and Science Forums.

The Industrial and Business Orientation Programme is focused on the need for academic training in tertiary institutions to be beneficially matched with some form of work experience in an industrial environment. This programme is planned and coordinated by a centralised body with the cooperation and support of government ministries and commercial and educational bodies.

A wide spectrum of other useful and interesting out-of-class Chemistry activities are also initiated and organised by professional and other bodies. Some could be due to the efforts of the students themselves especially through their own clubs and societies. These include talks, industrial visits, chemistry games, symposia, etc.

Another promising innovation in the Singapore education scene was the recent introduction of The Gifted Child Programme. Pupils are selected so that their special gifts may be developed to the fullest in special classes. From these students, one could expect future chemists of excellence, as well as those who, though not pursuing careers in Chemistry later on, may still make the most of the Chemistry classes they undergo in their secondary school days.

Finally, we in Singapore are encouraged by the formation of the Federation of Asian Chemical Societies and the establishment of its working committee on Chemical Education. Similarly, the Singapore Institute of Chemistry has established its own study group in which chemists from the University, the Institute of Education, the Curriculum Development Institute of Singapore, and some of our junior colleges and high schools are involved. We believe such a group will have tremendous impact on contributing towards not only the overall good of Chemical Education but also in areas which this conference emphasises.

ACKNOWLEDGEMENT

The author records his thanks to the Conference Organizers, the National University of Singapore, the JSPS, Miss Cecilia Lim, Mr Goh Kee Chor, Mr C Y Leow and the SNIC Chem Ed Working Group.

REFERENCES

1. Communications, Seventh International Conference on Chemical Education, Montpellier, (1983).
2. Proceedings, Sixth International Conference on Chemical Education, Maryland, (1981).
3. Chia, L H L, Chemists: Their Induction, Involvement And Influence, Bull of S'pore Nat'l Inst of Chem 4, 1 (1976).
4. Lippincott, W T, and Heikkinen, H, Source Book for Chemistry Teachers, Division of Chemical Education, American Chemical Society, (1981).
5. New Trends in Chemistry Teaching V, Unesco, (1981).
6. Frazer, M J, Trends in Chemical Education, Kemia-Kemi 6 4, 147-150 (1979).
7. Frazer, M J, and Sleet, R J, Resource Book on Chemical Education in the United Kingdom, London, Heyden, (1975).
8. Johnstone, A H, Research in Science Education at the University of Glasgow, European Journal of Sc. Educ. 1, 2 (1979).
9. Frazer, M J, Aims and Objectives of Programmes and Curricula in Chemistry, Meeting of the Sociedade Portuguesa de Quimica, Lisbon, (1978).
10. Chia, L H L, Chem Courses and Their Relevance in the S'pore Situation, Proc of a Symp on 'The Relevance of Secondary and Tertiary Chem Courses to Subsequent Employment', Aust., (1982).

# Non-conventional tertiary chemical education to foster chemists of excellence in Sri Lanka

J. N. Oleap Fernando

Professor of Chemistry, Open University, P.O.Box 1537, Colombo 10, Sri Lanka & President, Institute of Chemistry, Ceylon, Colombo 7, Sri Lanka.

Abstract - The Sri Lankan Government has sponsored the development within the state system of an Open University to cater for large numbers of students, the majority of whom are employed and of greater maturity. The professional Institute of Chemistry, Ceylon has also taken concrete steps to conduct professional graduateship courses and examinations with a view to meeting the shortage of chemistry honours graduates. The long term impact of these non-conventional programmes is of considerable importance and relevance in the production of chemists of excellence in Sri Lanka.

## INTRODUCTION

We have a situation in Sri Lanka, where a very widespread educational system has resulted in one of the highest literacy rates for developing countries. On the other hand, the relatively high cost of providing science education and the gradual neglect of English has produced a population whose general science literacy is not high. Adequate science educational opportunities have not spread widely to the rural areas. However the number aspiring for and qualifying to pursue science based courses in the Universities has been increasing. Despite increases both in the number of Universities as well as the number of students admitted for science based courses, a very severe competitive nature has entered at the point of admission to conventional Sri Lankan Universities. The alternate opportunities available for those students who are unable to pursue a conventional University based education have been grossly inadequate to meet the heavy demand.

The Government has increased the quantum of funds spent on higher education by an unprecedented amount. The irony however in that due to inflation the funds available to provide the minimum facilities have decreased in real terms. A question arises as to whether the astronomical increase in the number of university institutions and the number of undergraduates in such a constrained situation are on the long run causing too greater burden on the conventional university system. Will it in turn result in a lowering of the quality of graduates passing out? Are we succeeding in bringing tertiary science education within the reach of wider sections of students but by depriving them of the basic necessities for a successful career? In this context I am therefore happy that our government has in recent times accepted the policy of "consolidation & not proliferation" of universities in Sri Lanka.

It has to be recognized that the percentage of students admitted relative to the number qualifying for admission has been decreasing from year to year. The country continues to need more personnel for its development programmes, but in the context of growing demands on the national exchequer, a question arises as to whether a state sponsored free education system, can continue to meet all the graduate manpower needs of Sri Lanka.

It is within such a scenario that one has to observe the growth and expansion of non-conventional tertiary educational opportunities and this has happened in many instances with the active support of the Government.

## THE OPEN UNIVERSITY OF SRI LANKA

The Open University of Sri Lanka is a classic example of such an attempt to broaden the base of tertiary education though in a non-conventional manner. This institution was created in 1981 by the Government under the aegis the

University Grants Commission, like the other state financed conventional university institutions; staff recruitment and conditions of service are on a par with the conventional system. A considerable part of the funds required to run the Open University is received from the state. It is however non-conventional in three important aspects:-

(i) Part of the funds are collected from students in the form of tuition fees in contrast to the conventional unversity system.

(ii) The teaching is based on the distance education system, thus enabling the University to cater to very large numbers, many of whom are employed;

(iii) No formal educational qualifications are required from those seeking registration at the lowest (foundation) level.

The Open University commenced with courses at the lowest (foundation) level but now conducts the first year of the degree courses in Science & Law. Degree courses in Engineering will commence in two years' time.

The Open University is thus both a home-based University and a University of the later chance. It offers its students an opportunity to make up for chances which they did not have or take when they left school. It is also an opportunity to extend opportunities which they did take and add to them. Furthermore, here is an opportunity, not only for the formally qualified persons, but also for other students desiring to first obtain equivalent formal qualifications and then proceed to obtain a recognised university degree.

The multimedia study system developed by the British Open University with suitable modifications is the goal of the Open University of Sri Lanka. This presently includes printed material, practical classes, discussion classes and the submission of assignments.

The adoption of this type of system enables the Open University to enroll a much larger number of students than would be possible in a conventional university with face to face teaching. For example, over 1400 students were enrolled in our Open University for chemistry for the inaugural B.Sc. Degree Programme. This contrasts with about 800 students admitted to all the Science faculties in the seven other conventional universities in Sri Lanka.

The non-conventional Open University system has thus enabled the Government to throw open opportunities for tertiary education to large numbers in a manner, style and form that would be impossible under the conventional system. This has been made possible through a pragmatic realisation by the government that while it recognises its primary responsibility for providing free education at tertiary level, it is nevertheless hampered in its desire to provide such education to everyone who desires and deserves it. Contrary to the views expressed by some that such action represents a negation of the principles of free education, I hold the view that it represents the only feasible solution recognising stark realities and devoid of outdated dogma. I am convinced that, particularly in developing countries, a solution to widen the availability of education lay in provding education to the large number who desire it by levying at least part of the cost of that education.

## INSTITUTE OF CHEMISTRY, CEYLON

The employment potential of Chemistry honours graduates in Sri Lanka is such that for many decades the demand has been ahead of the supply. Relative to the other pure scientific disciplines the demand for Chemistry honours graduates is extremely high. Excellent job opportunities available for Chemistry honours graduates have increased the demand for providing courses that would lead to a production of such graduates. The aspiration of many Sri Lankan (science) undergraduates is to pursue honours degree courses in Chemistry. Unfortunately, despite the increase in the number of Universities over the past twenty years and the increase in the number of students selected to read for a Chemistry Honours Degree, the output of such graduates woefully lags behind demand. It is very unlikely to be remedied in the forseeable future through the traditional university system.

It was in such a context and with the desire to make a positive contribution to satisfy a much felt national need and demand that the Institute of Chemistry was founded. Ceylon took a big stepforward in 1979 to organise and sponsor courses of study at graduate level. This was a novel and non-conventional approach to tertiary chemical education. Our Institute thus became the first body, apart from the Universities, to provide tertiary level courses in science in Sri Lanka. Satisfactory completion of the relevant graduateship examinations conducted by the Institute enables admission to Graduate Membership of our Institute.

By providing facilities for the attainment of professional qualifications in Chemistry, our Institute was not trying to merely duplicate existing opportunities for university education. The courses and examinations conducted are primarily intended for persons employed at middle level to better their prospects. Late Developers as well as those who did not have the necessary resources to pursue higher studies immediately after their secondary education are now given a further opportunity. Persons possessing alternate qualifications are also given an opportunity they would never have under the traditional system. University undergraduates not selected to read for the much coveted Honours Degree course in Chemistry are also given a further opportunity to make good the loss.

Having been very closely associated with the conduct of these courses and examinations since their commencement I am fully convinced that we are not catering for a privileged class. The Institute has been able to provide this 4 year course leading to a professional qualification at tertiary level complete with practicals for an average monthly cost of 15 US $ or less than 750 US $ over 4 years.

Institute courses are conducted only in the English medium since the Institute has accepted as a matter of policy that the teaching and learning of science can be done satisfactorily only in an international language. Early last year four of our students became the first batch of professionally qualified chemists ever to be produced in Sri Lanka, and since then ten more have passed out. All of them have been able to obtain better employment and/or better their prospects in existing employment; we are confident that the Institute has been able to initiate a programme, which will play a key role in the task of national development in Sri Lanka.

Organization of a course of this magnitude was indeed an uphill task and a severe strain on the honorary services of Institute members. However with the ready co-operation of all the Institute is happy to have been able to satisfactorily complete six years of direct involvement in tertiary level education. A very interesting and salutary feature has been our ability to draw on the teaching personnel from most Universities in Sri Lanka and a number of research and other institutions.

The Institute wishes to place on record the support and encouragement received from the Royal Society of Chemistry, which assisted us to obtain the services of external examiners from the University of Nottingham. We started without a single book but thanks largely to generous donations of books from the British Government, the Asia Foundation and from the Royal Society of Chemistry we have been able to provide library facilities for our students. However, our courses are conducted in borrowed high school premises at week-ends while the library is housed at the premises of a leading research imstitute. The long term educational programmes of the Institute necessitates the early provision of permanent buildings of our own to house the lecture theatres, laboratories and a library and we look forward to institutional support and collaboration towards the 150,000 US $ needed for this purpose.

## CONCLUSION

Development and further expansion of tertiary educational opportunities in Sri Lanka particularly in Science appears to be not possible within the conventional (state) system unless it is going to be at the sacrifice of the high standards which Sri Lankan Universities have for long maintained. A rapid rise of university institutions to cater essentially to political and secretarian interests cannot take place in the context of a developing country except against such a background. It is however my belief that the only viable alternative available to obtain an accelerated expansion of higher educational opportunities without sacrifice of quality within the framework of a developing economy is to develop such opportunities in a non-conventional manner.

The Open University has opened up new vistas for the future to provide non-conventional avenues for Science education within the state system. The

Institute of Chemistry, Ceylon has pioneered the development of tertiary educational opportunities in chemistry outside the state system and without any state funding. However, in both these instances, students have to meet at least part of the expenditure required for their education. These Sri Lankan examples have shown in no uncertain terms that parents from many strata of Sri Lankan society are prepared to spend on their children's education, since they regard it as a sound investment for the future. The state as well as voluntary professional bodies must do all within their means to satisfy that demand. It is also a means of avoiding and minimising frustration in a situation where large numbers are denied admission to the conventional state institutions. This could also reduce ethnic conflict in multiracial societies such as Sri Lanka.

I welcome therefore the recent decision of the Sri Lankan government to encourage the establishment of degree awarding institutions. I am confident that devoid of conventional state bureaucracy these institutions will play a meaningful role in the tertiary educational scene in Sri Lanka in the future.

# Chemical education activities at Vikram A. Sarabhai Community Science Centre

Chandravadan J. Sanchorawala

Vikram A. Sarabhai Community Science Centre, Navrangpura, Ahmedabad 380 009

Dr. Vikram A. Sarabhai established a Group for Improvement of Science Education in 1963. This group consisted of research workers, administrators and faculty members from universities, colleges and schools. The group met regularly and discussed the programmes to be carried out to improve the science and mathematics education at all levels. Some of the group members carried out experimental activities in the colleges during the evening hours for school children. The group also studied new science and mathematics curricula and text books developed in USA and UK. The group decided to set up a central facility to coordinate the work of innovations in science and mathematics education. During 1966, Community Science Centre was established by Dr. Vikram A. Sarabhai. The Centre started its activities during 1967 from the premises of the local museum. The Centre decided to have its own premises during 1968 and the staff worked out a blueprint regarding the basic facilities required at the Centre. A competition was organised to design the Centre amongst the students of the School of Architecture, Ahmedabad whose first batch of students were graduating. Twelve entries were received from the students. Out of which one was selected. The building was ready in 1971 and the Centre shifted its activities to the new premises opposite Gujarat University. After the sad demise of Dr. Sarabhai in December 1971, the Centre was named Vikram A. Sarabhai Community Science Centre.

Various chemistry programmes carried out at the Centre have been:

1. Science learning through inquiry

This programme was carried out at the Centre for 11 years for college students, school students and school teachers. Initially for the first five years open-ended inquiry programmes were carried out. The programme consisted of all kinds of items with which the participants play and discuss amongst themselves various questions arising in their mind. After discussion for several hours, they choose topics to be investigated and form groups. The group devises experiments, to answer the questions raised by them, on the basis of the experiments carried out. The faculty of the Centre and the faculty from various schools and colleges act as catalyst who raise questions that lead to new experiments. After about 3 to 4 months, the group learned communication techniques and presented their work to all the participants and other invited guests using audio-visual aids, charts, models, experiments, etc.

Later on we found that the school teachers were not confident of running the open-ended inquiry in their classroom. The inquiries were directed to the concepts being taught in the classroom. A series of publications have been brought out on these inquiries by the Centre.

2. Motivated students programme

This programme consisted of students with a definite age or standard. The group consists of 25 to 30 students. The students learned the basic concepts as well as the various techniques in Chemistry, through lectures, discussions, experiments and audio-visual aids.

3. Motivated students programme at college level

The students from F.Y., S.Y., T.Y. or H.Sc. participated in this programme. This programme was carried out once a week for about 2 hours. We have more than 15 groups so far. For college students discussion of basic concepts in chemistry as well as project work was carried out. For M.Sc. students a series of programmes in chemical technology were carried out. This includes topics on:

1. Synthetic dyes
2. Petroleum refining and petro-chemicals
3. Natural products
4. Organic Polymers
5. Heavy Chemical Industries

Based on this programme we have published the following three books:

1. Outline of Synthetic Dyes
2. Outline of Petroleum Refining and Petro-chemicals
3. Outline of Organic Polymers

4. Chemistry projects

Earlier the National Science Talent Search examination had a compulsory project submission. Several students have carried out projects and obtained the Science Talent Scholarship. The students are encouraged to take up small projects in chemistry and carry them out by themselves.

5. Programmed learning in Chemistry

With the introduction of the higher secondary, many teachers found the basic concepts in chemistry quite difficult. About 15 higher secondary school teachers had undergone orientation programme in chemistry. We also brought out programmed learning material for teachers to improve their conceptual understanding. This material was later cyclostyled and made available to school children in the form of Publication on Programmed Learning in Chemistry in Gujarati.

6. Lecture demonstration in Chemistry

Lecture demonstration in several topics in chemistry have been carried out for large number of students and teachers at the Centre. The topics included chemical reactions, electrochemical series, energy, chemical magic show.

The Centre organised a Science Circus at the Centre as well as in other cities and rural areas. Interesting chemical experiments are demonstrated to the viewers.

Chemistry demonstrations

1. Change of rainbow colours of universal indicator by addition of base to acidic solution and vice versa
2. Precipitate formation and dissolution $(H_g(NO_3)_2$ + KI
3. Magic bottles
4. Rising and sinking balls
5. Sugar into carbon
6. Volcano
7. Catalytic combustion of glycerine
8. Chemiluminescence
9. Colour change on heating
10. Disappearance of colour on heating

7. Problem solving in chemistry

The Centre's staff is available for problem solving to teachers and students with previous appointment.

8. Exhibitions

The Centre develops exhibition on topics every year. Most of the exhibitions use integrated topics. The topics related to chemistry were molecules, salt, water, energy, measurement, pollution. These exhibitions are organised in 12 mt. x 12 mt. hall. The presentation is largely visual and participatory.

9. Teachers' training programme

The Centre from its inception has been involved in teacher training programmes every year. We have carried out teacher training programmes from primary to higher secondary level. We have done several refresher courses for teachers. The teachers were welcome to use the facility of the Centre and they bring students along with them most of the time. We have regular school visit programmes wherein teachers come with a batch of students and carry out experiments of their choice.

10. Computer

The Centre acquired two BBC microcomputers during 1984. Several computer training programmes have been conducted for students and teachers. We are planning to develop software in the areas of chemistry education.

11. Models in chemistry

The Centre has made several models, e.g., crystal structure of sodium chloride, cesium chloride, model of DNA. The Centre has developed the whole set of molecular orbital models made out of wood. They show sp, $sp^2$, $sp^3$, hybridization. We have made a set of models showing structure or methane, ethane, ethene, ethyne and benzene. The Centre has made models to explain close packing of atoms in metals.

12. Miscellaneous

The Centre has prepared slide shows on several topics related to chemistry.

1. Chromatography
2. Distillation of wood
3. Molecular models

The Centre had participated in developing video programmes for SITE broadcast through satellite to rural areas during 1975-76.

Participatory exhibits

The Centre's quadrangle has about 30 participatory exhibits. Some of them are based on chemistry.

1. Magic bottle
2. Change in colour of cobalt chloride solution on heating with a bulb
3. Electrolysis of water
4. Conductivity of solution of acid, baze, salt and sugar
5. Chemical models

13. Facility

The Centre has an excellent library with more than 10,000 books. The Centre subscribes to about 60 international journals in science and mathematics education.

The Centre has an excellent workshop where models, teaching aids and exhibits are designed and erected.

The Centre has an audio—visual department which helps with illustrations, book design and presentation. The audio—visual department also helps in designing exhibits.

# Biographical sketches of plenary and invited lecturers

**Professor Michinori Ōki**

Michinori Ōki was born in 1928 in Hyogo, Japan. After receiving his B. Sc. from The University of Tokyo, he joined the faculty of Tokyo Metropolitan University in 1950. He received his D. Sc. degree from the University of Tokyo and did postdoctoral work at University of Illinois in 1953-1955. He was promoted to an associate professor of Tokyo Metropolitan University in 1956 and then moved to The University of Tokyo as an associate professor in the same year where he has been professor of chemistry since 1962.

He is a recipient of The Chemical Society of Japan Award for Chemical Education, 1982. In addition to the domestic activity in chemical education, he was one of the first members of IUPAC Committee on Teaching of Chemistry in 1964-1972. He is the co-chairman of the 8th International Conference on Chemical Education.

He is well known as an organic chemist who first demonstrated the existence of stable rotational isomers of the ethane type.

**Professor Shin-ichi Sasaki**

Shin-ichi Sasaki received his Doctor's Degree of Science from the Tohoku University, 1959, Sendai, Japan. He has been a faculty member of Toyohashi University of Technology since 1979 where he is Vice President. He is a founder of the Division of Chemical Information and Computer Science, Chemical Society of Japan and was former President of the division.

His current research interests are in computer chemistry, particularly structure elucidation, spectral data base and molecular designing. He was awarded the Prize of Japan Society for Analytical Chemistry, 1974, and he received Niwa Medal of Japan Information Center of Science and Technology, 1980.

**Professor John W. Moore**

John Moore received his A.B. from Franklin and Marshall College and the Ph.D. from Northwestern University. He is currently Professor of Chemistry at Eastern Michigan University. In 1982 he received the Catalyst Award for Teaching of the Chemical Manufacturer's Association. He is co-author (with Elizabeth Moore) of Environmental Chemistry, (with W. G. Davies and R. W. Collins) of Chemistry and (with R. B. Pearson) of Kinetics and Mechanism, 3rd edition.

He edits the Computer Series in the Journal of Chemical Education and directs the NSF-sponsored Computer Project SERAPHIM. He is presently Secretary of the American Chemical Society Division of Chemical Education.

**Dr. Daniel Cabrol**

Daniel Cabrol studied in Nancy & Marseille and received his Doctorat de Specialite in Structural Chemistry in 1971 and the Doctorat es Sciences in 1985. He is currently Maitre de Conferences at the University of Nice, and heavily involved in teacher training and computer instruction.

His research interests has been in chemical solution kinetics and conformational analysis of biomolecules.

In 1978 he created and now directs a National Center of Information for Computer Use in Chemical Education. He is Co-Director of a National Network of research in this field and is actively engaged in the production of new chemical education software.

**Professor Geoffrey Norman Malcolm**

Geoffrey Norman Malcolm received his Master's degrees from Canterbury University in New Zealand and then went to Manchester University, U.K., for his 1956 doctorate in polymer solution chemistry.

He has taught at Massey University, New Zealand since 1969 and is currently Dean of Science there.

In addition to his continuing research in polymer and biopolymer chemistry, he has been much involved in chemical education, serving as Chief Examiner for a number of chemistry programs. He was the 1977 President of the N.Z. Institute of chemistry.

**Professor Reiko Isuyama**

Reiko Isuyama, a native of Sao Paulo, Brazil, received her D.Sc.from the University of Sao Paulo in 1975 and became Assistant Professor there the same year.

She has been on International Exchange Programs at Osaka & Tokyo, Japan, in Nice, France, in Norwich, U.K., at the University of Maryland, U.S.A., and in Yugoslavia and Peru. Her research interests are in organic adducts of transition and lanthanide metals and in chemical education.

She is a member of the Scientific Council of the Brazilian Foundation for Science Development. She is the Latin American Representative on the International Advisory Committee for 8-ICCE.

**Professor Peter Eric Childs**

Peter Eric Childs received his D.Phil. in Solid State Chemistry at Oxford in 1968 and taught at Makerere University, Uganda 1970-76.

He moved to Ireland where he is now Lecturer in Chemistry at Thomond College of Education. Since 1982 he has organized the annual conferences CHEM-ED IRELAND and edits Action-Chemistry.

He is secretary of the Ireland Region of the Education Division of the Royal Society of Chemistry.

**Professor Peter J. Kelly**

Peter J. Kelly is Professor of Education, Head of the Department of Education and Dean of the Faculty of Educational Studies, at the University of Southampton.

He has served as Director of the Nuffield, A-level Biological Sciences Project and was Chairman of the Commission for Biological Education of the International Union of Biological Sciences.

His degrees include the B.Sc., M.A., and Ph. D., degrees, and he is currently Chairman of the Committee on the Teaching of Science of the International Council of Scientific Unions.

**Professor Kui Wang**

Kui Wang received his chemical training at Yenching and Beijing Universities and is currently Professor at Beijing Medical College as well as Dean of the Faculty of Pharmaceutical Sciences and Head of its Research Laboratories.

Kui Wang has served as Vice Chairman of the Council of the Beijing Chemical Society, and of the Working Committee for the Popularization of Chemistry of the Chinese Chemical Society. His Research interests include the biochemical role of metals, the treatment of heavy metal poisoning, and anticancer complexes. He is seeking to reform the chemical education of students who plan to use chemical knowledge in the field of medicine.

**Mr. Isao Nakayama**

Isao Nakayama graduated from the Tokyo University of Education, Faculty of Science, Department of Biology. After studying animal physiology at the graduate school of the same university, he baecame a high school teacher of biology from 1957, and now he teaches at Azabu High School which is one of the most famous in Japan.

He has been very active in biological education at High School level, and co-authored several high school text books of biology and science I. He has also cooperated with many chemical educators in a project aiming to reform scientific education at the secondary level.

**Professor Wiero Jan Beek**

Wiero Jan Beek received his doctorate in chemical engineering from the Institute of Technology, Holland in 1962, and in 1963 joined the faculty at that Institute.

He has held a number of major industrial positions, in 1980 being made Chairman of the Board of the Unilever Research Laboratory in Vlaardingen. He has chaired the Dutch National Research Council for Agricultural Research and in 1983 was President of the Royal Dutch Chemical Society.

**Professor Avi Hofstein**

Avi Hofstein was educated at the Hebrew University, Jerusalem (Chemistry & Physics), Tel-Aviv University and the Weizmann Institute (Ph.D. Science Teaching 1975).

For three years he was Deputy Headmaster and chemistry teacher at the Technical High School of Tel-Aviv University while also beginning his association with the Chemistry Group of the Weizmann Institute's Department of Science Teaching.

Avi Hofstein is on the Board of Editors of the Journal of Research in Science Teaching having studied and published work on science education topics for many years. He has been working with others to introduce chemical industry into the secondary school chemistry curriculum.

**Professor Wha-Kuk Lee**

Wha-Kuk Lee received his B.S. from Taejon Presbyterian College, his M.S. from Chonbuk National University, and his Ph.D. from the University of East Anglia, UK (1983).

He taught high school for five years and since 1974 has been on the teaching staff at Chonbuk National University, currently being Associate Professor in its Department of Chemical Education.

He has published a General Chemistry text and is interested in Piaget's and other models for understanding how students learn science.

**Professor George C. Pimentel**

George Pimentel received his Ph.D. from the University of California in 1949 and was on the chemistry faculty there until 1977 when he became Deputy Director of the National Science Foundation. Since 1981 he has been Director of the University of California's Laboratory for Chemical Dynamics.

His research has been in spectroscopy, lasers, molecular structure, free radicals and hydrogen bonding. He was Editor of the CHEM-Study project.

He was elected a member of the U.S. National Academy of Sciences in 1966 and is head of its Commission to Study Chemistry's Future Opportunities and needs. He is President-Elect of the American Chemical Society.

His numerous honors in chemistry, chemical education and public service include the Linus Pauling Medal, the Wolf Prize in Chemistry, the Manufacturing Chemists Association College Chemistry Teacher Award, the National Science Foundation's Distinguished Service Gold Medal, and the National Medal of Science.

# List of participants

ACHARYA, Binod P.
P.O.B. 2102
Kathmandu, NEPAL

AISAWA, Toshio
Hachinohe Higashi High School
1-4-47 Ruike, Hachinohe-shi, Aomori 031

AIZAWA, Masuo
The University of Tsukuba
Sakura-mura, Niihari-gun, Ibaraki 305

ALDEN, Karl Inge G.
Swedish University of Agricultural
Science, Dept.of Chemistry and Molecular
Biology, Box 7015,75007, Uppsala SWEDEN

AMANO, Masatake
Tokai University
1117 Kitakaname, Hiratsuka-shi,
Kanagawa 259-12

ANDERSON, Frank E.
Department of Chemistry, Engineering
University of California, Berkeley,
CA 94720, U.S.A.

ANDERSON, Gordon J.
20 Prospect Hill Street,
Greenock, Renfrewshire, Scotland,
PA15 4DL, U.K.

ANDO, Takeshi
College of General Education,
Kyushu University
Ropponmatsu, Chuoo-ku, Fukuoka 810

ANRAKU, Mitsugi
Koishikawa High School
2-29-29 Honkomagome, Bunkyo-ku,
Tokyo 113

ANSARI, Salman
6148 Heppenheim-4
Odenwald Schule,
W. GERMANY

ARAI, Tomio
Yamasaki High School
Yanazaki-machi, Machida-shi, Tokyo 194

ARAI, Yasumasa
Daimon 1-13-9, Shiba, Minato-ku,
Tokyo 105

ARAKAWA, Hisao
Naruto University of Teacher Education
Takashima, Naruto-cho, Naruto-shi,
Tokushima 772

ARIKAWA, Hiroshi
Rikkyo High School
1-2-25 Kitano,
Niiza-shi, Saitama 352

ASADA, Hiroko
Tamagawa Primary School
1-12-1 Inaba, Higashiosaka-shi,
Osaka 578

ASAKA, Kiyotaka
Yanagihara Primary School
2-49-1 Yanagihara, Adachi-ku, Tokyo 120

ASAKAWA, Akira
Komae High School
3-9-1 Motoizumi, Komae-shi, Tokyo 201

ASHFORD, Theo A.
1832 Bearss Ave.
Tampa, FL 33612,
U.S.A.

ASHMORE, Anthony D.
The Royal Society of Chemistry
30, Russell Square, London,
WC1B 5DT, U.K.

ATOBE, Kiyoe
Chitose High School
3-8-1 Kasuya, Setagaya-ku, Tokyo 157

Al-Obadie, Mubarak S.
Department of Chemistry,
University of Kuwait
P. O. Box 5969, KUWAIT

BALOGH, Magdalena.
8200. Veszprem Szabad nep u,15
HUNGARY

BARGELLINI, Alberto
Via Bruno Buozzi, N.2
55049 Viareggio,
ITALY

BARON, Joseph A.
9144 Johnson Dr.
La Mesa CA 92041,
U.S.A.

BEEK, W.J.
P.O.Box 114, 3130 AC Vlaardingen,
The NETHERLAND

BEN-ZVI, Ruth
Department of Science Teaching,
Weizmann Institute of Science
Rehovot 76100, ISRAEL

BENFEY, Otto Theodor
801 Woodbook Drive
Greensboro NC 27410,
U.S.A.

BHANTHUMNAVIN, Pirawan
Dept. of Chemistry, Fact. of Science
Chulalongkorn University
Phya Thai Road, Bangkok 10500, THAILAND

BONTCHEV, Panayot R.
Fact. of Chemistry, University of Sofia
1, Anton Ivanov Str.,
Sofia 1126, BULGARIA

BOUGUERRA, Mohamed L.
Faculte des Sciences de Tunis
Campus Universitaire, 1016
Tunis, TUNISIA

BOWMAN, Leo H.
Russellville, AR 72801, U.S.A.

BRADLEY, John D.
Department of Chemistry,
University of the Witwatersrand
Johannesburg, SOUTH AFRICA

BRASTED, Robert C.
Dept. of Chemistry, Univ. of Minnesota,
207 Pleasant St S.E.,
Minneapolis, MN 55455, U.S.A.

BROCKLEHURST, Donald James
Mitaka High SChool
6-21-21 Shinkawa, Mitaka-shi, Tokyo 181

BURMEISTER, John L.
Department of Chemistry,
University of Delaware
Newark, DE 19716, U.S.A.

CAI, Jinfeng.
Dept. of Chemistry
South China Normal University
Guangzhou, China

CARDELLINI, Liberato L C.
Dipartimento Seienze Materiali Eterra
Via Montagnola 30
60128 Ancona, ITALY

CHANDAVIMOL, Manee
The Institute for the Promotion of
Teaching Science and Technology,
924 Sukhumvit Rd, Banqkok 10110 THAILAND

CHASTRETTE, Maurice
22, Chemin Combe Martin
F- 69300 Caluire,
FRANCE

CHENG, Mingrong
Curriculum and Teaching Materials
Research Institute
55 Sha Tan Hou Street, Beijing CHINA

BERG, Lillian D.
3319 Dauphine Drive,
Falls Church, VA 22042,
U.S.A.

BLUM, M.A.Gabrielle
320 State Rd.
Gt. Barrington, MA 01230,
U.S.A.

BOSCHMANN, Erwin
425 Agnes St, Indianapolis,
IN, 46202, U.S.A.

BOWMAN, Charles L.
2569 B Pheasant Run
Wexford, PA 15090,
U.S.A.

BRACE, Neal O.
Department of Chemistry
Wheaton College
Wheaton, IL 60187, U.S.A.

BRANDT, Ludo
Swertmolenstraat, 6
B-3020 Herent,
BELGIUM

BRIDGES, Bette Anne
221 Oak Street, Box 94A Brockton,
MA 02401, U.S.A.

BROWN, Carol B.
4827 Wesleyan
San Antonio, TX 78249,
U.S.A.

CABROL, Daniel
C.R.P.R.D.C.
University of Nice
F 06034 Nice Ceolex, FRANCE

CAINS, Isabel B.
Quenepa 108, Milaville
Rio Piedras, 00926,
PUERTO RICO

CEDER, Olof
Dept. of Organic Chemistry
Chalmers Univ. of Chem., Univ. of
Goteborg, S-412 96 Goteborg SWEDEN

CHAPMAN, Kenneth M.
American Chemical Society
1155 Sixteenth St., NW
Washington, DC 20036, U.S.A.

CHEN, Yi
Department of Chemistry,
Nanjing University
Nanjing, CHINA

CHIA, Lawrence, H.L.
Dept. of Chem., Faculty of Science,
National University of Singapore
Lower Kent Ridge 0511, SINGAPORE

CHIANG, Hung-Cheh
3rd. Fl., 5, Tai An St.,
Taipei 100 TAIWAN, ROC

CHIBA, Nobuyuki
Aomori High School
8-1-2 Sakuragawa, Aomori-shi, Aomori 030

CHIEH, Chung
Department of Chemistry,
University of Waterloo
Waterloo, Ontario, N2L 3G1, CANADA

CHIHARA, Hideaki
Fac. of Science, Osaka University
1-1 Machikaneyama-machi, Toyonaka-shi,
Osaka 560

CHILDS, Peter E.
Thomond College
Limerick, IRELAND (EIRE)

CHISMAN, Dennis G.
114 The Avenue
Sunbury on Thames,
Middx, TW16 SEA, U.K.

CLEVENGER. John V.
Truckee Meadows Community College
7000 Dandini Boulevard, Reno,
NV 89512, U.S.A.

CLINE, Susan J.
Ronnebarvej 136B
DK-2840 Holte
DENMARK

COLE, Andrew R.H.
School of Chemistry,
University of Western Australia
Nedlands 6009, AUSTRALIA

CROS, Daniele
10 Rue du Bosc Clapiers
34170 Castelnau Le Le2
FRANCE

DANIEL, David A.
American Chemical Society
1155 Sixteenth Street, N.W.
Washington D.C. 20036, U.S.A.

DEHNERT, Peter
Weingarten Str. 4
D-6109 Muhltal,
FRG W.GERMANY

DHOUBHADEL, Shiva P.
Royal Nepal Academy of Scienc. and Tec.,
Kathmandu, NEPAL

DOMINGUEZ, Ana Elizabeth
C. Beistegui 1706 Col. Narvarte
Mexico, D.F.03020, MEXICO

DOMINGUEZ, X.A.
Superiores de Monterrey
Succursal de Correos "J"
64849 Monterrey, N.L. MEXICO

DONGALA, E.B.
Fac. of Science,
University of Marien-Ngovabi
B.P. 69 Brazzaville, CONGO

DRAKE, Robert F.
Department of Chemistry,
Saint Mary's College
Winona, MN 55987, U.S.A.

EDA, Minoru
Chiba Prefectural General
Education Center
Wakaba, Chiba-shi, Chiba 260

EGUCHI, Hanako
Toshima High School
4-41 Chihaya-cho, Toshima-ku, Tokyo 171

ELOUADI, Brahim
Department of Chemistry
Oklahoma State Univ., Stillwater,
74078 Oklahoma, U.S.A.

ESHIET, Israel T.
Fac. of Education,
University of Cross River State
Uyo, NIGERIA

ESHIET, Israel Tompson
Faculty of Education,
University of Cross River State
P.M.B. 1017, UYO NIGERIA

EUBANKS, Dwaine
Department of Chemistry,
Oklahoma State University
Stillwater, OK 74078, U.S.A.

FANG, Tai-Shan
Department of Chemistry,
National Taiwan Normal University
Taipei 117, TAIWAN, ROC

FERNANDEZ, Dolores Esther
Alfredo Vicenti 32-5B
15004 La Coruna,
SPAIN

FERNANDO, J. N. Oleap
Open University,
P.O.B. 1537, Colombo 10
SRI LANKA

FIELDS, Ellis K.
Amoco Chemical Corp.
P.O.Box 400, Naperville,
IL, 60566, U.S.A.

FOGLIANI, Charles L.
Mitchell College of
Advanced Education
Bathurst, N.S.W. 2795, AUSTRALIA

FRIEDSTEIN, Harriet G.
41 Hilltop Dr.
Pittsford, NY 14534,
U.S.A.

FUJII, Kiyoshi
Tomakonai Technical College
443 Nishikioka, Tomakomai-shi,
Hokkaido 059-12

FUJITA, Michio
3-4-29 Asahigaoka
Ikeda-shi, Osaka 563 (Home)

FUJITA, Toshio
Dep. of Agricultural Chemistry,
Kyoto University
Sakyo-ku, Kyoto-shi, Kyoto 606

FUJIWARA, Hironobu
Joshi Seigakuin High School
3-12-2 Nakazato, Kita-ku, Tokyo 114

FUJIWARA, Shizuo
Dept. of Industry Chemistry,
Chiba University
1-33 Yayoi-cho, Chiba-shi, Chiba 260

FUKAMACHI, Moriyoshi
Tokorozawa-nishi High School
1649 Kitano, Tokorozawa-shi,
Saitama 359

FUKUCHI, Akiteru
Tokyo Gakugei University
4-1-1 Nukuikita-machi,
Koganei-shi, Tokyo 184

FUKUOKA, Hisao
Komaba Senior High School Attached to
The University of Tsukuba
4-7-1 Ikejiri, Setagaya-ku, Tokyo 154

FUKUOKA, Tatsuhiko
Kanazawa University
1-1 Marunouchi, Kanazawa-shi,
Ishikawa 920

FUKUSHI, Sachio
Chiba Institute of Technology
2-17-1 Tsudanuma,
Narashino-shi, Chiba 275

FUKUSHIMA, Hachiro
The Japan Association of Physics
and Chemistry
1-20-7 Kitaotsuka, Toshima-ku, Tokyo 170

FUNAKI, Makoto
Kitamoto High School
650 Miyauchi, Kitamoto-shi, Saitama 364

FURUHASHI, Akiko
Dept. of Chemistry, Fac. of Science,
& Engineering, Aoyama Gakuin University
Chitosedai, Setagaya-ku, Tokyo 157

FURUYA, Masao
Kanuma Agricultural Senior High School
8-73 Minami, Kanuma-shi, Tochigi 322-05

FUTAMI, Ayako
Tokyo Institute of Technology
2-12-1 Ookayama, Meguro-ku, Tokyo 152

GARDNER, Marjorie H.
Lawrence Hall of Science
University of California, Berkeley,
CA 94720, U.S.A.

GENDREAU, Harvey W.
1622 Massachusetts Ave.
Cambridge, MA 02138,
U.S.A.

GHOSH, Pradip K.
Chemistry Dept. Indian Institute of
Technology, Kanpur 208016, INDIA

GIESBRECHT, Ernesto
Univ. of Sao Paulo, Instituto de Quimica
Quimica, Caixa Postel 20780
01498, Sao Paulo, BRASIL

GLAZAR, Sasa
UNESCO International Center for Chemical
Studies, Vegova 4, pp 18/I
61001 Ljubljana, YUGOSLAVIA

GOODNEY, David E.
Dept. of Chemistry, Willamette Univ.
900 State Street
Salem, OR 97301, U.S.A.

GOTO, Atsushi
Rikkyo High School
1-2-25 Kitano, Niiza-shi, Saitama 352

GOTO, Hiroshi
RD & E Management Div.
Japan Management Association
3-1-22 Shibakoen, Minato-ku, Tokyo 105

GOTO, Yutaka
Josai High School
1-43, Chihaya-cho,
Toshima-ku, Tokyo 171

GROMKO, Mary S.
616 Becket Place
Colorado Springs,
Colorado 80906, U.S.A.

GRUNWALD, Peter
Universitat Hamburg, Institut fur
Physikallsche Chemle, Laufgraben 24,2
Hamburg, 13, FRG

HAAVISTO, Anja Elina
Ritokalliontie 8-16 D
00330 Helsinki 33
FINLAND

HADISUSILO, Susilowati
Department of Chemistry, F MIPA
Universitas Indonesia, Salemba Raya No.4
Jakarta Pusat, INDONESIA

HAIDER, S.Z.
Department of Chemistry,
Dhaka University
DHAKA-2, BANGLADESH

HAISHI, Nobuhiko
Shuchi High School
Toyota, Tamba-cho, Funai-gun,
Kyoto 629-04

HALSTED, Douglas A.
2140 Lincolnwood Drive,
Evanston, IL 60201,
U.S.A.

HAMADA, Shizuko
Tokyo Metropolitan
Bunkyo School for The Blind
1-7-6 Koraku, Bunkyo-ku, Tokyo 112

HAMPTON, Michael D.
Department of Chemistry,
University of Central Florida
Orlando, FL 32816, U.S.A.

HARA, Shigeyoshi
Chitose Senior High School
3-8-1 Kasuya, Setagaya-ku, Tokyo 157

HARRIS, William Charles
Chemistry Division of
National Science Fundation
1800"G"Street,Washington DC,20550,U.S.A.

HARUTA, Tokuhiro
Itoshima High School
380 Shinohara, Maebaru-cho, Itoshima-gun
Fukuoka 819-11

HASHIMOTO, Hisashi
Mie Prefectural Science Education Center
1506 Kitaura, Terakata-cho,
Yokkaichi-shi, Mie 510

HAYASHI, Yoshishige
Fac. of Education, Toyama University
3190 Gofuku, Toyama-shi, Toyama 930

HERNNANDEZ, Lilian M.de
CENAMEC. Apart. 75055, El Marques
Caracas 1070-A, VENEZUELA

HILL, John W.
Dept. of Chemistry, Univ. of Wisconsin
River Falls, WI 54022,
U.S.A.

HIRASAWA, Ryo
Dept. of Chemistry, College of Arts and
Sciences, The University of Tokyo
3-8-1 Komaba, Meguro-ku, Tokyo 153

HAGENAUER, Rosa
A-2500 Baben bei Wien
Frauengasse 3-5,
AUSTRIA

HAIGHT, Gilbent, P.
1706 Pleasant Street,
Urbana, IL, 61801,
U.S.A.

HAKKI, W.
Supreme Council of Sciences,
POB 4762, Damascus,
SYRIAN ARAB REPUBLIC

HAMADA, Hiroshi
The Chemical Society of Japan
1-5 Kanda Surugadai, Chiyoda-ku,
Tokyo 101

HAMADA, Shuichi
Science University of Tokyo
1-3 Kagurazaka, Shinjuku-ku,
Tokyo 162

HANAYA, Kaoru
Miyagi University of Education
Aoba, Aramaki, Sendai-shi,
Miyagi 980

HARIGAYA, Toshiharu
Kiyose High School
3-1-56 Matsuyama, Kiyose-shi, Tokyo 204

HARTOMO, Anton John
Jacan Embong Ploso 12
Surabaya 60271
East Java, INDONESIA

HASEGAWA, Shizuo
Shirako High School
4-17-1 Shirako-cho, Suzuka-shi,
Mie 510-02

HAWANG, Bao-Tyan
Department of Chemistry, National Taiwan
Normal University, 88 Sec.5,
Roosevelt Rd., Taipei 117, TAIWAN ROC

HENMI, Hiroshi
Tokai University
1117 Kitakaname, Hiratsuka-shi,
Kanagawa 259-12

HIKIME, Seiichiro
Hyogo University of Teacher Education
942-1 Shimokume, Yashiro-cho,
Hyogo 673-14

HIRANO, Tsuneo
Fac. of Engineering,
The University of Tokyo
7-3-1 Hongo, Bunkyo-ku, Tokyo 113

HIRATA, Takuro
Toyama Prefectural Consolidated
Education Center
2238 Gofuku, Toyama-shi, Toyama 930

HIRAYAMA, Toru
School of Engineering, Tokai University
Kitakaname, Hiratsuka-shi,
Kanagawa 259-12

HIROE, Saburo
3-2-24 Higashimotomachi,
Kokubunji-shi, Tokyo 185

HIROE, Toshihiko
Haijima Junior High School
2-2-12 Midori-cho, Akishima-shi,
Tokyo 196

HIROI, Tadashi
The Senior High School Attached to
The University of Tsukuba
1-9-1 Otsuka, Bunkyo-ku, Tokyo 112

HIROSE, Masakatsu
Senshu University
2-1-1 Higashimita, Tama-ku,
Kawasaki-shi, Kanagawa 214

HIROTA, Minoru
Dept. of Appl. Chem.,Fac. of Engineering
Yokohama National University
Tokiwa-dai, Hodogaya-ku. Yokohama 240

HISKIA, Achmad
Department Kimia-ITB
Julan Ganesa 10
Bandung 40132, INDONESIA

HITOI, Noboru
Katsuyama High School
3-10-75 Tatsumi-higashi, Ikuno-ku,
Osaka-shi, Osaka 544

HIYOSHI, Yoshiro
Monzen High School
5-3 Hirooka Monzen-machi,
Fugeshi-gun, Ishikawa 927-21

HOFSTEIN, Avi
Dept. of Science Teaching
The Weizmann Institute of Science
Rehovot 76100, ISRAEL

HOFSTEIN, Ronny
Dept. of Science Teaching
The Weizmann Institute of Science
Rehovot 76100, ISRAEL

HOLDSWORTH, David K.
Dept. of Chemistry,
University of Papua New Guinea
PAPUA NEW GUINEA

HORI, Fumio
Higashiyamato High School
3-954 Chuo Hogashiyamayo-shi, Tokyo 189

HORIUCHI, Akira Charles
Dept. of Chemistry, Fac. of Science,
St. Paul's University
Nishi-Ikebukuro, Toshima-ku, Tokyo 171

HOSOYA, Haruo
Dept. of Chemistry, Fac. of Science,
Ochanomizu University
Otsuka, Bunkyo-ku, Tokyo 112

HUA, Tongwen
Department of Chemistry, Peking
University, Beijing, CHINA

HUNMA, Rannoo R.
Mauritius Institute of Education,
Scuence Department
Reduit, MAURITIUS

HUNT, John Baker
1302 Madison
Chillum, MD 20782, U.S.A.

HUSS, Frank W.,III.
3658 Stettinius Ave.
Cincinnati, OH 45208,
U.S.A.

HYDE, Keith A.
11, Aysgarth Avenue,
Greave, Romiley Cheshire
SK6 4PX, U.K.

IDE, Koichiro
Fac. of Education,
Chiba University
Yayoi-cho, Chiba-shi, Chiba 260

IGARASHI, Hirokazu
Setagaya Primary School Attached to
Tokyo Gakugei University
4-10-1 Fukazawa, Setagaya-ku, Tokyo 158

IIJIMA, Takao
Fac. of Science, Gakushuin University
Mejiro, Toshima-ku, Tokyo 171

IIRI, Yuichi
The Ministry of Education, Japan
3-2-2 Kasumigaseki, Chiyoda-ku,
Tokyo 100

IKEGAMI, Yusaku
Chemical Research Institute of
Non-Aqueous Solutions, Tohoku University
2-1-1 Katahira, Sendai-shi, Miyagi 980

IKEO, Kazuko
Nara University of Education
Takabatake-cho, Nara-shi, Nara 630

IMAI, Nobuhisa
Dept. of Physics, Fac. of Science,
Nagoya University
Furo-cho, Chikusa-ku, Nagoya 464

INO, Takashi
Musashi University
1-26 Toyotama-kami, Nerima-ku, Tokyo 176

INO, Yoshitaka
Tokyo Gakugei University
4-1-1 Nukuikita-machi,
Koganei-shi, Tokyo 184

INOUE, Tomoaki
Zama High School
2-262 Iriya, Zama-shi, Kanagawa 228

ISHIBASHI, Humihide
Dept. of Chemistry,
Osaka University of Education,
Minamikawahori-cho,Tennoji-ku, Osaka 543

ISHIGE, Junichi
Setagaya Junior High School Attached to
Tokyo Gakugei University
4-3-1 Fukazawa, Setagaya-ku, Tokyo 158

ISHII, Aihiko
Joto Senior High School
22-1 Ojima 3-chome, Koto-ku, Tokyo 136

ISHIKAWA, Reiko
The 16th Adachi Junior High School
38-1 Senjuasahi-cho, Adachi-ku,
Tokyo 120

ISUYAMA, Reiko
Univ, de Sao Paulo, Instituto de Quimica
USP Caixa Postal 20780,
Cep 01498, Sao Paulo, BRASIL

ITO, Yasuzo
Hokkaido University of Education
Shiroyama, Kushiro-shi, Hokkaido 085

ITOH, Yoshinori
High School Attached to
Tsukuba University
Otsuka, Bunkyo-ku, Tokyo 112

IWASAKI, Hirosi
Miyazu High School
Hinode, Ine-cho, Yosa-gun, Kyoto 626-04

JACOBSON, Elaine M.
Box 80631
Fairbanks AK 99708,
U.S.A.

JAMES, June Marion
113 Shamrock Drive
Winnipeg, Manitoba
R2J 3W4, CANADA

JAMES, Sheryl J.
RR3 Box 260,
Lexington, IN 47138,
U.S.A.

JI, Yaown
Department of Chemistry,
Zhengzhou University
Henan Province, CHINA

INOUE, Syohei
Fac. of Engineering,
The University of Tokyo
7-3-1 Hongo, Bunkyo-ku, Tokyo 113

INUI, Yoshikazu
Nara Prefectural Board of Education
Noboriohji-cho, Nara-shi, Nara 631

ISHIBASHI, Susumu
Kwansei Gakuin High School
1-1-155 Uegahara, Nishinomiya-shi,
Osaka 662

ISHIGURO, Tetsuro
Chuo University
1-13-27 Kasuga, Bunkyo-ku, tokyo 112

ISHII, Tomoko
High School Attached to
Ochanomizu University
Otsuka, Bunkyo-ku, Tokyo 112

ISHISONE, Seiichi
Ryogoku High School
1-7-14 Kotobashi, Sumida-ku, Tokyo 130

ITO, Hiroko K.
2915 W 5th St.,
Brooklyn, NY 11224,
U.S.A.

ITOH, Seiichi
3-5-2-126 Kiyoshigaoka Kogane
Matsudo-shi, Chiba 270 (Home)

ITOYAMA, Tou-ichi
Fac. of Education, Kagawa University
1-1 Saiwai-cho, Takamatsu-shi,
Kagawa 760

IWATA, Atsuko
Aoyama High School
2-1-8 Jingumae, Shibuya-ku, Tokyo 150

JAMES, Frederick C.
Faculty of Applied Science
Royal Melbourne Institute of Technology
124 Latrobe St. Melbourne, AUSTRALIA

JAMES, Ralph E.
113 Shamrock Drive
Winnipeg, Manitoba
R2J 3W4, CANADA

JARISCH, Edith
Billrothstr. 39 A-1190
Vienna, AUSTRIA

JONES, Donald E.
1229 Woods Road
Westminster, Maryland 21157,
U.S.A.

JONES, Loretta L.
1211 W.Main St.,
Urbana, IL 61801,
U.S.A.

JOSEPH, Mungarulire
National University of Rwanda,
P.O.B. 117 Butare, Rwanda

JOSEPHSEN, Jens
Roskilde University
P.O.Box 260, DK 4000,
Roskilde, DENMARK

JUBBER, Peter Charles
Hilton College
P.O. Hilton Naral,
SOUTH AFRICA

KAGAWA, Shoji
Fac. of Engineering,
Kanto Gakuin University
Mutsuura, Kanazawa-ku, Yokohahama 236

KAJIYAMA, Masaaki
Mizuho Agricultural Senior High School
Ishihata, Mizuho-machi, Nishitama-gun,
Tokyo 190-12

KALLUS, Daniel J.
2213 Shell
Midland, TX 79705,
U.S.A.

KAMADA, Minoru
Kumagaya High School
1813 Kumagaya, Kumagaya-shi,
Saitama 360

KAMIYA, Isao
College of General Education,
Nagoya University
Chikusa-ku, Nagoya, Aichi 464

KANAME, Shinji
Suzurandai High School
9 Nakaichiri, Shimotanigami, Yamada-cho,
Kita-ku, Kobe-shi, Hyogo 651-11

KANATHARANA, Proespichaya
Chemistry Department
Prince of Songkla University
HAT-YAI, 90112, THAILAND

KANDA, Seiichi
Otsubojutaku 3-14, Hachimancho
Tokushima-shi, Tokushima 770 (Home)

KANEKO, Rokuro
Tokyo University of
Agriculture and Technology
2-24 Naka-cho, Koganei-shi, Tokyo 184

KANETSUNA, Hidenori
Kasai Minami High School
1-11-1 Minamikasai, Edogawa-ku,
Tokyo 132

KANZAKI, Natsuko
Tsurumi Senior High School
6-2-1 Shimosueyoshi, Tsurumi-ku,
Yokohama, Kanagawa 230

KAPOOR, K.L.
Dept. of Chemistry, Hindu College
University of Delhi,
Delhi 110007, INDIA

KASHIWAGI, Kenji
Yazu High School
Kunoji, Koge-cho, Yazu-gun,
Tottori 680-04

KASORI, Masao
3-11-6 Nagasaki, Toshima-ku, Tokyo 171
(Home)

KATABUA, Musanda
University of Kinshasa
B.P. 137
Kinshasa XI, ZAIRE

KATAE, Yasumi
Koishikawa High School
2-29-29 Honkomagome, Bunkyo-ku,
Tokyo 113

KATAHIRA, Katsuhiro
The University of Tsukuba
1-1-1 Tennoudai, Sakura-mura,
Niihari-gun, Ibaraki 305

KATO, Shunji
Kobe Gakuin University
Igawadani, Nishi-ku, Kobe-shi,
Hyogo 651-11

KATO, Yosisige
3-1 Asaoka-cho, Chikusa-ku, Nagoya-shi,
Aichi 464 (Home)

KATSUTA, Norio
Shiraoka High School
275-1 Takaiwa, Shiraoka,
Minamisaitama-gun, Saitama 349-02

KATZ, David A.
1621 Briar Hill Rd,
Gladwyne, PA 19035,
U.S.A.

KAZAMA, Tetsuya
Kodaira Minami High School
1500-1 Joosuihonmachi, Kodaira-shi,
Tokyo 187

KELLY, P.J.
Dept. of Education,
Southampton University
Southampton, S09 5NH, U.K.

KHODAIR, Ahmed I.A.
Vice President,
Suez Caual University
Ismailia, EGYPT

KHROMINE, Viatcheslav
Ministry of Higher Education, USSR
Lusinovskaja, 51, Moscow USSR

KIDOKORO, Tadahiko
Fac. of Science, Tokai University
1117 Kitakaname, Hiratsuka-shi,
Kanagawa 259-12

KIHARA, Hiroshi
Hyogo University of Teacher Education
942-1 Shimokume, Yashiro-cho, Kato-gun,
Hyogo 673-14

KIKUCHI, Mitsuko
College of Education, Nihon University
Tamura-machi, Koriyama-shi,
Fukushima 963

KIKUCHI, Osamu
Dept. of Chemistry, Fac. of Science,
The Tsukuba University
Sakuramura, Niihari-gun, Ibaraki 305

KILKER, Richard
Apt. 1616, Birchwood Court
North Brunswick, NJ 08902,
U.S.A.

KIM, Chang Hwan
College of Science
Yonsei University
Seoul, KOREA

KIM, Chang Yol
Dept, of Chemical Education,
Chonbuk University
Chinju KOREA

KIM, Si-Joong
Dept. of Chemistry,
Korea University
Seoul, KOREA

KINJO, Yoshiaki
1 Senbaru Nishihara-cho, Okinawa

KIRSCH, Albert S.
175 Freeman Street,
Brookline, MA 02146,
U.S.A.

KIRSCHNER, Stanley
Department of Chemistry,
Wayne State University
Detroit, MI 48202, U.S.A.

KISHIMOTO, Takao
School of Science and Engineering
Waseda University
3-4-1 Okubo, Shinjuku-ku, Tokyo 106

KITAHAMA, Hachiro
Wakayama Technical College
Noshima 464, Nada, Gobo-shi,
Wakayama 649-15

KITAHARA, Ayao
Science University of Tokyo
1-3 Kagurazaka, Shinjuku-ku,
Tokyo 162

KITAHARA, Takio
Fac. of Science, Tokai University
1117 Kitakaname, Hiratsuka-shi,
Kanagawa 259-12

KITAMURA, Masaharu
Nagashima High School
Higashinagashima, Kiinagashima-cho,
Kitamuro-gun, Mie 519-32

KITO, Akio
Hibiya High School
Nagatacho Chiyoda-ku, Tokyo 100

KOGA, Gen
Ibaraki University
Bunkyo 2-1-1, Nito-shi, Ibaraki 310

KOIDE, Tsutomu
Osaka University of Education
Minami-kawahori, Tennoji-ku, Osaka-shi,
Osaka 543

KOJIMA, Kazuo
Tachikawa High School
2-13-5 Nishiki-cho, Tachikawa-shi,
Tokyo 190

KOLB, Doris K.
3130 Chartwell Rd.,
Peoia, IL 61614
U.S.A.

KOLB, Kenneth E.
3130 Chartwell Rd
Peoria, IL 61614,
U.S.A.

KOMATSU, Masakichi
Hongo 3-42-6, Bunkyo-ku, Tokyo 113
Nankodo, Limited

KONNO, Hiroshi
Keimei Gakuen School
5-11 Haijima, Akishima-shi, Tokyo 196

KOSHIO, Genya
Fac. of Agriculture,
Tamagawa University
Tamagawagakuen, Machida-shi, Tokyo 194

KOTANI, Masahiro
Fac. of Science, Gakushuin University
1-5 Mejiro Toshima-ku, Tokyo 171

KOYANO, Kinko
Apt. 1-13, Myodaiji-cho, Sakashita 11-72
Okazaki-shi, Aichi 444 (Home)

KULKARNI, Ram Anant.
Excel Estate, S V Road, Goreraon (West).
Bombay, 400 062, INDIA

KURAHASHI, Masamichi
587-3 Okuboryoke
Urawa-shi, Saitama 338

KURODA, Eiko
Kwansei Gakuin High School
1-1-155 Uegahara, Nishinomiya-shi,
Hyogo 662

KUWAHARA, Minoru
The Institute of Vocational Training
1960 Aihara, Sagamihara-shi,
Kanagawa 229

KYZYAKOV, Juvie
Fac. of Chemistry,
Moscow State University
117234, Moscow, USSR

LANGLEY, William F.
1 Short St.,
Mafikeng 8670
BOPHUTHATSWANA

LERMAN, Zafra M.
Science Department, Columbia College
600 S. Michigan Avenue,
Chicago, IL 60605, U.S.A.

LIU, Zhi-Xin
Department of Chemistry,
Beijing Normal University
Beijing, CHINA

LOW, W.K.
Physical Education Unit,
University of Hong Kong
HONG KONG

MAKELA, Marja-Leena Mirjami
Hiidenranta 8, 03400 Vihti,
FINLAND

MALCOLM, Geoffrey N.
Department of Chemistry and Biochemistry
Massey University
Palmerston North, NEW ZEALAND

MANUNAPICHU, Kamchorn
Dept. of Chemistry, Fac. of Science,
Mahidol University
Rama VI Road, Bangkok, THAILAND

MASSIE, Samuel P.
12203 Brittany Place
Laurel, MO 20708,
U.S.A.

MASUI, Yukio
Science Education Institute of
Osaka Prefecture
Karita,Sumiyoshi-ku,Osaka-shi, Osaka 558

KUNUGI, Akira
Tokyo College of Pharmacy
1432-1 Horinouchi,
Hachioji-shi, Tokyo 192-03

KURATA, Takeo
Meiji University
1-1-1 Higashimita, Tama-ku,
Kawasaki-shi, Kanagawa 214

KUROISHI, Yoshinobu
Dept. of Chemistry, Fac. of Science
Saitama University
Urawa-shi, Saitama 338

KUZUMI, Eiko
Tokyo Gakugei University
Nukui-kitamachi, Koganei-shi, Tokyo 184

LAGOWSKI, J.J.
Department of Chemistry,
The University of Texas
Austin,TX, 78712, U.S.A.

LEE, Wha-Kuk
Dept, of Chemical Education,
Chonbuk National University
ChonjU 520, KOREA

LIU, Yi-Lun
Department of Chemistry,
Northwest University
Xian CHINA

LOEFFLER, William R.
Benchmark Technologies Corp.
1995 Tremainsville Rd.
Toledo, OH 43613, U.S.A.

MAARSCHALK, Jan
6 Dartmoor Road,
Florida Hills 1710,
SOUTH AFRICA

MAKINOSE, Kiyokazu
Nima High School
Yunotu, Yunotu-chi, Nima-gun,
Shimane 699-25

MANALO, Juanita A.
The Philippine Women's University
Taft Avenue, Manila
PHILIPPINES

MARUYAMA, Masao
Miyagi University of Education
Aoba, Aramaki, Sendai-shi, Miyagi 980

MASUDA, Shigeru
Dept. of Chemistry, College of Arts
and Sciences, The University of Tokyo
3-8-1 Komaba, Meguro-ku, Tokyo 153

MATSUBARA, Shizuo
The National Institute for
Educational Research
6-5-22 Shimomeguro, Meguro-ku, Tokyo 153

MATSUDA, Keizo
Fac. of Science, Tokai University
Kitakaname, Hiratsuka-shi,
Kanagawa 259-12

MATSUI, Kichinosuke
Chofu The Third Junior High School
3-2-7 Somechi, Chofu-shi, Tokyo 182

MATSUMOTO, Shoji
Fukaya High School
Shukune 315, Fukaya-shi, Saitama 366

MATSUURA, Ryohei
Fukuoka Women's University
Kasumigaoka 1-1-1, Higashi-ku,
Fukuoka-shi, Fukuoka 813

MATSUURA, Shu
Dept. of Physics, Fac. of Science,
Nagoya University
Furo-cho, Chikusa-ku, Nagoya 464

MATSUURA, Tatsuo
Institute for Atomic Energy,
Rikkyo University
Nagasaka Yokosuka-shi, Kanagawa 240-01

MERA, Seiji
Shinjuku High School
6-2-1 Sendagaya, Shibuya-ku, Tokyo 151

MIKAN, Francis M.
2103 Tower Drive
Austin, TX 78703,
U.S.A.

MIRANDA, S.R.
University of the Philippins
Quezon City, PHILIPPINS

MISAWA, Katsumi
Musashino Girls' Senior High School
1-1-20 Shinmacho Hoya-sho, Tokyo 202

MITOMO, Shun-ichi
Kawagoe Girl's High School
Rokken-cho Kawagoe-shi, Saitama 350

MITSUI, Sumio
Urawa Senior High School
5-3-3 Ryoke, Urawa-shi, Saitama 336

MITSUZAWA, Shummei
Fac. of Science, Tokai University
Kitakaname, Hiratsuka-shi,
Kanagawa 259-12

MIURA, Shigeru
782-8 Shizunami, Haibara-cho,
Haibara-gun, Shizuoka 421-04 (Home)

MIYAGI, Masaaki
High School Attached to
Tokyo Gakugei University
4-1-5 Shimouma, Setagaya-ku, Tokyo 154

MIYATA, Haruo
Shujitsu Joshi University
1-6-1 Nishigawara,
Okayama-shi, Okayama 703

MOCHIZUKI, Kazuyuki
Koishikawa High School
2-29-29 Honkomagome, Bunkyo-ku,
Tokyo 113

MOERWANI, Patimah
Department of Chemistry, F MIPA
Universitas Indonesia, Salemba Raya No.4
Jakarta Pusat, INDONESIA

MOORE, Elizabeth A.
Department of Chemistry,
Eastern Michigan University
Ypsilanti, MI 48197, U.S.A.

MOORE, John W.
Department of Chemistry,
Eastern Michigan University
Ypsilanti, MI 48197, U.S.A.

MORIGUCHI, Yoshiki
Dept. of Chemistry, Fac. of Science,
Fukuoka University of Education
Akama, Munakata-shi, Fukuoka 811-41

MUKAIYAMA, Teruaki
Dept. of Chemistry, Fac. of Science,
The University of Tokyo
7-3-1 Hongo, Bunkyo-ku, Tokyo 113

MUNGARULIRE, Joseph
National University of Rwanda,
P.O.B. 117 Butare, Rwanda

MURAKAMI, Mitsuhiro
Nara University of Education
Takabatake-cho, Nara-shi, Nara 630

MURAKAMI, Yoshikazu
Fac. of Education, Ehime University
1-5-22 Mochida-chi Marsuyama-shi,
Ehime 790

MURAOKA, Motomu
Dept. of Chemistry, Fac. of Science,
Josai University
1-1 Keyaki-dai, Sakado, Saitama 350-02

MURASE, Norio
Fac. of Science and Engineering,
Tokyo Denki University
Hatoyama-cho, Hiki-gun, Saitama 350-03

MURATA, Kenji
Okazakihigashi High School
Ushiroyama, Ryusenji-cho, Okazaki-shi,
Aichi 444-35

MUTAI, Kiyoshi
College of Arts and Science,
The University of Tokyo
Komaba, Meguro-ku, Tokyo 153

McDONALD, David W.
P.O.Box 265
Sidney MT 59270,
U.S.A.

McDONALD, Joy-Lyn L.
P.O.Box 265
Sidney MT 59270,
U.S.A.

McKELVEY, Ronald D.
RT2 Box 70E
La Crosse,
W1 54601, U.S.A.

NAGASATO, Mitsuaki
Korigaoka Senior High School
2-18-1 Higashinakaburi, Hirakata-shi,
Osaka 573

NAGATA, Shinichi
Higashimurayama High School
4-26-1 Onta-cho, Higashimurayama-shi,
Tokyo 189

NAKAMURA, Daiyu
Dept. of Chemistry, Fac. of Science,
Nagoya University
Furocho-cho, Chikusa-ku, Nagoya 464

NAKAMURA, Takeo
Nikko Rica Corporation
22 Oosanuki, Meiwamura, Oura-gun,
Gunma 370-07

NAKAMURA, Yoshinobu
Kasukabe Girl's High School
6-1-1 Kasukabe-higashi, Kasukabe-shi,
Saitama 344

NAKANISHI, Keiji
Shizyonawate Gakuen Women's College
Gakuencho Daito-shi, Osaka 574

NAKANO, Hidehiko
Dept. of Applied Chemistry,
Himeji Institute of Technology
2167 Shosha, Himeji-shi, Hyogo 671-22

NAKAYAMA, Isao
Azabu High School
2-3-29 Motoazabu, Minato-ku, Tokyo 106

NAKAYAMA, Minoru
Tokyo Institute of Technology
2-12-1 Ookayama, Meguro-ku, Tokyo 152

NAKAYAMA, Yuichi
Hachijo Senior High School
3020 Okago, Hachijo-cho,
Hachijo-jima, Tokyo 100-14

NAPARI, Pirjo Helena
Pyorrekuja 4 G 95
01600 Vantaa,
FINLAND

NARA, Sakari
Hokkaido University of Education
Hachiman-cho, Hakodate-shi, Hokkaido 040

NARASAKA, Koichi
Dept. of Chemistry, Fac. of Science,
The University of Tokyo
7-3-1 Hongo, Bunkyo-ku, Tokyo 113

NATSUME, Kiyohisa
The University of Tokyo
7-3-1 Hongo, Bunkyo-ku, Tokyo 113

NAWA, Nagayasu
Takeo High School
5540-2 Takeo, Takeo-machi, Takeo-shi,
Saga 843

NDIAYE, Abdourahim
Dept. of Physical Chemistry,
Fac. of Science, University of Dakar
Dakar, SENEGAL

NEWBOLD, Brian T.
Dept. of Chemistry, Fact. of Science,
Saitama University, Urawa 338
JAPAN

NG, Yin Ling
Physical Education Unit,
University of Hong Kong, HONG KONG

NICKON, Alex
Department of Chemistry,
Johns Hopkins University
Baltimore MD 21218, U.S.A.

NIGORIKAWA, Tomio
Kiyose High School
3-1-56 Matsuyama, Kiyose-shi, Tokyo 204

NIIDA, Satoshi.
The Senior High School Attached to
Tokyo Gakugei University
4-1-5 Shimouma, Setagaya-ku, Tokyo 154

NIITSUMA, Takuitsu
5-11-3 Shoran, Izumi-shi, Miyagi 981-31

NISHIHIRA, Teruko
Jounan High Schooli
6-16-36 Roppongi, Minato-ku, Tokyo 106

NISHIKAWA, Tomonari
Yamatogawa High School
4-1-72 Karita, Sumiyoshi-ku, Osaka 558

NISHIO, Shigekazu
Sakai Senior High School Attached to
The Osaka University of Commerce
358 Horiage Sakai-shi, Osaka 593

NISHITA, Kishiko
Susaki High School
391-2 Shimobun-ko, Susaki-shi, Kochi 785

NODA, Shiro
Sapporo Hiragishi High School
5-18 Hiragishi, Toyohira-ku,
Sapporo-shi, Hokkaido 062

NODA, Tamehisa
Tachikawa High School
2-13-5 Nishiki-cho, Tachikawa-shi,
Tokyo 190

NODA, Toyoko

NOMURA, Yujiro
National Center for University
Entrance Examination
2-19-23 Komaba, Meguro-ku, Tokyo 153

NOZAWA, Naomi
Kumagaya High School
1813 Kumagaya, Kumagaya-shi,
Saitama 360

O'BRIEN, James F.
2252 E. Latoka
Springfield MI 65804,
U.S.A.

OCHIAI, Masae
Minami High School
3-11-10 Nakamagome, Ota-ku, Tokyo 143

ODAKA, Yoshio
Kogakuin University
2665-1 Nakano-machi, Hachioji-shi,
Tokyo 192

OGAWA, Keiichiro
Dept. of Chem., College of Arts and
Sciences, The Univerisity of Tokyo
3-8-1 Komaba, Meguro-ku, Tokyo 153

OGINO, Kazuko
College of Medical Sciences,
Tohoko University
Seiryo-machi, Sendai-shi, Miyagi 980

OGUCHI, Shoshichi
Fac. of Education, Bunkyo University
3337 Minamiogishima, Koshigaya-shi,
Saitama 343

OHIRA, Kenji
Mitaka High School
6-21-21 Shinkawa, Mitaka-shi, Tokyo 181

OHKAWA, Takashi
Koyo Gakuin High School
3-138 Sumiishi-cho,
Nishinomiya-shi, Hyogo, 662

OHKI, Yumi
Nishiarai Junior High School
7-22-1 Nishiarai,
Adachi-ku, Tokyo 123

OHMACHI, Tadatoshi
Mitaka High School
6-21-21 Shinkawa, Mitaka-shi, Tokyo 181

OHOKA, Minoru
Soken Chemical & Engineering Co. Ltd.
132 Kamihirose, Sayama-shi,
Saitama 350-13

OHSUGI, Naoyuki
NHK Gakuen High School
2-36 Fujimidai, Kunitachi-shi,
Tokyo 186

OKAMOTO, Hajime
13 Ikenouchi-cho, Fukakusa,
Fushimi-ku, Kyoto 612

OKI, Michinori
Dept. of Chemistry, Fac. of Science,
The University of Tokyo
Hongo, Bunkyo-ku, Tokyo 113

OLIVEIRA, Elisabeth
Universidade de Sao Paulo Instituto de
Quimica, B2S USP C.Postal 20780,
CEP 01000 Sao Paulo, BRASIL

OLVER, Norman H.
4A Victor Ave., Kew,
Victoria 3101,
AUSTRALIA

OMICHI, Hiroaki
3-12-6 Umezono, Kiyose-shi, Tokyo 204
(Home)

ONDA, Masao
Sophia University
Kioicho Chiyoda-ku, Tokyo 102

ONG, Henry K.O.
Dept. of Chemistry, Fac. of Science,
National University of Singapore
Kent Ridge, 0511, SINGAPORE

ONO, Hiroshi
Higashimurayama High School
4-26-1 Onta-cho, Higashimurayama-shi,
Tokyo 189

OOTUKA, Haruo
Institute for Applied Optics
20 Kanda-nishikityo 3 Tiyoda-ku,
Tokyo 101

ORNA, Mary Virginia
Department of Chemistry,
College of New Rochelle
New Rochelle, NY 10801, U.S.A.

OSAKI, Kenji
6-11-23 Tsukahara
Takatsuki-shi, Osaka 606 (Home)

OSAKI, Tomoe
Fukuoka Institute of Technology
Shimowajiro, Higashi-ku, Fukuoka-shi,
Fukuoka 811-02

OSAWA. Masumi
Dept.of Chemistry,
Tokyo Gakugei University
Nukuikita-machi, Koganei-shi, Tokyo 184

OTANI, Yoshihisa
Komaba High School Attached to
The University of Tsukuba
4-7-1 Ikejiri, Setagaya-ku, Tokyo 154

OTSUKI, Isamu
Miyagi Second Girl's High School
4-1 Renbo 1-chome, Sendai-shi,
Miyagi 980

PEREIRA, Mariana P.B.A.
Av. Combatentes da Grande Guerra
15-1-D, P 2700 Amadora,
PORTUGAL

PIMENTEL, George C.
Laboratory of Chemical Biodynamics,
University of California
Berkeley, CA 94720, U.S.A.

PIRSON, Pierre
Facultes Notre-Dame de la Paix
Department de Chimie, 61, Rue de
Bruxelles, B-5000 Namur, BELGIUM

POKROVSKY, A.
7, Place de Fontenoy, 75700 Paris,
FRANCE

PROTOPAPAS, Nicos
5 Lohn Metaxas Str Nicosia P,S 164,
CYPRUS

PRYDE, Lucy T.
Dept. of Chemistry, Southwestern College
900 Otay Lakes Road,
Chula Vista, CA 92010, U.S.A.

PSHEZHETSKII, Valeri S.
117234 Moscow University
USSR

RAM, Nayan
B-14, Gandhi Nagar
Moradabad-244001,
INDIA

RAMETTE, Richard W.
805 Highland Ave,
Northfield, MN 55057,
U.S.A.

RAUT, Kamalakar B.
708 Penwaller Rd.
Savannah, GA 31410,
U.S.A.

RAUT, Mudhuri
2000 E 95th St.,
Kansas City, MO 64131,
U.S.A.

ROBERTSON, Brock
Dept. of Chemistry, Univ. of Calgary
Calgary Alberta T2N 1N4
CANADA

ROCHA-FILHO, R.C.
Dep. of Chemistry, UFS. Car
C.P.676, Sao Carlos -S.P.
13560, BRAZIL

ROGERS, Barbara A.
425 Ena Road #607-C
Honolulu, HI 96815,
U.S.A.

RONQUILLO, Rebecca D.
Oikoshi 19-302, Tsukuba University
Sakura-mura, Niihari-gun
Ibaraki-ken 305 JAPAN

SAIJYO, Motoyasu
Nihon University Buzan Girl's
Senior High School
3-15-1 Nakadai, Itabashi-ku, Tokyo 174

SAITO, Jyo
1054 Sugekari, Kani-shi, Gifu, 509-02
(Home)

SAITO, Koichi
Komaba Toho High School
4-5-1 Ikegiri, Setagaya-ku, Tokyo 154

SAITOU, Akira
Fac. of Science, Tokai University
Kitakaname, Hiratsuka-shi,
Kanagawa 259-12

SAKAI, Takayoshi
3-11 Miyuki-cho, Hirosaki-shi,
Aomori 036

SAKAKI, Tomohiko
Ikezono 2-60, Chikusa-ku, Nagoya-shi,
Aichi 464

SAKANOUE, Masanobu
Fac. of Science, Kanazawa University
Waki Tatsunokuchi-machi,
Nomi-gun, Ishikawa 923-12

SAKAO, Takashi
Fac. of Education,
Kagoshima University
Korimoto, Kagoshima-shi, Kagoshima 890

SAKURAGI, Hirochika
Dept. of Chemistry,
The University of Tsukuba
Sakura-mura, Niihari-gun, Ibaraki 305

SAMPEI, Takao
Joto High School
3-22-1 Ohjima, Koto-ku, Tokyo 136

SANCHORAWALA, Chandravadan J.
Navrangpura, Ahmedabad 380 009
Gujarat State, INDIA

SANGEN, Osamu
Dept. of Applied Chemistry,
Himeji Institute of Technology
2167 Shosha, Himeji-shi, Hyogo 671-22

SANKAWA, Ushio
Fac. of Pharmaceutical Sciences
The University of Tokyo
7-3-1 Hongo, Bunkyo-ku, Tokyo 113

SANO, Shunsuke
Sakaide High School
2-1-5 Bunkyo-cho, Sakaide-shi,
Kagawa 762

SANWO, Norio
6165 Hikio-cho, Fukuyama-shi,
Hiroshima 721

SARQUIS, Arlyne M.
Miami University-Middletown
4200 E University Blvd
Middletown, OH 45042, U.S.A.

SARQUIS, Jerry L.
Chemistry Department,
Miaimi University
Oxford OH 45056, U.S.A.

SASAKI, Ken-ichi
Dept. of Chemistry, Fac. of Science,
Nagoya University
Furo-cho, Chikusa-ku, Nagoya, Aichi 464

SASAKI, Shin-ichi
Toyohashi University of Technology
Temauk-cho, Toyohashi-shi, Aichi 440

SASAKI, Tsunetaka
Tokai University
1117 Kitakaname, Hiratsuka-shi,
Kanagawa 259-12

SASAKI, Yoshiko
9-1 Hirose-cho, Sendai-shi,
Mitygi 980

SASAMURA, Yasuaki
Tomakomai Technical College
443 Nishikioka, Tomakomai-shi,
Hokkaido 059-12

SATO, Sanae
25-52 Minamiyawata-cho,
Shizuoka-shi, Shizuoka 422 (Home)

SATO, Tomohisa
Hikarigaoka Senior High School
1 Hikarigaoka, Nerima-ku, Tokyo 176

SAWADA, Kingo
Morioka Dai-ichi Senior High School
Ueda, Morioka-shi, Iwate 020

SCHAFFER, Claus E.
Ronnebarvej 136B
DK-2840 Holte
DENMARK

SCHWARTZ, A.Truman
Department of Chemistry,
Macalester College
Saint Paul, MS 55105, U.S.A.

SEDDON, G. Malcolm
School of Chemical Sciences,
University of East Anglia
Norwich NR4 7TJ, U.K.

SEKIGUCHI, Katsuyasu
1-6-9 Harayama, Urawa-shi,
Saitama 336 (Home)

SEKIYA, Takuzo
Komaba Toho High School
4-5-1 Ikejiri Setagaya-ku, Tokyo 154

SENGOKU, Toshihiko
Hokkaido Science Education Center
Miyanomori, 4-jo 7-chome, Chuo-ku,
Sapporo-shi, Hokkaido 064

SENVAR, Cemil Bedri
Marmara Universitesi
At. Egitim Fakultesi
Ziverbey-Istanbul, TURKEY

SETA, Keiichi
Hirakata Junior High School
2115 Hirakata, Koshigaya-shi,
Saitama 343

SETH, Chand K.
Dept. of Chemistry
Hindu College, Delhi-110007
INDIA

SEYDEL, Judith C.
2815 W.Morningside Dr.
Idaho Falls, ID 83402,
U.S.A.

SHAGAWA, Hirozi
Shirakawa High School
54 Minaminobori-machi, Shirakawa-shi,
Fukushima 961

SHARMA, R. C.
Rameshwar Ji Ka Nohra
Rampura Kota-324006 (Raj.)
INDIA

SHIBA, Tetsuo
Fac. of Science, Osaka University
1-1 Machikaneyama-cho, Toyonaka-shi,
Osaka 560

SHIBATA, Shoji
Meiji College of Pharmacy
1-35-23 Nozawa, Setagaya-ku, Tokyo 154

SHIMADA, Toshiro
Hikarigaoka High School
2-1-35 Asahi-machi, Nerima-ku, Tokyo 176

SHIMIZU, Sakie
Hachinohe Institute of Technology
88-1 Obiraki, Myo, Hachinohe-shi,
Aomori 031

SHIMOZAWA, John T.
Dept. of Chemistry, Fac. of Science,
Saitama University
Urawa-shi, Saitama 338

SHIOHAMA, Takashi
Arai Senior High School
1-10-1 Tamachi, Arai-shi, Nigata 944

SHOJI, Keiko
Shiroishi Girl's High School
2-1 Chorenjo, Shiroishi-shi,
Miyagi 989-02

SILBERSTEIN, Judith
Dept. of Science Teaching
The Weizmann Institute of Science
Rehovot 76100, ISRAEL

SONE, Kozo
Dept. of Chemistry, Fac. of Science,
Ochanomizu University
Otsuka Bunkyo-ku, Tokyo 112

SUEFUJI, Yoshimasa
Hakuo High School
1-6-22 Motoasakusa, Taito-ku, Tokyo 111

SUGA, Takayuki
Fac. of Science, Hiroshima University
1 Higashisenda-machi, Naka-ku,
Hiroshima-shi, Hroshima 730

SUGITA, Ichirou
Konosu High School
1020 Ooma, Konosu-shi, Saitama 365

SHAPIRO, Stanley Jay
Apt. 19E, 400 Second Ave.
N.Y.C NY 10010,
U.S.A.

SHIAU, George T.
Dept. of Chemistry, National Taiwan
Normal University, 88, Sec.5,
Roosevelt Rd, Taipei, TAIWAN ROC

SHIBASAKI, Takehiro
2-9-9 Uemine, Yono-shi, Saitama 338
(Home)

SHIMADA, Hiroshi
Fac. of Science, Tokai University
Kitakaname, Hiratsuka-shi,
Kanagawa 259-12

SHIMIZU, Kazuyuki
Minamino Senior High School
2-11-1 Minamino Tama-shi, Tokyo 192-02

SHIMODA, Yoshio
Hibiya High School
2-16-1 Nagata-cho, Chiyoda-ku, Tokyo 100

SHIN, Hyonsohb
Korea University
1-700 Ogawa-cho, Kodaira-shi, Tokyo 187

SHIOTA, Michio
Dept. of Chemistry,
Ochanomizu University
Otsuka, Bunkyo-ku, Tokyo 112

SHUKLA, Raj Kishor
Head of The Dept. of Chemistry
c/o Naraini Chauraha Naraini
Banda U.P. 210129, INDIA

SMITH, Douglas D.
648 Valley Forge Trail
P.O.Box 206, Rockton, IL
61072, U.S.A.

STOGRYN, Daniel E.
1826 Bara Rd
Glendale, CA 91208,
U.S.A.

SUGA, Shoiti
Miyashiro High School
Higashi 611, Miyashiro-machi,
Minamisaitama-gun, Saitama 345

SUGIHARA, Gohsuke
Fukuoka University
Jonan-ku, Fukuoka-shi, Fukuoka 814-01

SUGIYAMA, Niichi
Osaka Meisei Gakuen
5-44 Esashi, Tennoji-ku,
Osaka-shi, Osaka 543

SUSSMAN, Martin
Dept. of Chemical Engineering,
Tufts University
Medford, MA, 02155, U.S.A.

SUURING, Dora
27 Silverstream Crafton Downs,
Wellington 4,
NEW ZEALAND

SUZUKI, Akira
Funabashi-asahi High School
Asahi-cho, Funabashi-shi, Chiba 273

SUZUKI, Chieko
Osaka Kunei Women's College
1-4-1 Shojaku, Settsu-shi, Osaka 564

SUZUKI, Kazuo
Tokyo Metropolitan Institute for
Educ. Resear. and In-Service Training
1-1-14 Meguro Meguro-ku Tokyo 153

SUZUKI, Shuichi
Saitama Institute of Technology
1690 Fusaiji, Okabe-machi,
Saitama 369-02

SWARTNEY I. Joyce
Buffalo State College
1300 Elmwood Ave.
Buffalo, N.Y. 14222, U.S.A.

TAGA, Mitsuhiko
Dept. of Chemistry, Fac. of Science,
Hokkaido University
Kita 10-jo,Nishi 8-Chome, Sapporo 060

TAKAESU, Akira
Futaba High School
14-1 Rokuban-cho, Chiyoda-ku, Tokyo 102

TAKAMIYA, Nobuo
Dept. of Chemistry,
Waseda University
3-4-1 Okubo Shinjiku-ku, Toyko 160

TAKANASHI, Masahide
Keiogijuku Yochisha
2-35-1 Ebisu, Shibuya-ku, Toyko 150

TAKANO, Bokuichiro
Dept. of Chem., The College of Arts and
Sciences, The University of Tokyo
3-8-1 Komaba, Meguro-ku, Tokyo 153

TAKANO, Jiro
Fac. of Science, Tokai University
1117 Kitakaname, Hiratsuka-shi,
Kanagawa 259-12

TAKANO, Takeko
School of Science and Engineering
Waseda University
3-4-1 Okubo, Shinjuku-ku, Tokyo 160

TAKANO, Tsutomu
Shumei-Yachiyo High School
803 Sohnohashi, Yachiyo-shi, Chiba 276

TAKAYANAGI, Masao
Dept. of Chemistry, Fac. of Science,
The University of Tokyo
Hongo Bunkyo-ku, Tokyo 113

TAKEBAYASHI, Matsuji
1-122 Osakaba, Yao-shi, Osaka 581 (Home)

TAKEBAYASHI, Yasutsugu
Shumei Ageo High School
1012 Ueno, Ageo-shi, Saitama 362

TAKEDA, Chikayuki
Koshigaya Minami High School
6-220 Kawayanagi-cho, Koshigaya-shi,
Saitama 343

TAKEDA, Kazuyoshi
Science University of Tokyo
1-3 Kagurazaka, Shinjuku-ku, Tokyo 162

TAKEMURA, Hiroshi
Chukyo Women's University
55 Nadakayama, Yokone-cho, Obu-shi,
Aichi 474

TAKEOKA, Yoshiko
Otsuma Women's University
12 Sanban-cho, Chiyoda-ku, Tokyo 102

TAKEUCHI, Yoshito
Dept. of Chemistry, College of Arts
and Sciences, The University of Tokyo
3-8-1 Komaba, Meguro-ku, Tokyo 153

TAKIZAWA, Yasuomi
Dept. of Chemistry
Tokyo Gakugei University
Nukuikita-machi, Koganei-shi, Tokyo 184

TAM, Patrick
Department of Chemistry,
Hong Kong Baptist College, 244 Waterloo
Road, Kowloon, HONG KONG

TANAKA, Fumio
Chigusa High School
Meito-ku, Nagoya-shi, Aichi 465

TANAKA, Motoharu
Fac. of Science, Nagoya University
Furo-cho, Chikusa-ku,
Nagoya-shi, Aichi 464

TANAKA, Sinji
Yokkaichi-Yogo High School
Hachioji-cho, Yokkaichi-shi, Mie 510

TASAKA, Koa
International Christian University
3-10-2 Osawa, Mitaka-shi, Tokyo 181

TERAO, Hiromitsu
Fac. of Education, Tokushima University
1-1 Minamijosanjima-cho, Tokushima-shi,
Tokushima 770

THULSTRUP, Erik W.
Chem. Department. Royal Danish School of
Educational Studies, Emdrupvej 115 B
DK-2400 Copenhagen NV, DENMARK

TOGARI, Susumu
Meijo University
Yagoto, Tempaku-cho, Tempaku-ku,
Nagoya-shi, Aichi 468

TOKORO, Roberto
Universidade de Sao Paulo, Instituto de
Quimica USP-B85, C.Postal 20780,
CEP 0100, Sao Paulo, BRASIL

TORIMOTO, Noboru
Science Education Institute of
Osaka Prefecture
Karita Sumiyoshi-ku Osaka-shi, Osaka 558

TOROP, William
162 Hunters Run.
Newtown Square, PA
19073 U.S.A.

TOWNSEND, Iain T.
Mitchell College of Advanced Education
Bathurst N.S.W. 2795 AUSTRALIA

TSUJI, Masahiro
Dept. of Chemistry, Fac. of Science,
Furo-cho, Chikusa-ku, Nagoya-shi,
Aichi 464

TSUKAHARA, Togo
1-31-1 Akatsuka, Itabashi-ku, Tokyo 175
(Home)

TSUNAKAWA, Sukenari
Mitsubishi Che. Ind. Ltd.
1000 Kamoshida-cho, Yokohama-shi,
Midori-ku, Kanagawa 227

TSUNETSUGU, Josuke
Dept. of Chemistry, Fac. of Science,
Saitama University
Urawa-shi, Saitama 338

UEHARA, Hiromichi
Fac. of Science, Josai University
1 Keyakidai, Sakado-shi, Saitama 350-02

UMEKI, Matsusuke
Higashimurayama High School
4-26-1 Onta-cho, Higashimurayama-shi,
Tokyo 189

TAYA, Kazuo
Dept. of Chemistry,
Tokyo Gakugei University
Nukuikita-machi, Koganei-shi, Tokyo 184

TERATANI, Shousuke
Tokyo Gakugei University
4-1-1 Nukuikita-machi,
Koganei-shi, Tokyo 184

TODA, Shozou
Dept. of Agricultural Chemistry,
The University of Tokyo
1-1-1 Yayoi, Bunkyo-ku, Tokyo 113

TOKITA, Sumio
Dept. Applied Chemistry,
Fac. of Engineering, Saitama University
Urawa-shi, Saitama 338

TOMIZAWA, Kenzo
Honjo Kita High School
Nitte 2167-1, Honjo-shi,
Saitama 367

TORIYAMA, Yoshiko
School for The Blind Attached to
The University of Tsukuba
3-27-6 Mejiro-dai, Bunkyo-ku, Tokyo 112

TOSHIMA, Naoki
Fac. of Engineering,
The University of Tokyo
7-3-1 Hongo Bunkyo-ku, Tokyo 113

TSUCHIYA, Masako
St. Dominic's Inst. Senior High School
1-10-1 Okamoto, Setagaya-ku, Tokyo 157

TSUKAGOSHI, Hiroshi
Tadao High School
18-2276 Kiso-machi, Machida-shi,
Tokyo 194

TSUMAKI, Takao
Koishikawa High School
29-29 Honkomagome 2-chome, Bunkyo-ku,
Tokyo 113

TSUNAKAWA, Sukenari
1000 Kamoshida-cho, Midori-ku,
Yokohama-shi, Kanagawa 227
Mitsubishi Che. Ind. Lta.

UCHIKOBA, Hiroyuki
Fuji High School
5-21-1 Yayoi, Nakano-ku, Tokyo 164

UENO, Masasi
Adachi High School
1-3-9 Chuohon-cho Adachi-ku, Tokyo 120

UNO, Katsushi
Tomakomai Technical College
443 Nishikioka, Tomakomai-shi,
Hokkaido 059-12

UYEMURA, Masakatsu
Tokyo Institute of Polytechnics
1583 Iiyama, Atsugi-shi, Kanagawa 243-02

VAERNEWIJCK, Jan A.
Boswegel 8
B-9210 Heusden,
BELGIUM

VAN BRANDT, P.
Cathoric University of Louvatn
BELGIUM

VRTACNIK, Margareta
UNESCO International Center for Chemical Studies, Vegova 4, 61001 Ljubljana, pp 18/I, YUGOSLAVIA

WADDINGTON, D.J.
Department of Chemistry,
University of York,
Heslington, York YO1 5DD, U.K.

WAKAIRO, Kinzi
469-17 Shimomakuri, Koshigaya-shi,
Saitama 343 (Home)

WAKI, Takeshi
Fac. of Education, Tokushima University
1-1 Minamijosanjima-machi,
Tokushima-shi, Tokushima 770

WAKIHARA, Masataka
Tokyo Institute of Technology
2-12-1 Ookayama
Meguro-ku, Tokyo 152

WALTERS, Schalk Willem
Department of Education
P.O.Box 13, CAPE TOWN 8000
SOUTH AFRICA

WANG, Kui
Faculty of Pharmaceutical Sciences
Beijing Medical College
Beijing 100083, CHINA (PRC)

WATABE, Masatoshi
Kogakuin University
2665-1 Nakano, Hachioji-shi, Tokyo 192

WATANABE, Hiroshi
Dept. of Chemistry, College of Arts and Sciences, The University of Toyko
Komaba, Meguro-ku, Toyko 153

WATANABE, Mikio
Fac. of Science, Tokai University
1117 Kitakaname, Hiratsuka-shi,
Kanagawa 259-12

WATANABE, Tokuko
2-443-45 Amanuma-cho, Omiya-shi,
Saitama 330 (Home)

WATANABE, Tomohiro
Rikkyo High School
1-2-25 Kitano, Niiza-shi, Saitama 352

WATANABE, Yutaka
Hiratsuka Konan High School
5-1 Suwa-cho, Hiratsuka-shi,
Kanagawa 254

WATANUKI, Kunihiko
Dept. of Chemistry, College of Arts and Sciences, The University of Tokyo
3-8-1 Komaba, Meguro-ku, Tokyo 153

WATERMAN, Edward L.
700 Blue Mesa Avenue
Fort Collins, CO 80526,
U.S.A.

WAYBORN, Hart R.
Department of Chemistry,
University of Manitoba
Winnipeg, Manitoba R3T 2N2, CANADA

WEGEMAR, Borje R.V.
Norra Allegatan 32B
S-72219 Vasteras,
SWEDEN

WEILL III, David R.
Department of Science
Shady Side Academy, 423 Fox Chapel Road
Pittsburgh, PA, 15238, U.S.A.

WEISSMANN, Katherine E.
15585 West Baldwin Road
Chesaning, MI 48616,
U.S.A.

WELLER, Bernard E.
Dept. of Physical Sciences Politec. of the South Bank, Borough Rd.,
London SEIOAA, London

WEY, Ming-tong
Science Education Center, National Taiwan Normal University, 88 Sec.5
Rooseveit Rd, Taipei, 117, TAIWAN, ROO

WHYTE, Helena M.
100 El Morro
Los Alanos, NM 87544,
U.S.A.

WOODS, Gordon T.
Monmouth School,
Monmouth, Gwent,
NP5 3XP, U.K.

XIA, Yan
Department of Chemistry,
East China Normal University
Shanghai 200062, CHINA

YACHI, Tadayoshi
Tokai Uniersity,
2-28-4 Tomigaya,
Shibuya-ku, Tokyo 151

YAGODIN, Gennadij A.
D. I. Mendeleev Inst. of Chem. Tech.
Miusskaia Sq. 9, Moscow
125047, USSR.

YAGUCHI, Hirosi
Tottori Commercial High School
401 Kita 2-chome, Koyama, Tottori-shi,
Tottori 680

YAKUBOV, Khamid M.
Tadjik State University, Lenin Street,
17, 734016, Dushanbe,
USSR

YAMADA, Kazutoshi
Dept. of Industrial Chemistry
Chiba University
Yayoi-cho 1 Chiba-shi, Chiba 260

YAMADA, Shoichiro
Institute of Chemistry, College of
General Education, Osaka University
Machikaneyama, Toyonaka-shi, Osaka 560

YAMAGUCHI, Kazumi
Fac. of Engineering, Hokkaido University
Kiya 13-jo, Nishi 8-chome, Kita-ku,
Sapporo-shi, Hokkaido 060

YAMAGUCHI, Ryohei
Dept. of Industrial Chemistry,
Kyoto University
Yoshidahon-machi, Sakyo-ku, Kyoto 606

YAMAJI, Susumu
Jyohoku Saitama High School
585-1 Yanagitsubo, Furuichiba,
Kawagoe-shi, Saitama 356

YAMANA, Shukichi
Fac. of General Education, Kinki Univ.
3-4-1 Kowakae, Higashiosaka-shi,
Osaka 577

YAMASHITA, Tokuko
Kakegawa Technical High School
400 Kuzukawa, Kakegawa-shi, Shizuoka 436

YANAGISAWA, Atsuhiro
Hokkaido Tokai University
224 Chyuwa, Asahikawa-shi,
Hokkaido 070

YASUDA, Takeo
Shumei Ageo High School
1012 Ueno, Ageo-shi, Saitama 326

YONEKAWA, Mitsuyuki
Tsu Nishi High School
2210-2 Kobe, Tsu-shi, Mie 514

YOSHIDA, Ryushi
Hibiya High School
2-16-1 Nagata-cho, Chiyoda-ku,
Tokyo 100

YAGOH, Kazuo
1002 Fujinoki
Toyama-shi, Toyama 930 (Home)

YAGUCHI, Masaharu
Media Co., LTD
9-3, Dogenzaka 1-Chome
Shibuya-ku, Tokyo 150

YAMADA, Katsuzo
3-36-16 Tamagawa, Setagaya-ku, Tokyo 158

YAMADA, Ryuichi
Dept. of Chemistry,
National Defence Academy
Hashirimizu, Yokosuka-shi, Kanagawa 239

YAMADA, Tetsuya
Sosa High School
A-1630 Yokaichiba-shi, Chiba 289-21

YAMAGUCHI, Masafumi
Kobe Kaisei Girl's Senior High School
2-7-1 Aotani-cho, Nada-ku, Kobe-shi,
Hyogo 657

YAMAGUCHI, Shigeo
Fac. of Engineering,
Kanto Gakuin University
Mutsuura, Kanazawa-ku, Yokohama 236

YAMAMOTO, Yasuro
Meijo Gakuin High School
3-15-34 Huminosato, Abeno-ku, Osaka-shi,
Osaka 545

YAMASAKI, Akira
The University of Electro-Communications
1-5-1 Chofu-ga-oka, Chofu-shi,
Tokyo 182

YAMAUCHI, Sinpei
11 Moroki-cho, Uetakano, Sakyo-ku,
Kyoto-shi, Kyoto 606

YANO, Hisao
Kinuta Technical High School
2-9-1 Okamoto, Setagaya-ku,
Tokyo 157

YASUOKA, Takashi
Fac. of Science, Tokai University
1117 Kitakaname, Hiratsuka-shi,
Kanagawa 259-12

YOSHIDA, Hiroshi
c/o Mr. Tada, 1-9-13 Daizawa,
Setagaya-ku, Tokyo 155

YOSHIDA, Toshihisa
Dept. of Chemistry, Fac. of Education,
Saitama University
Urawa-shi, Saitama 338

YOSHIDA, Yoshio
High School Attached to
The University of Tsukuba
1-9-1 Otauk Bunkyo-ku, Tokyo 112

YOSHIKAWA, Hisaya
Hakusan High School
678 Minami-Ieki, Hakusan-cho,
Ichishi-gun, Mie 515-31

YOSHIMURA, Tadayoshi
Toyohashi University of Technology
Tempaku-cho, Toyohashi-shi, Aichi 440

YOSHIOKA, Tamio
Tokai University
2-28-4 Tomigaya
Shibuya-ku, Tokyo 151

YUKAWA, Yasuhiko
Futaba High School
14-1 Rokuban-cho, Chiyoda-ku,
Tokyo 102

ZALKOW, Vera B.
Kennesaw College
Marietta, GA 30061,
U.S.A.

ZHU, Ming-hua
East China Institute of Chemical
Technology, 130 Meilong Road,
Shanghai 201107, CHINA

ZOOK, Rosslyn T.
David's Lane
R.R. 3, Box 61
Pound Ridge, NY 10576, U.S.A.

YOSHIHIRO, Yoshiro
Meiji University
Higashimita, Tama-ku, Kawasaki 214

YOSHIMURA, Shin
Dept. of Chemistry, College of Arts
and Sciences, The University of Tokyo
3-8-1 Komaba, Meguro-ku, Tokyo 153

YOSHINO, Yukichi
Fac. of Science,
Toho University
2-2-1 Miyama Funabashi-shi, Chiba 274

YU, Qing-Sheng
Department of Chemistry,
Zhejiang University,
Hangzhou, CHINA

YUKI, Masanari
Toyama Technical High School
2238 Gofuku, Toyama-shi, Toyama 930

ZECHMANN, Heiner
Tafernerstr 22
A-9500 Villach,
AUSTRIA

ZOLLER, Uri.
Division of Chemical Studies
Haifa University Oranim,
P.O. Kiryat Tivon, 36910, ISRAEL

ZUUR, Aad P.
Gorlaeus Laboratorium,
P.Box 9502, 2300 Ra Leiden,
THE NETHERLANDS

# List of foundations which provided support for the conference

The Organizing Committee wishes to express its sincere gratitude to the following Organizations for their financial support of this Conference.

1. The Commemorative Association for the Japan World Exposition (1970)
2. Inoue Foundation for Science
3. The Kajima Foundation
4. The Naito Foundation
5. Sankyo Foundation of Life Science
6. Shimadzu Science Foundation
7. Toray Science Foundation
8. Yoshida Foundation for Science and Technology